Lecture Notes in Mathematics

Edited by A. Dold and B. Eckmann

806

Burnside Groups

Proceedings of a Workshop Held at the
University of Bielefeld, Germany
June–July 1977

Edited by J. L. Mennicke

Springer-Verlag
Berlin Heidelberg New York 1980

Editor

Jens L. Mennicke
Universität Bielefeld, 4800 Bielefeld, Federal Republic of Germany

Assisted by

F. J. Grunewald
Sonderforschungsbereich, Theoretische Mathematik
Universität Bonn, 5300 Bonn, Federal Republic of Germany

G. Havas,
M. F. Newman
Department of Mathematics, Institute of Advanced Studies
Australian National University, Canberra, ACT 2600, Australia

AMS Subject Classifications (1980): 20-02, 20-04, 20 E 10, 20 F 05,
20 F 10, 20 F 50

ISBN 3-540-10006-7 Springer-Verlag Berlin Heidelberg New York
ISBN 0-387-10006-7 Springer-Verlag New York Heidelberg Berlin

Printing and binding: Beltz Offsetdruck, Hemsbach/Bergstr.
2141/3140-543210

PREFACE

The present notes arose out of a workshop which was held at the University of Bielefeld in summer 1977. The main purpose of the workshop was to survey the present knowledge on Burnside groups, in particular the work of Novikov-Adian. The technical difficulties of this work are such that communication becomes a serious problem. The editors hope that the notes of Professor Adian's lectures will help a prospective reader to find access to the work which is now available in book form. The original Russian book was translated into English by Wiegold and Lennox and has appeared in the Ergebnisse series.

A first attack on the finiteness problem for $B(2, 8)$ is a second part of these notes. The authors are well aware of the incompleteness of the present results. They hope, however, that some of the methods and techniques may be helpful for future progress.

The workers in the field seem to agree that the structure of $B(2, 2^k)$ should become stable for some k .. However, it is not even clear whether one should expect that for large k these groups are finite or infinite. It seems clear, however, that $B(2, 8)$ should be an important test case.

M.F. Newman has compiled a list of problems which we hope will stimulate interest.

It is a pleasure to acknowledge financial support from Deutsche Forschungsgemeinschaft, Heinrich-Hertz-Stiftung, and University of Bielefeld. Our thanks go to the participants, in particular to Professor Sergej I. Adian who did the bulk of the lecturing. Our thanks also go to the technical staff for their valuable help: to the secretaries of the Department of Mathematics, and to the staff of ZiF in Bielefeld, and in particular to Mrs B.M. Geary who did a splendid job in typing the notes.

My personal thanks go to all the cooperators for their patience and efficiency. All the incompleteness of the present volume is my responsibility.

Bielefeld, January 1980   J. Mennicke

TABLE OF CONTENTS

PROCEEDINGS OF THE BURNSIDE WORKSHOP
BIELEFELD, June-July 1977, 1-40.

20F05

20F10, 20F50

CLASSIFICATIONS OF PERIODIC WORDS AND THEIR
APPLICATION IN GROUP THEORY

S.I. ADIAN

1. Introduction

Recently Springer-Verlag published the English translation of my book *The Burnside Problem and Identities in Groups*. The original Russian version was published in Moscow in 1975. In this book we outlined a method for studying periodic groups, which was first introduced in the joint work of P.S. Novikov and the author, "Infinite periodic groups I, II, III", *Izv. Akad. Nauk SSSR Ser. Mat.* 32 No. 1, 2, 3 (1968). The book contains a solution of the well-known Burnside problem for large odd exponent as well as other applications of the method and its modifications to construct groups with some interesting properties.

In 1977 Professor J. Mennicke invited the author to give a series of lectures at Bielefeld University. These lectures were meant as a help for a prospective reader to find his way through the book. The main difficulty which the reader faces is a collection of some 50 notions and more than 100 basic properties of these notions. The proof of the main result then consists of a simultaneous induction step for all the notions and their properties. Because there are too many cross-references to the inductive hypothesis, for nonformal understanding of any of these definitions or properties, depending on the inductive parameter α, the reader must be familiar with the meaning of all considered notions and with many of their properties for lower values of the inductive parameter α.

In order to avoid these difficulties, a rigid exposition requires an axiomatic approach. For $\alpha = 0$, all the relevant properties are more or less obvious. But in the inductive step from α to $\alpha + 1$, we ask the reader to consider all occurring notions for rank $\leq \alpha$ in the beginning as notions defined axiomatically, satisfying some properties named axioms. So in the beginning of this step, starting from the inductive hypothesis we have to give formal deductions of all relevant properties of rank $\alpha + 1$. Then as the reader becomes more and more familiar with the meaning of

the notions our exposition will become less formal and we can omit some details, which
by then will be clear. After the inductive step is completed (see Adian (1979),
V. 2.5, p. 236), all our notions will become meaningful for the reader and all the
properties of these notions will be verified for all values of α .

The aim of the introductory lectures given by the author in 1977 at Bielefeld
University was to help the nonformal understanding of a prospective reader of the book
by introducing the basic ideas of our theory. In the lectures there was given a
precise definition of all notions with some necessary comments and demonstrations for
rank 1 and 2 . Sometimes the lectures led to informal discussions of questions
raised by the audience.

The present article presents more or less the subject of these lectures. We have
left out only long precise formal definitions of all notions for rank α , taking into
account that in the meantime the book has become available in English, hence the
reader may look up the definitions in §4 of Chapter 1. Instead, we introduce an
approximative version of the most basic notions of our theory. This version is free
from many technical details and demonstrates the basic ideas of our method. At the
end, we give a summary of results obtained by our method.

I would like to thank the Mathematische Fakultät der Universität Bielefeld and in
particular Professor J. Mennicke for his invitation and for his help during my visit
to Bielefeld and the preparation of these notes.

I would like also to thank the Deutsche Forschungsgemeinschaft for their
financial support.

2. Free periodic groups

As is well-known, any group can be presented by generators and defining
relations. In particular, the free group $F_m = \langle a_1, a_2, \ldots, a_m \rangle$ of rank m can be
introduced in the following way. Consider an alphabet

$$(1) \qquad a_1, a_2, \ldots, a_m, a_1^{-1}, a_2^{-1}, \ldots, a_m^{-1} .$$

The defining relations of F_m are

$$(2) \qquad a_i a_i^{-1} = 1 , \quad a_i^{-1} a_i = 1 \ (i = 1, 2, \ldots, m) ,$$

where 1 denotes the empty word. From (2) we obtain the following consequences

$$(3) \qquad P a_i a_i^{-1} Q = PQ , \quad P a_i^{-1} a_i Q = PQ ,$$

where P, Q are arbitrary words in the alphabet (1). Two words X and Y are
called equal in F_m (we write $X = Y$ in F_m or $X \equiv Y$) if there exists a finite
sequence of words $X_1, X_2, \ldots, X_\lambda$ such that $X = X_1$, $X_i = X_{i+1}$ for $1 \leq i < \lambda$

and $X_\lambda = Y$ are of the form (3), possibly after permuting the left and right hand
sides. This is an equivalence relation in the set of all words in the alphabet (1).
Let $\{A\}$ denote the equivalence class containing A . The set of all classes with
multiplication $\{X\}\{Y\} = \{XY\}$ is the free group of rank m , that is F_m ; the unit
element of F_m is $\{1\}$, and the inverse element of $\{X\}$ is the class $\{X^{-1}\}$, where
X^{-1} is the result of writing the word X in reverse order and replacing all letters
a_i (or a_i^{-1}) by their inverses a_i^{-1} (or a_i).

The presentation of a group

$$G = \langle a_1, a_2, \ldots, a_m; A_j = 1 \rangle$$

with defining relations $A_j = 1$, where j runs through some index set J , is
obtained from the presentation of F_m by adding the new equations

(4) $PA_jQ = PQ$ for all $j \in J$.

In particular the free periodic group of exponent n ,

$$B(m, n) = \left\langle a_1, a_2, \ldots, a_m; x^n = 1 \right\rangle$$

is obtained from F_m by adding the identical relation $x^n = 1$, or, what is the same,
by adding defining relations $A^n = 1$ for all words A in the alphabet (1).

Let a group G be presented by defining relations. If two words A and B are
equal in this group, then by scanning all finite sequences of equations (3) and (4),
beginning with the word A , we can find a sequence ending with B . But such a
presentation of a group G does not give a method for proving that two words are not
equal in G . Moreover, the unsolvability of the word problem for finitely presented
groups in the general case means that there does not exist an algorithm which
enumerates all pairs of words (x, y) which are not equal in G . To prove that A
and B are not equal one has to find a property, which holds for all words equal to
A and does not hold for B . To find such properties for the free periodic groups
$B(m, n)$ turned out a difficult problem. For this reason the Burnside problem raised
in 1902 remained open for a long time (see Burnside (1902)). In the work (Novikov and
Adian (1968)) it was first shown that for all $m > 1$ and all odd $n \geq 4381$, the
groups $B(m, n)$ are infinite. For the proof of this result the authors introduced a
new way of describing the groups $B(m, n)$, based on a classification of periodic
words and their transformations depending on a natural parameter, called rank. Below
we shall describe the basic ideas of the method for studying periodic groups first
introduced in Novikov and Adian (1968) and later improved and refined in Adian (1979).

Unless otherwise stated in what follows n is odd and $n \geq 665$, $q = 90$.

3. Words and Occurrences

We shall consider words in the alphabet (1). A word X is called *uncancellable* if it has not the form $Pa_i a_i^{-1}Q$ or $Pa_i^{-1}a_iQ$. Letter-for-letter equality of two words X and Y is denoted by $X \overline{\circ} Y$. We denote the length of the word A by $\partial(A)$, that is $\partial(A)$ is the number of letters comprising A .

3.1. If $E \overline{\circ} PQ$, then the word P is called a *start* of E and Q is an *end* of E . If $E \overline{\circ} PQ$, then QP is called a *cyclic shift* of E .

We say that the word E *occurs* in the word X if there are words P and Q such that $X \overline{\circ} PEQ$. One and the same word can occur in a given word X in different places. In order to distinguish between two different occurrences of E in X we use an extra symbol $*$. If $X \overline{\circ} PEQ$ then we call $P*E*Q$ an *occurrence of* E *in* X . Thus an occurrence of E in X is a triple (P, E, Q) of subwords of X . The word E is called the *base* of the occurrence $P*E*Q$, and is denoted by $\mathrm{Bas}(P*E*Q)$. We shall consider only occurrences with non-empty bases. The length of an occurrence is the length of its base. If $W = P*E*Q$, then we denote by W^{-1} the occurrence $Q^{-1}*E^{-1}*P^{-1}$.

One can think of an occurrence in a given word as a segment of this word. Let $X \overline{\circ} PEQ \overline{\circ} RDS$, $V \overset{\sim}{\div} P*E*Q$, $U \overset{\sim}{\div} R*D*S$. Two occurrences V and U in a given word X can be situated in the following ways.

1. $\partial(PE) \leq \partial(R)$

2. $\partial(P) < \partial(R)$, $\partial(S) < \partial(Q)$

3. $\partial(P) < \partial(R)$, $\partial(Q) < \partial(S)$

4. $\partial(P) < \partial(R)$, $\partial(Q) = \partial(S)$

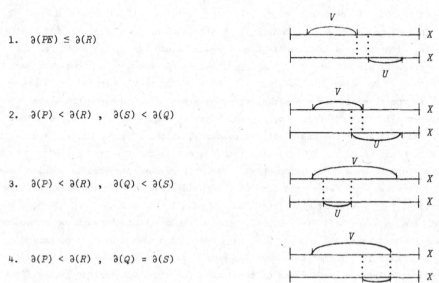

5. $\partial(P) = \partial(R)$, $\partial(Q) < \partial(S)$

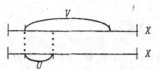

In case 1) we say U and V do not intersect, in the other cases they intersect. In cases 1) and 2) we say that V lies to the left of U and write $V < U$. In case 1) we also say that V lies strictly to the left of U and write $V \ll U$. In cases 3), 4) and 5), U occurs in V , thus U is an intersection of U and V . In case 4) we say that U is an end of V , in case 5), U is a start of V .

3.2.   The *union* of two occurrences is the occurrence containing both of them that has base of shortest length. In cases 3), 4) and 5), V is the union of V and U . In cases 1) and 2) the union of V and U has the form $P*H*S$ for some H .

3.3.   Let $P*E*Q$ and $R*E*S$ be occurrences in words X and Y and V an occurrence in X contained in $P*E*Q$. Then there exist words B, D and C such that $E \; \overline{o} \; BDC$ and $V = PB*D*CQ$. We can show this situation geometrically in the following way.

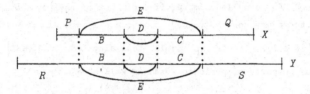

In such a case we denote the occurrence $RB*D*CS$ in Y by

$$\phi(V; \; P*E*Q, \; R*E*S) \; .$$

For any two occurrences W and W_1 with the same base, the function $V_1 = \phi(V; W, W_1)$ sets up a one-to-one mapping of the set of occurrences in W onto the set of all occurrences in W_1 .

Clearly, the function $V_1 = \phi(V; W, W_1)$ preserves the relations $<, \ll$ and carries the common part (union) of two occurrences contained in W to the common part (union) of their images in W_1 .

If $V_1 = \phi(V; W, W_1)$, then $V = \phi(V_1; W_1, W)$.

Let W_1, W_2, W_3 be occurrences with the same base and suppose that V_1 is contained in W_1 . If $V_2 = \phi(V_1; W_1, W_2)$ and $V_3 = \phi(V_2; W_2, W_3)$ then $V_3 = \phi(V_1; W_1, W_3)$.

4. Periodic Words

For any integer $t > 0$ we denote by A^t the word $AA \ldots A$, with A repeated t times. For any word A we set

$$A^{-t} \doteq \left(A^{-1}\right)^t \quad \text{and} \quad A^0 \doteq 1 \; .$$

4.1. We call an uncancellable word of the form $A_1 A^t A_2$, where A_1 is an end of A, A_2 is a start of A and $\partial\left(A_1 A^t A_2\right) > 2\partial(A)$, a *periodic word* of rank 1 with period A. For empty A_1 (or A_2) we call A the *left* (or *right*) *period* of $A_1 A^t A_2$. The set of all periodic words of rank 1 with period A is denoted by Per(A) or Per(1, A).

Clearly, if B is a cyclic shift of A, then Per(A) = Per(B). If $X \in$ Per(A), B occurs in X and $\partial(B) = \partial(A)$, then B is a cyclic shift of A.

Clearly, if $A \; \overline{\underline{\circ}} \; B^r$ for $r > 0$, then Per(A) $\subset$ Per(B).

4.2. The word A is called *simple*, if it is not of the form D^r for any $r > 1$. Clearly, for every non-empty word A there exists a simple word B such that $A = B^t$ for some $t \geq 1$. If AB is simple, then also BA is simple.

The following property of periodic words is well-known.

4.3. If $A^t A' \; \overline{\underline{\circ}} \; B^r B'$, where A' is a start of A, B' is a start of B and $\partial\left(A^t A'\right) \geq \partial(AB)$, there exists a word D such that $A \; \overline{\underline{\circ}} \; D^k$ and $B \; \overline{\underline{\circ}} \; D^s$ for some k and s. In particular if A is simple, then $D \; \overline{\underline{\circ}} \; A$.

If $X \in$ Per(A) for some A, then X has a unique left period, which is also a left period of minimal length of the word X. Similarly for right periods.

5. Basic Notions for Rank 0

We denote by R_0 the set of all uncancellable words in the alphabet (1). These words also will be called *reduced* words of rank 0.

We call an occurrence $P*E*Q$ in the word $X \in R_0$ a *kernel of rank* 0 of the word X, if $\partial(E) = 1$, that is if E is a letter. We denote by Ker(0, X) the set of all kernels of rank 0 of the word X. By Reg(0, X) we denote the set of all occurrences in the word X with a non-empty base. It is clear that any occurrence $W \in$ Reg(0, X) begins (and ends) with some kernel of rank 0. By $\partial_0(W)$ we denote the number of kernels of rank 0 contained in W. Clearly, we have

$$\partial_0(W) = \partial\left(\text{Bas}(W)\right) \; .$$

Two words X, $Y \in R_0$ are called *equivalent in rank* 0 , if $X \overline{\underset{0}{}} Y$. For this we also write $X \underset{0}{Q} Y$. If $X \underset{0}{Q} Y$ and $V \in \text{Reg}(0, X)$, then we set by definition $V = f_0(V; X, Y)$.

We call two occurrences V and W in the words $X \in R_0$ and $Y \in R_0$ *mutually normalized* if $\text{Bas}(V) \overline{\underset{0}{}} \text{Bas}(W)$.

If X, $Y \in R_0$, then there exists a word T such that $X \overline{\underset{0}{}} X_1 T$, $Y \overline{\underset{0}{}} T^{-1} Y_1$ and $X_1 Y_1 \in R_0$. We say that this word $X_1 Y_1$ is the result of *coupling* of the words X and Y in rank 0 , and we write

$$[X, Y]_0 = X_1 Y_1 .$$

This operation is clearly unique and associative.

The set R_0 with the operation $[X, Y]_0$ is the simplest form to describe the group F_m with m generators $a_1, a_2, \ldots, a_m$. Our aim will be to find a similar description for the free periodic groups $B(m, n)$.

6. Elementary words of rank 1

The cyclically uncancellable simple words (no cyclic shift is cancellable) are precisely the *minimal periods of rank* 1 . From these periods we select elementary periods of rank 1 .

6.1. A minimal period A of rank 1 is called an *elementary period of rank* 1 if no periodic word E with period B of rank 1 and $\partial(E) > 8\partial(B)$ occurs in A^8 . For instance the words $a_1 a_2$ and $a_2 a_3 (a_2 a_1)^7$ are elementary periods of rank 1 . The words $a_1^{10} a_2$ and $a_2 a_1 a_2 a_3 (a_2 a_1)^7$ are minimal periods of rank 1 but they are not elementary periods of rank 1 , because in the second case the word $(a_2 a_1)^8 a_2$ with the period $a_2 a_1$ occurs in the word $(a_2 a_1 a_2 a_3 (a_2 a_1)^7)^8$. Notice that if A is a minimal period, then A^{-1} also is. If A is an elementary period of rank 1 , then A^{-1} also is.

6.2. If A is an elementary period of rank 1 and E is a periodic word with period A , then E will be called an *elementary word of rank* 1 . By 4.1 and 4.3, for any elementary word E of rank 1 one can find a corresponding elementary period A that is unique up to a cyclic shift.

6.3. For an arbitrary elementary word E of rank 1 with period A we call the number

$$\left\lceil \frac{\partial(E)}{\partial(A)} \right\rceil = \text{the smallest integer} \quad r \quad \text{such that} \quad \frac{\partial(E)}{\partial(A)} \le r \ ,$$

the number of segments of rank 1 of the word E and denote it by $l_1(E)$. We call
$l_1(E)$ also the number of segments of an arbitrary occurrence $P*E*Q$ and we write
$l_1(P*E*Q) = l_1(E)$.

It follows easily from our definition of $l_1(E)$, that for an arbitrary elemen
elementary word E_1E_2 we have the inequalities

$$l_1(E_1) + l_1(E_2) - 1 \le l_1(E_1E_2) \le l_1(E_1) + l_1(E_2) \ .$$

If $l_1(E) \ge r$, then we say that E is an elementary r-power of rank 1 . Any
occurrence of the elementary word E of rank 1 will be called a *normalized*
occurrence of E . For a given word $X \in R_0$ and given number $r \ge 2$ we denote by

$$\text{Norm}(1,\ X,\ r)$$

the set of all occurrences of elementary r-powers of rank 1 in the word X .

6.4. Let an occurrence

(5) $P*A^tA_1*Q$

in some word $X \in R_0$ be given, where A^tA_1 is an elementary word of rank 1 with
period $A \overline{\underline{o}} A_1A_2$. We say that the occurrence (5) can be continued to the left if P
is nonempty and if after adding the last letter of P to the word A^tA_1 we obtain
also a periodic word with period A . In other words, the occurrence (5) can be
continued to the left if there exists a nonempty word H such that $P \overline{\underline{o}} P_1H$ and
$HA^tA_1 \in \text{Per}(A)$. In this case we call the occurrence $P_1*HA^tA_1*Q$ a *continuation* of
(5) *to the left*. Similarly we define *continuations to the right*. If (5) can be
continued to the left or to the right then the corresponding occurrence $P_1*HA^tA_1F*Q_1$
is called a *continuation* of (5). It is convenient also to call (5) a continuation of
itself.

6.5. Obviously the occurrence (5) can be continued to the left if and only if P
and A have the same last letter. Similarly for a continuation to the right, A_2A_1
and Q must have the same first letter.

We call (5) a *maximal occurrence* if it has no proper continuation to the left or
to the right. A continuation of (5) is called a *maximal continuation* of (5) if it is
a maximal occurrence. Similarly, we introduce the maximal continuation to the left

(or to the right) of (5).

We denote by

$$\text{Max Norm}(1, X, r)$$

the set of all maximal occurrences of r-powers of rank 1 in X .

EXAMPLE.　The occurrence $a_1^{-1} a_2 * \left(a_1^3 a_2 \right)^{10} * a_2^5$ can be continued to the left and

cannot be continued to the right.　Its maximal continuation is $a_1^{-1} * a_2 \left(a_1^3 a_2 \right)^{10} * a_2^5$.

6.6.　Two occurrences in the same word $X \in R_0$,

(6)　　　　　　　　　　　　　　$P*E*Q$ and $R*D*S$

of the elementary words E and D of rank 1 we call *compatible* if there exists an occurrence which contains both occurrences (6) and is a continuation of both of them.

It is easy to see that the occurrences (6) are compatible if and only if their union is a continuation of both of them (see 4.3), and consequently E and D have the same periods.　If the occurrences (6) are compatible then we write

(7)　　　　　　　　　　　　　　$\text{Comp}(P*E*Q, R*D*S)$.

Of course in (7) we assume that $RDS \; \overline{\circ} \; PEQ$.　Obviously (7) is a symmetric and transitive relation.

EXAMPLES.　The occurrences

(8)　　　　　　　$a_1 * \left(a_2 a_1 \right)^{15} a_2 * a_1 a_2 \left(a_1 a_2 \right)^7 a_1^3 a_2$

and

(9)　　　　　　　$a_1 \left(a_2 a_1 \right)^{15} a_2 a_1 a_2 * \left(a_1 a_2 \right)^7 * a_1^3 a_2$

are compatible.　The occurrences

$$a_1 * \left(a_2 a_1 \right)^{15} a_2 * a_1 a_2^{-1} \left(a_1 a_2 \right)^7 a_1^3 a_2$$

and

$$a_1 \left(a_2 a_1 \right)^{15} a_2 a_1 a_2^{-1} * \left(a_1 a_2 \right)^7 * a_1^3 a_2$$

are not compatible, because the base of their union is not a periodic word with the same period.

Notice that any occurrence has a unique maximal continuation.　If two occurrences are compatible then they have the same maximal continuation.

The occurrences (8) and (9) have the maximal continuation

$$* (a_1 a_2)^{24} a_1 * a_1^2 a_2 \ .$$

6.7. Two elementary words E and D of rank 1 are called *related* if their periods coincide. In this case any two occurrences $P*E*Q$ and $R*D*S$ are also called related. In this case we shall write

$$\text{Rel}(E, D) \quad \text{and} \quad \text{Rel}(P*E*Q, R*D*S) \ .$$

6.8 LEMMA. *Suppose* $V \in \text{Norm}(1, Z, 9)$ *and* $W \in \text{Norm}(1, Z, r)$. *If* V *is contained in* W , *then* W *is a continuation of* V *and therefore* $\text{Comp}(V, W)$. *If* $\text{Rel}(V, W)$ *or* $\text{Rel}(V, W^{-1})$, *then the same is true already when* $l_1(V) > 2$.

Proof. Let A be an elementary period of $\text{Bas}(V)$ and C be an elementary period of $\text{Bas}(W)$. By the assumption of our lemma, we have

$$(10) \qquad \qquad \text{Bas}(V) = A^t A' \overline{\ \circ\ } D^r D' \ ,$$

where A' is a start of A , D is a cyclic shift of C , D' is a start of D , and $\partial(A^t A') > 8\partial(A)$. By 6.1 we have $\partial(D^r D') > 8\partial(D)$, because D is an elementary period of rank 1 . Then by 4.3 we have $A \overline{\ \circ\ } D$. Therefore W is a continuation of V .

The second part of the lemma is easier since by the condition of $\text{Rel}(V, W)$ we have $\partial(A) = \partial(C)$. Then it follows from $l_1(V) \geq 2$ and (10) that $\partial(A^t A') > \partial(A)$ and $A \overline{\ \circ\ } D$. Consequently we have $\text{Comp}(V, W)$.

6.9. It follows from 6.8 that if two occurrences $V \in \text{Norm}(1, X, 9)$ and $W \in \text{Norm}(1, Z, 9)$ are not compatible, then the common part of them contains less than 9 segments of each.

7. Reversals of Rank 1

7.1. Suppose that $r \geq 9$, $PA^t A_1 Q \in R_0$ and $P*A^t A_1 *Q$ is a maximal occurrence of an elementary r-power $A^t A_1$ with the period $A \overline{\ \circ\ } A_1 A_2$. Then the transition

$$(11) \qquad \qquad PA^t A_1 Q \to P(A^{-1})^{n-t-1} A_2^{-1} Q$$

is said to be an *r-reversal of rank* 1 of the occurrence $P*A^t A_1 *Q$ if $(A^{-1})^{n-t-1} A_2^{-1}$ is also r-power with the elementary period A^{-1} of rank 1 .

7.2. Notice, that by 6.4 it follows easily from $PA^t A_1 Q \in R_0$ that $P*(A^{-1})^{n-t-1} A_2^{-1} *Q$ is a maximal occurrence. Then we have $P(A^{-1})^{n-t-1} A_2^{-1} Q \in R_0$

because $P*A^t A_1 *Q$ is a maximal occurrence.

Therefore, if (11) is an r-reversal of $P*A^t A_1 *Q$, then the transition

$$P\left(A^{-1}\right)^{n-t-1} A_2^{-1} Q \rightarrow PA^t A_1 Q$$

is an r-reversal of $P*\left(A^{-1}\right)^{n-t-1} A_2^{-1} *Q$.

EXAMPLE. The transition

$$a_1^4 \left(a_2 a_1^{-2}\right)^q a_2 a_1^{-1} a_2^3 \rightarrow a_1^4 \left(a_1^2 a_2^{-1}\right)^{n-q-1} a_1 a_2^3$$

is a q-reversal of rank 1 of the maximal occurrence $a_1^4 * \left(a_2 a_1^{-2}\right)^q a_2 a_1^{-1} * a_2^3$ (see

2.4). Here we have $A = a_2 a_1^{-2}$, $A_1 = a_2 a_1^{-1}$ and $A_2 = a_1^{-1}$. It is impossible to

carry out even a 9-reserval of rank 1 of the occurrence

$$a_1^4 * \left(a_2 a_1^{-2}\right)^{n-7} a_2 a_1^{-1} * a_2^3 .$$

7.3. Any reversal (11) of the occurrence $P*A^t A_1 *Q$ will also be called a

reversal of an arbitrary occurrence W , compatible with $P*A^t A_1 *Q$.

7.4. It follows easily from 6.3 that for an arbitrary elementary period
$A \overline{\underline{o}} A_1 A_2$ we have

$$n \le l_1 \left(A^t A_1\right) + l_1 \left(\left(A^{-1}\right)^{n-t-1} A_2^{-1}\right) \le n + 1 .$$

Then for an arbitrary maximal occurrence $P*A^t A_1 *Q$ of the elementary word $A^t A_1$ with

the period $A \overline{\underline{o}} A_1 A_2$ in $X \in R_0$ we can carry out an r-reversal (11) of rank 1 if

the inequalities

$$r \le l_1 \left(A^t A_1\right) \le n - r - 1$$

hold.

7.5. Let P_1 be the set of all words $X \in R_0$ which contain no occurrences of

elementary $(n-q)$-powers of rank 1 . Then for any $X \in P_1$ and any maximal

occurrence $P*A^t A_1 *Q$ in X one can carry out the q-reversal (11) of rank 1 if

$l_1 \left(A^t A_1\right) \ge q$.

7.6. Consider a 9-reversal (11) of the occurrence

$P * A_1^t A_1 * Q \in$ Max Norm(1, X, 9) . An occurrence $F * E * G$ in the word $PA^t A_1 Q$ is said to be *stable* in the reversal (11) if either $P \overline{\underline{\circ}} FEP_1$ or $Q \overline{\underline{\circ}} Q_1 EG$. In such a case the occurrence

$$F * E * P_1 \left(A^{-1} \right)^{n-t-1} A_2^{-1} Q \quad \text{or} \quad P \left(A^{-1} \right)^{n-t-1} A_2^{-1} Q_1 * E * G$$

will be called the *image* of the occurrence $F * E * G$ in (11) and will be denoted by

$$f_{X \to Y}(F * E * G) .$$

7.7. An occurrence V in the word $X \in R_0$ is stable in a given reversal of an occurrence $W \in$ Norm(1, X, 9) if and only if V does not intersect the occurrence W .

This follows from 7.6.

7.8. We shall consider q-reversals of rank 1 (see 2.4). Some of the q-reversals of rank 1 we call *real reversals*. The precise definition of real reversals of rank 1 is based on the notion of cascades of rank 1 (see I.4.22 and I.4.23), and essentially has been used in Adian (1979) in proofs of some important properties of our notions in rank 2 . The following properties of real reversals are important.

(a) Any real reversal of rank 1 is a q-reversal of rank 1 .

(b) If there exists a real reversal of an occurrence $W \in$ Norm(1, X, 9) , then there exists an occurrence $V \in$ Norm(1, X, q) such that Comp(W, V) and V is stable in the real reversal of an arbitrary occurrence of rank 1 that is not compatible with V (see I.4.23).

(c) If $W \in$ Max Norm(1, X, $q+34$) and $l_1(W) < n - q - 52$, then one can carry out some real reversal $X \to Y$ of the occurrence W (see III.3.24).

(d) If $X \to Y$ is a real reversal of rank 1 , then $Y \to X$ also is a real reversal of rank 1 (see III.3.24).

If one is interested only in rank 1 , then one can call any q-reversal of rank 1 a real reversal of rank 1 .

We call an occurrence V an *active occurrence of rank* 1 if $V \in$ Norm(1, X, 9) and if it is possible to find some real reversal of it. We denote by

$$\text{Act}(1, X)$$

the set of all active occurrences of rank 1 in a given word $X \in P_1$.

Obviously, if $V, U \in$ Norm(1, X, 9) , Comp(V, U) and $V \in$ Act(1, X) , then $U \in$ Act(1, X) .

(e) Suppose that $X \to Y$ is a real reversal of an occurrence V and an occurrence $U \in \text{Norm}(1, X, 9)$ is stable in $X \to Y$. Then

$$U \in \text{Act}(1, X) \iff f_{X \to Y}(U) \in \text{Act}(1, Y)$$

(see III.3.31).

EXAMPLE. In the word $\left(a_1 a_2\right)^{2q} a_1^{81} a_2^{3q}$ we have two active maximal occurrences (see (c))

$$*\left(a_1 a_2\right)^{2q} a_1 * a_1^{80} a_2^{3q} \quad \text{and} \quad \left(a_1 a_2\right)^{2q} a_1^{81} * a_2^{3q} * \, ,$$

but the occurrence $\left(a_1 a_2\right)^{2q} * a_1^{81} * a_2^{3q}$ is not active (see (a) and 7.1).

7.9. We shall call an occurrence $V \in \text{Norm}(1, X, 9)$ a *kernel of rank* 1 in the word $X \in P_1$ if it is stable in a reversal of an arbitrary active occurrence of rank 1 that is not compatible with it and if no proper continuation of it has this property. Furthermore, we shall call a kernel V an *active kernel of rank* 1 if $V \in \text{Act}(1, X)$. The set of all kernels of rank 1 in a word $X \in P_1$ is denoted by $\text{Ker}(1, X)$. We denote by $\partial_1(X)$ the number of kernels of rank 1 in X.

EXAMPLES. The word $\left(a_1 a_2\right)^{2q} a_1^{81} a_2^{3q}$ has 3 kernels of rank 1:

$$*\left(a_1 a_2\right)^{2q} a_1 * a_1^{80} a_2^{3q} \, , \quad \left(a_1 a_2\right)^{2q} a_1^{81} * a_2^{3q} *$$

and

$$\left(a_1 a_2\right)^{2q} * a_1^{81} * a_2^{3q} \, .$$

The first two kernels are active and third is not active.

The word

(12)
$$\left(a_1^{10} a_2^{2q} a_1^{10} a_2^{-n+2q}\right)^n$$

has $2n$ inactive kernels of rank 1 with base a_1^{10} and n active kernels of rank 1 with each of the bases a_2^{2q} and a_2^{-n+2q}.

7.10. We list some properties of real reversals and kernels of rank 1 that follow easily from our definitions.

(a) If U and V are distinct kernels of rank 1 of a word $X \in P_1$, then they are not compatible and neither of them is contained in the other, that is either $U < V$ or $V < U$.

Suppose that $\text{Comp}(U, V)$. Then U and V have the same maximal continuation W and by 7.9 each of them is a maximal part of W that is stable in a reversal of an arbitrary active occurrence of rank 1 that is not compatible with W. Therefore $U \overline{\circ} V$. The rest of (a) follows from 6.8.

(b) If a kernel V of rank 1 of a word X is active, then a maximal continuation of it does not intersect any other kernel of rank 1 and $l_1(V) \geq q$.

Follows from 7.9, 7.7 and 7.8 (b).

(c) If $X \in P_1$ and $W \in \text{Norm}(1, X, 25)$, then W is compatible with some kernel of rank 1 of X.

Proof. Let $\overline{W}$ be the maximal continuation of W. By 6.9 and 7.7 for the maximal end W_1 of $\overline{W}$ that is stable in the real reversal of the closest active neighbour of W to the left we have

$$\partial\big(\text{Bas}(W_1)\big) \geq \partial\big(\text{Bas}(\overline{W})\big) - 8\partial(A) > 16\partial(A) \ ,$$

where A is a period of $\text{Bas}(W)$. Similarly, for the maximal start W_2 of W_1 that is stable in the real reversal of the closest active neighbour of W to the right we have

$$\partial\big(\text{Bas}(W_2)\big) \geq \partial\big(\text{Bas}(W_1)\big) - 8\partial(A) > 8\partial(A) \ .$$

Therefore $l_1(W_2) \geq 9$. Then by 7.7, W_2 is a kernel of rank 1 and it is compatible with W.

(d) If $V \in \text{Ker}(1, X)$ and W is the maximal continuation of V, then $l_1(W) \leq l_1(V) + 16$.

This follows from 7.7 and 6.9.

(e) If $W \in \text{Act}(1, X)$, then there exists an active kernel $V \in \text{Ker}(1, X)$ such that $\text{Comp}(W, V)$.

This follows from 7.8 (b) and 7.9.

(f) Suppose that the transition $X \to Y$ is a real reversal of rank 1. For each kernel $V \in \text{Ker}(1, X)$ there exists one and only one kernel $W \in \text{Ker}(1, Y)$ that is the image of V in the reversal $X \to Y$. This one-to-one correspondence of the set $\text{Ker}(1, X)$ with $\text{Ker}(1, Y)$ we denote by

$$f_1(V; X, Y) \ .$$

It preserves the relations $<$ and $\ll$. If $V \in \text{Act}(1, X)$, then $f_1(V; X, Y) \in \text{Act}(1, Y)$.

Proof. Let $X \to Y$ be the real reversal of $U \in \text{Norm}(1, X, 9)$. Then it has the form (11), where $X \; \overline{\underline{\circ}} \; PA^t A_1 Q$, $Y \; \overline{\underline{\circ}} \; P(A^{-1})^{n-t-1} A_2^{-1} Q$ and $\mathrm{Comp}\left[U, \; P*A^t A_1 *Q\right]$.

Let V be any kernel of rank 1 of X . If $\mathrm{Comp}\left[V, \; P*A^t A_1 *Q\right]$ does not hold, then by 7.9 and 7.6, V is stable in the reversal $X \to Y$. Then by 7.8 (e) and 7.9,

$$f_{X \to Y}(V) \in \mathrm{Ker}(1, Y) \; .$$

Therefore in this case we can set

$$f_1(V; \; X, \; Y) = f_{X \to Y}(V) \; .$$

Let $\mathrm{Comp}\left[V, \; P*A^t A_1 *Q\right]$ hold. By 7.2, $P*(A^{-1})^{n-t-1} A_2^{-1} *Q \in \mathrm{Act}(1, Y)$. By (e) there exists an active kernel $U \in \mathrm{Ker}(1, Y)$ such that $\mathrm{Comp}\left[V, \; P*(A^{-1})^{n-t-1} A_2^{-1} *Q\right]$. Then we set in this case

$$f_1(V; \; X, \; Y) = U \; .$$

The remaining assertions of (f) are obvious.

8. Reduced Words and Equivalence in Rank 1

8.1. We define the set R_1 as follows:

(13) $X \in R_1 \iff X \in N_1$ and $l_1(W) \le n - 176$ for any $W \in \mathrm{Ker}(1, X)$.

We define the sets K_1, L_1 and M_1 in a similar way by substituting in (13) the parameters $n - 218$, $\frac{n+1}{2} + 21$ and $\frac{n+1}{2}$, respectively, for $n - 176$.

Since $n \ge 665$ we have the following relations:

$$M_1 \subset L_1 \subset K_1 \subset R_1 \subset P_1 \subset R_0 \; .$$

EXAMPLES. It is easy to see that the word (12) belongs to R_1 , but not to M_1 .

The word

(14) $a_1^{n-170}\left(a_2 a_1^8\right)^{2q}$

belongs to P_1 and has two kernels

$$*a_1^{n-178} *a_1^8\left(a_2 a_1^8\right)^{2q} \quad \text{and} \quad a_1^{n-170} *\left(a_2 a_1^8\right)^{2q} * \; .$$

Clearly it belongs to R_1 .

8.1. It is not true that if $X \in R_1$ and $X \; \overline{\underline{\circ}} \; PDQ$, then $D \in R_1$. For example, the word (14) has a subword a_1^{n-170} which does not belong to R_1 , because

$*a_1^{n-170}* \in \mathrm{Ker}\left(1, a_1^{n-170}\right)$. But the following implications are true.

If $PDQ \in K_1$, then $D \in R_1$.

If $PDQ \in L_1$, then $D \in K_1$.

If $PDQ \in M_1$, then $D \in L_1$.

Suppose that $PDQ \in K_1$. By 8.1 and 7.10 (d) no elementary $\left((n-218)+17\right)$-power

occurs in PDQ . Then for an arbitrary kernel $V \in \mathrm{Ker}(1, D)$, we have

$l_1(V) < n - 201$. Therefore $D \in R_1$. The proofs of the last two implications are

similar.

The proof of the following proposition also is not difficult.

8.3. If $X \in R_1$ and $\partial(X) > 1$, then there exist words $P \in R_1$ and $Q \in R_1$

such that $X \overline{\underline{o}} PQ$ (see IV.1.23).

8.4. We shall say that two words $X, Y \in P_1$ are *equivalent in rank* 1 if

either $X \overline{\underline{o}} Y$ or there is a sequence of real reversals of rank 1 of the form:

(15) $X \to X_1 \to X_2 \to \ldots \to X_i \to X_{i+1} \to \ldots \to X_\lambda \to Y$.

By $X \underset{\sim}{1} Y$ we shall denote the combined assertion $X \in R_1$, $Y \in R_1$ and X is

equivalent to Y in rank 1 .

Obviously, the relation $X \underset{\sim}{1} Y$ is reflexive, symmetric and transitive (see

7.8 (d)).

EXAMPLE. On carrying out on the word (12) a sequence of n real reversals of

every fourth kernel of rank 1 , we get

$$\left(a_1^{10} a_2^{2q} a_1^{10} a_2^{-n+2q}\right)^n \to a_1^{10} a_2^{2q} a_1^{10} a_2^{2q} \left(a_1^{10} a_2^{2q} a_1^{10} a_2^{-n+2q}\right)^{n-1} \to$$

$$\to \left(a_1^{10} a_2^{2q} a_1^{10} a_2^{2q}\right)^2 \left(a_1^{10} a_2^{2q} a_1^{10} a_2^{-n+2q}\right)^{n-2} \to \ldots \to \left(a_1^{10} a_2^{2q} a_1^{10} a_2^{2q}\right)^n \overline{\underline{o}} \left(a_1^{10} a_2^{2q}\right)^{2n} .$$

Clearly, $\left(a_1^{10} a_2^{2q}\right)^{2n} \in M_1$.

8.5. Suppose $X \underset{\sim}{1} Y$. If $X \overline{\underline{o}} Y$, then we set $f_1(V; X, Y) = V$. If there

exists a sequence (15) of real reversals of rank 1 , then we define $f_1(V; X, Y)$ to

be the composition of the uniquely defined mappings for the reversals in the sequence

(15) indicated in 7.10 (f).

It follows clearly from 7.10 (f) that the mapping $f_1(V; X, Y)$ preserves the linear ordering relation between kernels of rank 1 and that

$$W = f_1(V; X, Y) \Longleftrightarrow V = f_1(W; Y, X) .$$

8.6. For an arbitrary occurrence W in $X \in R_1$ we denote by $\partial_1(W)$ the number of all kernels of rank 1 that occur in W .

We shall call an occurrence W in a word $X \in R_1$ a *regular occurrence of rank 1* if some start and some end of it belong to Ker(1, X) .

We denote the set of all regular occurrences of rank 1 in a word X by Reg(1, X) .

8.7. We extend the mapping $f_1(V; X, Y)$ for $X \perp Y$ introduced in 8.5 to the set Reg(1, X) . Suppose that $X \perp Y$ and $W \in$ Reg(1, X) . We look at kernels $U, V \in$ Reg(1, X) , where U is a start of W and V is an end of W . By 8.5 there are uniquely defined kernels $f_1(U; X, Y)$ and $f_1(V; X, Y)$ of Y whose union belongs to Reg(1, Y) . We denote this union by $f_1(W; X, Y)$. In this way we obtain a function $W_1 = f_1(W; X, Y)$ that maps the set Reg(1, X) one-to-one onto the set Reg(1, Y) .

8.8. Suppose that $X \in R_1$ and $X \to Y$ is a real reversal of a kernel $V \in$ Ker(1, X) . If $l_1(V) \geq r$, then $l_1(f_1(V; X, Y)) \leq n - r + 1$. If a kernel $U \in$ Ker(1, X) is not compatible with V , then

$$l_1(f_1(V; X, Y)) = l_1(V) .$$

The proof follows from 7.4 and 7.10 (f).

Therefore, if $X \in R_1$ and $l_1(V) \geq 177$, then $Y \in R_1$.

8.9. For an arbitrary word $X \in R_1$ one can find a word $Z \in M_1$, such that $X \perp Z$.

Proof. We shall proceed by induction on the number $m(X)$ of kernels $V \in$ Ker(1, X) such that $l_1(V) > \frac{n+1}{2}$. If $m(X) = 0$, then $X \in M_1$. If $m(X) > 0$, then we have $l_1(V) > \frac{n+1}{2}$ for some kernel $V \in$ Ker(1, X) . By 7.8 (c), $V \in$ Act(1, X) , that is one can carry out a real reversal $X \to Y$ of V .

By 8.8 we have $l_1(f_1(V; X, Y)) < \frac{n+1}{2}$ and $l_1(U) = l_1(f_1(U; X, Y))$ for an arbitrary $U \in$ Ker(1, X) , that is distinct from V . Then $Y \in R_1$, $X \perp Y$ and $m(Y) = m(X) - 1$. By inductive hypothesis there exists a word $Z \in M_1$ such that

$Y \downarrow Z$. Therefore $X \downarrow Z$.

8.10. If no elementary q-power of rank 1 occurs in a word $X \in R_1$, then for an arbitrary $Y \in R_1$ we have

$$X \mathrel{\substack{\perp \\ \sim}} Y \Longleftrightarrow X \mathrel{\overline{\circ}} Y .$$

In fact, by 7.8 (a) no sequence of real reversals of the form (15) can exist for such a word X . Thus if $X \mathrel{\substack{\perp \\ \sim}} Y$, then by 8.4 we have $X \mathrel{\overline{\circ}} Y$.

8.11. Suppose that $X, Y \in R_1$. We say that two occurrences $V \in \mathrm{Reg}(1, X)$ and $U \in \mathrm{Reg}(1, Y)$ are *mutually normalised in rank* 1 if the following conditions are fulfilled:

(a) If $X \mathrel{\substack{\perp \\ \sim}} Z$ for some Z , then a word Z_1 with $Y \mathrel{\substack{\perp \\ \sim}} Z$ can be found such that

(16) $$\mathrm{Bas}\bigl(f_1(V;\, X,\, Z)\bigr) \mathrel{\overline{\circ}} \mathrm{Bas}\bigl(f_1(W;\, Y,\, Z_1)\bigr) ,$$

and moreover for an arbitrary occurrence U contained in $f_1(V;\, X,\, Z)$ we have

(17) $$U \in \mathrm{Ker}(1,\, Z) \Longleftrightarrow \phi\bigl(U;\, f_1(V;\, X,\, Z),\, f_1(W;\, Y,\, Z_1)\bigr) \in \mathrm{Ker}\bigl(1,\, Z_1\bigr) .$$

(For the definition of ϕ see 3.3.)

(b) Conversely, if Z_1 is an arbitrary word such that $Y \mathrel{\substack{\perp \\ \sim}} Z_1$, then a word Z with $X \mathrel{\substack{\perp \\ \sim}} Z$ can be found satisfying (16) and (17).

To denote the fact that V and W are mutually normalised in rank 1 , we write

$$\mathrm{Mut\ Norm}_1(V,\, W) .$$

8.12. The following properties of the relation $\mathrm{Mut\ Norm}_1(V,\, W)$ are more or less obvious:

(a) The relation $\mathrm{Mut\ Norm}_1(V,\, W)$ is reflexive, symmetric and transitive.

(b) $U = f_1(V;\, X,\, Y) \Rightarrow \mathrm{Mut\ Norm}_1(V,\, U)$.

(c) If $\mathrm{Mut\ Norm}_1(V,\, W)$, then $\partial_1(V) = \partial_1(W)$.

8.13. We denote $\overline{M}_0 = R_0$ and define the set $\overline{M}_1$ as the set of all words $X \in M_1$ such that the implication

$$\mathrm{Mut\ Norm}_1(U,\, V) \Rightarrow \mathrm{Rel}(U,\, V)$$

holds for arbitrary kernels $U, V \in \mathrm{Ker}(1, X)$.

For an arbitrary word $X \in R_1$ a word $Y \in \overline{M}_1$ can be found such that $X \mathrel{\substack{\perp \\ \sim}} Y$

(see IV.3.12).

9. Coupling in Rank 1

9.1. For arbitrary words $X, Y \in R_1$ we define a binary operation $[X, Y]_1 = Z$, called *coupling of rank* 1, as follows:

$$[X, Y]_1 = PQ \Longleftrightarrow X \underset{\sim}{\downarrow} PT , \quad Y \underset{\sim}{\downarrow} T^{-1}Q \quad \text{and} \quad PQ \in R_1 \quad \text{for some} \quad T .$$

We prove that this operation is defined for an arbitrary pair of words $X, Y \in R_1$ and it is associative.

9.2. Suppose $n \geq 779$. If $X, Y \in L_1$, then there is a word $Z \in R_1$ such that $X \underset{\sim}{\downarrow} Z$, $Z \; \overline{\underline{\circ}} \; PT$, $Y \; \overline{\underline{\circ}} \; T^{-1}Q$ and $PQ \in R_1$ for some word T.

Proof. Clearly we can find a word T such that $X \; \overline{\underline{\circ}} \; PT$, $Y \; \overline{\underline{\circ}} \; T^{-1}Q$ and $PQ \in R_0$. We can write $X \underset{\sim}{\downarrow} PT$. By our assumption for an arbitrary kernel $V \in \mathrm{Ker}(1, PT)$ we have $l_1(V) \leq \frac{n+1}{2} + 21$. Let us look for all the words $PT \in R_1$ such that

$$X \underset{\sim}{\downarrow} PT , \quad Y \; \overline{\underline{\circ}} \; T^{-1}Q , \quad PQ \in R_0 \quad \text{for some} \quad T \quad \text{and} \quad l_1(V) \leq \frac{n+1}{2} + 21$$

for an arbitrary kernel $W \in \mathrm{Ker}(1, PT)$, that has nonempty intersection V with $*P*T$.

Suppose that we have chosen among these words PT a word with T of maximal possible length (notice that by (18), T^{-1} is a start of Y).

If $PQ \in R_1$, then we have found the word $Z \; \overline{\underline{\circ}} \; PT$ that we need.

Suppose that $PQ \notin R_1$. Then by 8.1 some elementary word E occurs in PQ, such that $l_1(E) \geq n - 175$. By 7.10 (d) and our assumption no elementary $\left(\frac{n+1}{2} + 37\right)$-power of rank 1 occurs in the words P and Q. Then for some words E_1 and E_2 we have

$$E \; \overline{\underline{\circ}} \; E_1 E_2 , \quad P \; \overline{\underline{\circ}} \; P_1 E_1 , \quad \text{and} \quad Q \; \overline{\underline{\circ}} \; E_2 Q_1 .$$

By 6.3 we have

$$l_1(E_1) \geq n - 175 - l_1(E_2) - 1 \geq \frac{n-1}{2} - 212 .$$

Then by the assumption $n \geq 779$ we have $l_1(E_1) \geq 177 > q + 34$. By 7.8 (c),

$P_1 * E_1 * T$ is an active occurrence of rank 1 . Let $E \; \overline{\underline{o}} \; A^t A_1$, where $A \; \overline{\underline{o}} \; A_1 A_2$ is

an elementary period of rank 1 . Obviously we may assume that $P_1 * E_1 * T$ cannot be

be continued to the left. By 6.5 it cannot be continued also to the right, because

otherwise T and E_2 must have the same first letter. But that is not possible

because $T^{-1} E_2 Q_1$ is uncancellable. Therefore $P_1 * A^t A_1 * T$ is a maximal occurrence.

Then we have the following real reversal of it

$$P_1 A^t A_1 T \to P_1 \left(A^{-1}\right)^{n-t-1} A_2^{-1} T ,$$

where $P_1 * \left(A^{-1}\right)^{n-t-1} A_2^{-1} * T$ is a maximal occurrence of the elementary word

$\left(A^{-1}\right)^{n-t-1} A_2^{-1}$ of rank 1 . Let us denote

$$Z = P_1 \left(A^{-1}\right)^{n-t-1} A_2^{-1} T \quad \text{and} \quad U = P_1 * \left(A^{-1}\right)^{n-t-1} A_2^{-1} * T .$$

By 8.8 we have $Z \in R_1$ and $X \not{\perp} Z$. Let

$$Z \; \overline{\underline{o}} \; P_2 T_1 , \quad Y \; \overline{\underline{o}} \; T_1^{-1} Q_2 \quad \text{and} \quad P_2 Q_2 \in R_0 .$$

Clearly T is an end of T_1 . Then P_2 is a start of $P_1 \left(A^{-1}\right)^{n-t-1} A_2^{-1}$ and Q_2 is

an end of $E_2 Q_1$. One of the two words E_2 and $A_2 A^{n-t-1}$ is a start of the other,

because they both are periodic words with the same left period $A_2 A_1$.

If $\partial\left[A_2 A^{n-t-1}\right] \leq \partial\left(E_2\right)$, then P_2 is a start of P_1 .

If $\partial\left[A_2 A^{n-t-1}\right] > \partial\left(E_2\right)$, then for some E_0 we have $\left(A^{-1}\right)^{n-t-1} A_2^{-1} \; \overline{\underline{o}} \; E_0 E_2^{-1}$ and

$P_2 \; \overline{\underline{o}} \; P_1 E_0$. It follows from 7.4 and 6.3 that

$$l_1\left(E_0\right) \leq l_1\left(E_0 E_2^{-1}\right) - l_1\left(E_2\right) + 1 \leq n - l_1\left(E_1\right) + 1 - l_1\left(E_2\right) + 1 \leq n - l_1\left(E_1 E_2\right) + 2 \leq 177 .$$

In both cases we have $\partial\left(T_1\right) > \partial(T)$ and the kernel $W \in \text{Ker}(1, Z)$ that is compatible

with U has an intersection V with $* P_2 * T_1$ such that $l_1(V) \leq 177$. For the other

kernels $W \in \text{Ker}(1, Z)$ the same property follows by 8.8 from the corresponding

property of the occurrence $* P * T$. Thus we have a contradiction to our assumption of

maximality of the length of T . It proves that $PQ \in R_1$ and therefore

$$PQ = [X, Y]_1 .$$

9.3. If $n \geq 779$, then for arbitrary words $X, Y \in R_1$ there are words

$X_1 T \in R_1$ and $T^{-1} Y_1 \in R_1$ such that $X \downarrow X_1 T$, $Y \downarrow T^{-1} Y_1$ and $X_1 Y_1 \in R_1$, that is $X_1 Y_1 = [X, Y]_1$.

This follows from 9.2, 8.9 and 8.1.

We have proved that the coupling operation of rank 1 is defined for arbitrary $X, Y \in R_1$ if $n \geq 779$. One can prove the same also for odd $n \geq 665$, but it needs a slightly more complicated consideration (see V.1.4).

9.4. The result of the coupling operation $[X, Y]_1 = Z$ is not unique in general for given words $X, Y \in R_1$. Of course it depends on the choice of the words PT and $T^{-1} Q$ (see 9.1). But one can prove that the result of this operation is uniquely defined up to equivalence in rank 1 (see V.1.7) and it is an associative operation (see V.1.8).

9.5. Let B_1 be the set of all equivalence classes of R_1 under the relation $\downarrow$.

If $X \in R_1$, then we denote the element of B_1 containing X by $\{X\}_1$. We define the binary operation o_1 on B_1 as follows:

$$\{X\}_1 o_1 \{Y\}_1 = \{[X, Y]_1\}_1 .$$

It follows easily from 9.3 and 9.4 that the set B_1 is a group under o_1 . The class $\{1\}_1$ of the empty word 1 is the identity element of this group, and for arbitrary $X \in R_1$ the element $\{X^{-1}\}_1$ is a inverse for $\{X\}_1$. We denote this group by $\Gamma(m, n, 1)$.

9.6. The group $\Gamma(m, n, 1)$ is isomorphic to the group $B(m, n, 1)$, presented by the generators

(18) $a_1, a_2, \ldots, a_m$

and the defining relations

(19) $A^n = 1$

for all elementary periods A of rank 1 .

Proof. For an arbitrary elementary period A of rank 1 , the relation (19) holds in the group $\Gamma(m, n, 1)$, because for such A clearly we have $A^{3q} \downarrow A^{-n+3q}$ and hence $\{A^{3q}\}_1 = \{A^{-n+3q}\}_1$. On the other hand for an arbitrary real reversal (11) clearly we have $PA^t A_1 Q = P(A^{-1})^{n-t-1} A_2^{-1} Q$ in $B(m, n, 1)$.

Thus for arbitrary words X, $Y \in R_1$ it follows from $X \underset{\lambda}{\sim} Y$ that $X = Y$ in $B(m, n, 1)$. Therefore if $[E, D]_1 = C$, then

$$ED = C \text{ in } B(m, n, 1) .$$

9.7. The word problem for the group $B(m, n, 1)$ is solvable.

For an arbitrary word X in the alphabet (1) we can easily find a word Y such that $Y \in R_1$ and $X = Y$ in $B(m, n, 1)$. For this we substitute A^{-n+t} for an arbitrary occurrence of A^t in X , where $t > \frac{n+3}{2}$, and cancel the result of the substitution. We obtain a shorter word X_1 that is equal to X in $B(m, n, 1)$. After a finite number of these transformations we obtain a word $Y \in R_0$ which does not contain elementary $\left((n+5)/2\right)$-powers of rank 1 . By 8.1 and 7.10 (d), $Y \in R_1$. Then by 8.10 we have $X = 1$ in $B(m, n, 1)$ if and only if $Y \underset{\circ}{=} 1$.

9.8. Infiniteness of the group $B(m, n, 1)$ for $m > 1$ follows easily from the existence of an infinite sequence of Aršon in the alphabet (18) (see §3 of Chapter I in Adian (1979)). In fact, by 8.10 such a sequence establishes an infinite sequence of elements of $\Gamma(m, n, 1)$ which are pairwise distinct.

10. Elementary Words of Rank 2

10.1. The occurrence $R*E*S$ in the word $X \in \text{Per}(A)$ is said to be interior relative to the period A if $\partial(R) \geq 8\partial(A)$ and $\partial(S) \geq 8\partial(A)$. We denote the set of all occurrences in X of this sort by $\text{Inn}(X, A)$.

Two occurrences $P*E*Q$ and $R*E*S$ in the same word $X \in \text{Per}(A)$ are said to *correspond in phase* relative to the period A if there is an integer r such that $\partial(R) - \partial(P) = r\partial(A)$. We denote this binary relation by $\text{Corr}_A(P*E*Q, R*E*S)$. For $r = 0$ we have that $P*E*Q \underset{\circ}{=} R*E*S$. If $r > 0$ (or $r < 0$) we shall say that $R*E*S$ is the result of shifting $P*E*Q$ to the right (to the left) by r periods A .

10.2. A minimal period A of rank 1 that is not an elementary period of rank 1 is said to be a *period of rank* 2 , if $A^n \in R_1$. We denote the set of all periodic words with period A of rank 2 by $\text{Per}(2, A)$.

Suppose that A is a period of rank 2 . Then by 6.1 some periodic word E with a minimal period B of rank 1 occurs in A^8 such that $\partial(E) > 8\partial(B)$. Clearly, we have $\partial(B) < \partial(A)$. Consider such a period B of minimal length. This B must be an elementary period of rank 1 . Therefore if a minimal period A of rank 1 is not elementary, then some elementary 9-power $E \underset{\circ}{=} B^t B_1$ of rank 1 occurs in A^8 . By 4.3 for any such subword E we have $\partial(E) < 2\partial(A)$ that is E occurs already in A^3 . Let $A^3 = PB^tB_1Q$. Clearly we may suppose that the occurrence

$$(20) \qquad\qquad A^8 P * B^t B_1 * Q A^{n-11}$$

in the word A^n is maximal. For any integer $i < n - 19$ one can consider the occurrence

$$(21) \qquad\qquad A^{8+i} P * B^t B_1 * Q A^{n-11-i} .$$

Clearly, any occurrence (21) is a maximal occurrence and it is interior relative to A . Any two occurrences (21) for $i = t$ and $i = j$ are not compatible, if $t \neq j$. By 6.8 the common part of two such occurrences has a base shorter than B^2 .

10.3. For any period A of rank 2 there is a kernel $V \in \mathrm{Ker}\left(1, A^n\right)$ of the form (20) where $\partial\left[B^tB_1\right] < 2\partial(A)$ (see I.4.3).

Let A be a period of rank 2 . If the set $\mathrm{Act}\left(1, A^n\right)$ is empty, then by 7.9 any occurrence (21) is a kernel of rank 1 . If not, then some elementary q-power of rank 1 occurs in A^n , by 7.10 (b). In this case we can suppose that $l_1\left[B^tB_1\right] \geq q$ in the maximal occurrence (20). Then by 7.10 (c) some kernel V of rank 1 is compatible with (20) and it occurs in it.

10.4. One can prove for any $X \in \mathrm{Per}(2, A)$ that if $U, V \in \mathrm{Inn}(X, A)$, $\mathrm{Corr}(U, V)$ and $U \in \mathrm{Reg}(1, X)$, then $V \in \mathrm{Reg}(1, X)$ and $\mathrm{Mut\ Norm}_1(U, V)$ (see II.1.9). Therefore by 8.12 (c) we have in this case $\partial_1(U) = \partial_1(V)$. For instance if $U \in \mathrm{Ker}(1, X)$, then $V \in \mathrm{Ker}(1, X)$.

That is the reason for the following definition.

10.5. Let $X \in \mathrm{Per}(2, A)$, let $V \in \mathrm{Ker}(1, X)$, $V, U \in \mathrm{Inn}(X, A)$ and let U be the result of shifting V to the right by one period A . If W is the union of U and V , then we call $\partial_1(W) - 1$ the density of X in rank 1 relative to the period A and denote it by

$$\rho_{1,A}(X) = \partial_1(W) - 1 .$$

It follows easily from 10.4 that $\rho_{1,A}(X)$ does not depend on the choice of the kernels U and V . The density $\rho_{1,A}(X)$ indicates the number of kernels of rank 1 which lie in one period A of the word X . It is some kind of length of A in rank 1 . We may also denote it by $\rho_1(A) = \rho_{1,A}(X)$, because it does not depend on the choice of the word $X \in \mathrm{Per}(2, A)$ (see II.1.18). If B is a cyclic shift of A , then $\rho_1(B) = \rho_1(A)$.

EXAMPLES. Let $X = A^n$. For $A = a_1^{10} a_2^{2q} a_1^{10} a_2^{-n+2q}$ we have $\rho_1(A) = \rho_{1,A}(X) = 4$.

For $A = \left[a_1^8 (a_2 a_3)^{23} \right]^q a_2$ we have $\rho_1(A) = \rho_{1,A}(X) = q$, and for $A = a_1^8 (a_2 a_1)^{23}$ we have $\rho_1(A) = 2$.

10.6. We shall call an occurrence W in a word $X \in \mathrm{Per}(2, A)$ a *normal generating occurrence of rank* 2 in X if $W \in \mathrm{Reg}(1, X) \cap \mathrm{Inn}(A, X)$.

If W is a normal generating occurrence of rank 2 in $X \in \mathrm{Per}(2, A)$ and $X \downarrow Y$, then the occurrence

$$f_1(W; X, Y)$$

will also be called a normal generating occurrence of rank 2 in the word Y .

If $X \in \mathrm{Per}(2, A)$, we denote by $\mathrm{Int}(X, 2, A)$ the set of all words Y such that $X \downarrow Y$.

10.7. We call a period A of rank 2 a *minimal period of rank* 2 if there exists no word $Y \downarrow A^n$ with the following property: for some normal generating occurrence W in the word A^n we have $\partial_1(W) > 2\rho_1(A)$ and

$$f_1(W; A^n, Y) = R * C^t C_1 * S ,$$ where $C \; \overline{\circ} \; C_1 C_2$ is a period of rank 2 with the density $\rho_1(C) < \rho_1(A)$.

EXAMPLES. $a_1^8 (a_2 a_1)^{23}$ is a minimal period of rank 2 because the word $\left[a_1^8 (a_2 a_1)^{23} \right]^n$ has no active kernels of rank 1 and hence

$$\left[a_1^8 (a_2 a_1)^{23} \right]^n \downarrow Y \Rightarrow Y \; \overline{\circ} \; \left[a_1^8 (a_2 a_1)^{23} \right]^n .$$

By the same reason $\left[a_1^8 (a_2 a_3)^{23} \right]^q a_2$ also is a minimal period of rank 2 .

The period $a_1^{10} a_2^{2q} a_1^{10} a_2^{-n+2q}$ is not a minimal period of rank 2 , because we have

$$X \; \overline{\circ} \; a_1^{10} a_2^{2q} a_1^{10} a_2^{-n+2q} \downarrow \left[a_1^{10} a_2^{2q} \right]^{2n}$$

(see example in 8.4) and for the generating occurrence

$$W = \left[a_1^{10} a_2^{2q} a_1^{10} a_2^{-n+2q} \right]^8 * \left[a_1^{10} a_2^{2q} a_1^{10} a_2^{-n+2q} \right]^9 * \left[a_1^{10} a_2^{2q} a_1^{10} a_2^{-n+2q} \right]^{n-17}$$

we have

$$f_1\left(W;\ X,\ \left(a_1^{10}a_2^{2q}\right)^{2n}\right) = \left(a_1^{10}a_2^{2q}\right)^{16} * \left(a_1^{10}a_2^{2q}\right)^{18} * \left(a_1^{10}a_2^{2q}\right)^{n-34}\ ,$$

where obviously $a_1^{10}a_2^{2q}$ is a period of rank 2 and

$$\rho_1\left(a_1^{10}a_2^{2q}\right) = 2 < \rho_1\left(a_1^{10}a_2^{2q}a_1^{10}a_1^{-n+2q}\right)\ .$$

10.8. We call a minimal period A of rank 2 an *elementary period of rank* 2 if no word $Y \downarrow A^n$ exists with the following property: For some normal generating occurrence in the word A^n we have

$$f_1\left(W;\ A^n,\ Y\right) = R*C^tC_1*S\ ,$$

where $C \overset{\circ}{=} C_1C_2$ is a period of rank 2 with the density $\rho_1(C) < \rho_1(A)$ and $\partial\left(R*C^tC_1*S\right) > 8\rho_1(C)\ .$

EXAMPLES. The words $a_1^8(a_2a_1)^{23}$ and $a_1^{10}a_2^{2q}$ are elementary periods of rank 2 . The minimal period of rank 2 , $\left(a_1^8(a_2a_3)^{23}\right)^q a_2$ is not an elementary period of rank 2 because denoting $A = \left(a_1^8(a_2a_3)^{23}\right)^q a_2$, we have a normal generating occurrence in the word A^n :

$$W = A^8 a_1^8 * \left((a_2a_3)^{23}a_1^8\right)^{q-1}(a_2a_3)^{23} * a_2 A^{n-9}\ ,$$

where $C = (a_2a_3)^{23}a_1^8$ is an elementary period of rank 2 with the density $\rho_1(C) = 1$ and $\partial_1(W) = q > 8\rho_1(C)\ .$

10.9. If A is an elementary period of rank 2 and $P*E*Q$ is a normal generating occurrence in some word $Y \downarrow X$, where $X \in \text{Per}(2, A)$, then E is called a *normal elementary word of rank* 2 , generated by the occurrence $P*E*Q$. Any subword D of a normal elementary word of rank 2 , $E \overset{\circ}{=} E_1DE_2$, generated by $P*E*Q$, is called an *elementary word of rank* 2 , generated by the occurrence PE_1*D*E_2Q in the word $Y \in \text{Int}(X, 2, A)$ (see 10.6).

10.10. For an arbitrary generating occurrence W of rank 2 in $Y \in \text{Int}(X, 2, A)$ we call the number

$$\left|\frac{\partial_1(W)}{\partial_1(A)}\right| = \text{the smallest integer}\ \ r\ \ \text{such that}\ \ \frac{\partial_1(W)}{\partial_1(A)} \leq r\ ,$$

the number of segments of rank 2 of the generating occurrence W and denote it by $l_2(W)$. If A is an elementary period and $E = \text{Bas}(W)$, then we call $l_2(W)$ also the number of segments of the word E and the number of segments of rank 2 of an arbitrary occurrence $R*E*S$ in a word $Z \in R_1$ and we write

$$l_2(R*E*S) = l_2(E) = l_2(W) .$$

It follows from our definition of $l_2(E)$, that for an arbitrary elementary word E_1E_2 we have the inequalities (II.5.2):

$$l_2(E_1) + l_2(E_2) - 1 \leq l_2(E_1E_2) \leq l_2(E_1) + l_2(E_2) + 2 .$$

If $l_2(E) \geq r$, then we say that E is an *elementary r-power of rank* 2 .

In general the number $l_2(E) = l_2(W)$ can depend on the choice of the corresponding period A of rank 2 . But for an elementary 9-power E any two generating occurrences $P*E*Q$ and $R*E*S$ have the same number of segments, that is $l_2(E)$ is independent of the choice of the original elementary period A and the corresponding generating occurrence $W = P*E*Q$ (see II.5.1).

Notice that an elementary word of rank 2 does not need to be a periodic word. But it is homogeneous in some sense, because it is a subword of a word $Y \in \text{Int}(X, 2, A)$ that is equal in rank 1 to the periodic word X with period A .

When one transforms a word $PEQ \in R_1$ with $P*E*Q \in \text{Norm}(2, PEQ, r)$ into a word $Z \underset{\sim}{\downarrow} PEQ$, where $Z \in \overline{M}_1$ (see 8.13), then the base of the image $f_1(P*E*Q; PEQ, Z)$ is a periodic elementary r-power of rank 2 (see II.4.5).

10.11. Let E be an elementary word of rank 2 , generated by a generating occurrence $P*E*Q$ in $Y \in \text{Int}(X, 2, A)$. We call an arbitrary occurrence $R*E*S$ in a word $Z \in R_1$ a *normalised* occurrence of E if

$$\text{Mut Norm}_1(R*E*S, P*E*Q) .$$

For a given word $X \in R_1$ and $r \geq 2$ we denote by $\text{Norm}(2, X, r)$ the set of all normalised occurrences of elementary r-powers of rank 2 in the word X .

10.12. Let an elementary word E of rank 2 be generated by an occurrence $R*E*S$ in $Y \in \text{Int}(X, 2, A)$. We say that a given occurrence $P*E*Q$ in a word $Z \in R_1$ can be continued to the left if there is a non-empty word H such that $P \overset{\text{o}}{=} P_1H$ and HE is an elementary word of rank 2 generated by some generating occurrence R_1*HE*S_1 in some $Y_1 \in \text{Int}(X_1, 2, A)$ with the same period A . In this case we call the occurrence P_1*HE*Q a *continuation* of $P*E*Q$ *to the left.*

Similarly we define continuations to the right, maximal continuations and maximal occurrences of elementary words of rank 2 (see 6.4 and 6.5).

We note by $\text{Max Norm}(2, Z, r)$ the set of all maximal normalized occurrences of elementary r-powers of rank 2 in the given word $Z \in R_1$.

10.13. Two occurrences in the same word $Z \in R_1$, $P{*}E{*}Q$ and $R{*}D{*}S$, of the elementary words E and D of rank 2 will be called *compatible* if their union is a continuation of either of them. In this case we write $\text{Comp}(P{*}E{*}Q, R{*}D{*}S)$.

If $\text{Comp}(P{*}E{*}Q, R{*}D{*}S)$, then E and D can be generated by some generating occurrences in the same word $Y \in \text{Int}(X, 2, A)$.

The relation Comp is reflexive, symmetric and transitive. The mapping ϕ (see 3.3) preserves this relation (see II.5.8).

It follows from the definition of elementary periods of rank 2 (see 10.8) that if two occurrences $V \in \text{Norm}(2, X, 9)$ and $U \in \text{Norm}(2, X, 9)$ are not compatible, then the common part of them contains less than 9 segments of each (see II.5.7).

10.14. The following properties of normalized occurrences of elementary words of rank 2 are most important.

(a) If $P{*}E{*}Q \in \text{Norm}(2, Z, r)$ and $r \geq 9$, then for an arbitrary occurrence $F{*}E{*}H$ in $Z_1 \in R_1$,

$$\text{Mut Norm}_1(P{*}E{*}Q, F{*}E{*}H) \iff F{*}E{*}H \in \text{Norm}\left(2, Z_1, r\right)$$

(see II.5.5).

(b) If $V \in \text{Norm}(2, Z, r)$ and $Z \gtrless Y$, then

$$f_1(V; Z, Y) \in \text{Norm}(2, Y, r)$$

(see II.5.16).

(c) Suppose that $R{*}E{*}S$ is a generating occurrence in $Y \in \text{Int}(X, 2, A)$ of an elementary 9-power E of rank 2 . If $E \overline{\underline{\circ}} HDF$,

$$\partial_1(R{*}E{*}S) = \partial_1(RH{*}DF{*}S) - 4\rho_1(A) = \partial_1(RH{*}D{*}FS) - 8\rho_1(A) ,$$

then for an arbitrary occurrence $P{*}E{*}Q$ in $Z \in R_1$ we have

$$\text{Mut Norm}_1(RH{*}D{*}FS, PH{*}D{*}FQ) .$$

(see II.6.6).

Hence an arbitrary occurrence $P{*}E{*}Q$ in $Z \in R_1$ of an elementary r-power E of rank 2 contains an occurrence $PH{*}D{*}FQ \in \text{Norm}(2, Z, r-8)$ that is compatible with $P{*}E{*}Q$.

11. Basic Concepts for Rank 2

We formulate here a definition of simple reversals of rank 2 that may be used for $n \geq 1003$ (see I.4.18).

11.1. Suppose that $r \geq 9$, $A \overset{\circ}{=} A_1 A_2$ is an elementary period of rank 2 ,

$PA^t A_1 Q \in K_1$ and $P\left(A^{-1}\right)^{n-t-1} A_2^{-1} Q \in K_1$. The transition

$$(22) \qquad\qquad PA^t A_1 Q \to P\left(A^{-1}\right)^{n-t-1} A_2^{-1} Q$$

is said to be a *simple r-reversal* of an occurrence $W \in \mathrm{Norm}\left(2,\ PA^t A_1 Q,\ 9\right)$ if

$\mathrm{Comp}\left(W,\ P*A^t A_1 *Q\right)$ and the occurrences $P*A^t A_1 *Q$ and $P*\left(A^{-1}\right)^{n-t-1} A_2^{-1} *Q$ are

compatible with some normalized occurrences of elementary r-powers of rank 2 .

If $r \geq s \geq 9$, then obviously any simple r-reversal of W is a simple s-reversal of W .

An occurrence $R*E*G \in \mathrm{Reg}\left(1,\ PA^t A_1 Q\right)$ is said to be *stable* in the simple reversal (22) of rank 2 if either $P \overset{\circ}{=} FEP_1$ and

$$\mathrm{Mut\ Norm}_1\left(F*E*P_1 A^t A_1 Q,\ F*E*P_1 \left(A^{-1}\right)^{n-t-1} A_2^{-1} Q\right)\ ,$$

or $Q \overset{\circ}{=} Q_1 EG$ and

$$\mathrm{Mut\ Norm}_1\left(PA^t A_1 Q_1 *E*G,\ P\left(A^{-1}\right)^{n-t-1} A_2^{-1} Q_1 *E*G\right)\ .$$

In such a case the occurrence

$$F*E*P_1 \left(A^{-1}\right)^{n-t-1} A_2^{-1} Q \quad \left(\text{or}\ \ P\left(A^{-1}\right)^{n-t-1} A_2^{-1} Q_1 *E*G\ \right)$$

will be called an *image* of the occurrence $F*E*G$ in (22).

For the simple r-reversal (22) of W the uniquely defined occurrence $U \in \mathrm{Max\ Norm}\left(2,\ P\left(A^{-1}\right)^{n-t-1} A_2^{-1} Q,\ r\right)$ that is compatible with $P*\left(Q^{-1}\right)^{n-t-1} A_2^{-1} *Q$ will be called the *maximal image* of W in the reversal (22).

11.2. Suppose that $X \in R_1$ and $W_1 \in \mathrm{Norm}(2,\ X,\ 9)$. A transition $X \to Y$ is said to be an *r-reversal* of W_1 if words $PA^t A_1 Q,\ P\left(A^{-1}\right)^{n-t-1} A_2^{-1} Q \in K_1$ can be found

such that $X \nmid PA^t A_1 Q$, $Y \nmid P\left(A^{-1}\right)^{n-t-1} A_2^{-1} Q$ and the transition (22) is a simple r-reversal of the occurrence $W = f_1\left(W_1;\ X,\ PA^t A_1 Q\right)$. We shall call an arbitrary 9-reversal of an occurrence $W \in \mathrm{Norm}(2,\ X,\ 9)$ a *reversal of rank 2* .

We shall say that an occurrence $V \in \text{Reg}(1, X)$ is *stable* in the reversal $X \to Y$ of rank 2 if the occurrence $V_1 = f_1\left(V; X, PA^t A_1 Q\right)$ is stable in the simple reversal (22). In this case if V_2 is the image of V_1 in (22), then the occurrence $f_1\left(V_2; P(A^{-1})^{n-t-1} A_2^{-1} Q, Y\right)$ is said to be the *image* of V in the reversal $X \to Y$. This image of a stable occurrence V in $X \to Y$ is uniquely defined. We shall denote it by $f_{X \to Y}(V)$.

If U is the maximal image of $W = f_1\left(W_1; X, PA^t A_2 Q\right)$ in a simple reversal (22) of W, then we call the occurrence

$$W' = f_1\left(U; P(A^{-1})^{n-t-1} A_2^{-1} Q, Y\right)$$

the *maximal image* of W_1 in the reversal $X \to Y$ of W_1.

11.3. It is clear that for an arbitrary 9-reversal of rank 2, $X \to Y$ of an occurrence $W \in \text{Norm}(2, X, 9)$ the transition $Y \to X$ is a reversal of the maximal image $W' \in \text{Norm}(2, Y, 9)$ of W in $X \to Y$. If $V' = f_{X \to Y}(V)$, then $V = f_{Y \to X}(V')$.

11.4. If an occurrence $V \in \text{Reg}(1, X)$ is stable in a given reversal of an occurrence $W \in \text{Norm}(2, X, 9)$, then V does not intersect the maximal normal continuation of W (see III.2.2). But the converse statement in 7.7 is not true for rank 2.

11.5. Suppose that $X \to Y$ is a 9-reversal of an occurrence $U \in \text{Norm}(2, X, 9)$ and the occurrence $W \in \text{Norm}(2, X, 21)$ is not compatible with U.

If $U < W$, then there is an end W_1 of W that is stable in $X \to Y$, and $l_2\left(W_1\right) \geq l_2(W) - 17$.

If $W < U$, then there is a start W_2 of W that is stable in $X \to Y$, and $l_2\left(W_2\right) \geq l_2(W) - 17$ (see III.2.10).

11.6. Notice also the following main properties of the reversals of rank 2.

(a) If $W \in \text{Max Norm}(2, X, 9)$ and $18 \leq r \leq l_2(W) \leq n-r-18$, then there exists a r-reversal of W (see III.1.9).

(b) If $X \to Y$ is a reversal of the occurrence $P*B^t B_1 *Q \in \text{Norm}(2, X, 9)$, where $X \in K_1$, $B \overline{\underline{\circ}} B_1 B_2$ is an elementary period of rank 2, then

$$Y \underset{\downarrow}{\lambda} \left[P, B^{-n+t}, B_1 Q\right]_1$$

(see III.1.10).

(c) If $X \to Y$ is a q-reversal of rank 2 , then X is not equivalent in rank 1 either to Y or to Y^{-1} .

11.7. Some of the q-reversals of rank 2 will be called real reversals of rank 2 . For the precise definition of real reversals of rank 2 we use the notion of cascades of rank 2 (see I.4.22).

The basic properties of the real reversals of rank 2 are similar to the properties (a)-(e) of the real reversals of rank 1 which have been given in 7.8.

We call an occurrence V an *active occurrence of rank* 2 if $V \in \mathrm{Norm}(2, X, 9)$ and if there exists some real reversal of it. We denote by $\mathrm{Act}(2, X)$ the set of all active occurrences of rank 2 in a given word X .

11.8. Let P_2 be the set of all words $X \in R_1$, such that $\mathrm{Norm}(2, X, n-88) = \emptyset$.

We call an occurrence $V \in \mathrm{Norm}(2, X, 9)$ a *kernel of rank* 2 in the word $X \in P_2$ if it is stable in a reversal of any active occurrence of rank 2 that is not compatible with it and if no proper continuation of it has this property. The set of all kernels of rank 2 in a word $X \in P_2$ is denoted by $\mathrm{Ker}(2, X)$. We denote by $\partial_2(X)$ the number of kernels of rank 2 in $X \in P_2$.

The basic properties of real reversals and kernels of rank 2 are similar to the properties (a)-(f) of the same notions for rank 1 which have been given in 7.10.

11.9. Similar to 8.1 we define the sets

$$M_2 \subset L_2 \subset K_2 \subset R_2 \subset P_2 \subset R_1 \ ,$$

according to I.4.26.

We say that two words $X, Y \in P_2$ are *equivalent in rank* 2 if either $X \downarrow Y$ or there is a sequence of real reversals of rank 2 of the form (15).

By $X \overset{2}{\downarrow} Y$ we denote the combined assertion: $X, Y \in R_2$ and X is equivalent to Y in rank 2 . This is a reflexive, symmetric and transitive relation.

Then for any two given words $X, Y \in R_2$, such that $X \overset{2}{\downarrow} Y$, we define a unique mapping $f_2(V; X, Y)$ of the set $\mathrm{Ker}(2, X)$ onto the set $\mathrm{Ker}(2, Y)$.

Obviously any kernel $V \in \mathrm{Ker}(2, X)$ is a regular occurrence of rank 1 . If $X, Y \in R_2$ and $X \downarrow Y$, then we have

$$f_2(V; X, Y) = f_1(V; X, Y) \ .$$

11.10. Similar to 8.10 we prove that if no elementary q-power occurs in a word

$X \in R_2$, then for an arbitrary $Y \in R_2$ we have

$$X \underset{2}{\sim} Y \Longleftrightarrow X \underset{1}{\sim} Y .$$

Any kernel of rank 2 contains a kernel of rank 1 , which is an occurrence of some periodic elementary 9-power of rank 1 . Hence from 11.10 and 8.10 we obtain the following lemma.

11.11. If no periodic 9-power of rank 1 occurs in a word $X \in R_1$, then for an arbitrary $Y \in R_2$ we have

$$X \underset{2}{\sim} Y \Longleftrightarrow X \ \overline{\circ} \ Y .$$

11.12. According to I.4.36, for arbitrary words $X, Y \in R_2$ we define the binary operation $[X, Y]_2 = Z$, called *coupling of rank* 2 . This operation is uniquely defined up to equivalence in rank 2 , and it is an associative operation. Also it has the following property:

If $X, Y \in R_2$, $[X, Y]_1 = Z$ and $Z \in R_2$, then $[X, Y]_2 = Z$.

11.13. Considering the set B_2 of all equivalence classes $\{X\}_2$ of R_2 under the relation $\underset{2}{\sim}$, we define the binary operation

$$\{X\}_2 \circ_2 \{Y\}_2 = \{[X, Y]_2\}_2 .$$

It is associative. The set B_2 is a group under $\circ_2$. We denote this group by $\Gamma(m, n, 2)$.

Similar to 9.6 we can prove the following lemma.

11.14. The group $\Gamma(m, n, 2)$ is isomorphic to the group $B(m, n, 2)$, presented by the generators (16) and the defining relations (17) for all elementary periods A of rank 1 or 2 .

Infiniteness of the group $\Gamma(m, n, 2)$ follows from 11.11. The solvability of the word problem for $B(m, n, 2)$ follows from the effectiveness of all our notions in rank 2 .

Obviously the group $B(m, n, 2)$ is a factor group of $B(m, n, 1)$.

11.15. Notice that the group $B(m, n, 2)$ does not depend on the value of the parameter q that we have used in the definition of real reversals. Remember that by 7.8 (a) and (c) any real reversal is a q-reversal and any $(q+34)$-reversal is a real reversal. If one considers for the same alphabet (1) and the same large parameter n $2q$-reversals or $3q$-reversals instead of q-reversals, then finally one obtains in rank 2 the same group $\Gamma(m, n, 2)$ that is isomorphic to $B(m, n, 2)$. This indicates that we have a lot of freedom in the choice of the class of real reversals of rank 2 .

12. The Induction Step from α to $\alpha + 1$

Now the reader can see what we need to do for the induction step from α to $\alpha + 1$.

12.1. Suppose that all our notions as defined in §4 of Chapter I in Adian (1979) have all the necessary properties in rank α . These properties have been proved for ranks 0 and 1 .

We look for a minimal period A of rank α that is not an elementary period of rank α . If $A^n \in R_\alpha$, then we see, that A^n has many kernels of rank α , and they are distributed in A^n homogeneously in all the periods A .

Then we consider the generating occurrences of rank $\alpha + 1$ in any word $Y \wr A^n$. We first choose the minimal periods of rank $\alpha + 1$ among all periods of rank $\alpha + 1$, then we choose the elementary periods of rank $\alpha + 1$ among all the minimal periods of rank $\alpha + 1$.

The elementary words of rank $\alpha + 1$, generated by generating occurrences in words $Y \in \mathrm{Int}(X, \alpha, A)$, where A is an elementary period of rank $\alpha + 1$, have many important properties which were considered in Chapter II of Adian (1979).

On the basis of these properties we introduce the notion of an r-reversal of rank $\alpha + 1$ of an occurrence $V \in \mathrm{Norm}(\alpha, X, 9)$ for $r \geq 9$ and $X \in R_\alpha$. Some of the q-reversals will be called real reversals of rank $\alpha + 1$ or, in other words, some occurrences $V \in \mathrm{Norm}(\alpha, X, 9)$ will be called active occurrences of rank $\alpha + 1$. Having established the set $\mathrm{Act}(\alpha+1, X)$, we can define the basic notions of a kernel of rank $\alpha + 1$, the set $R_{\alpha+1}$ and the relation $X \underset{\sim}{\alpha+1} Y$. Then as in I.4.36 we introduce the operation of coupling in rank $\alpha + 1$,

$$[X, Y]_{\alpha+1} = Z ,$$

which gives us a group operation $\circ_{\alpha+1}$ on the set $B_{\alpha+1}$ of all equivalence classes $\{X\}_{\alpha+1}$ of $R_{\alpha+1}$ under the relation $\underset{\sim}{\alpha+1}$. We denote this group by $\Gamma(m, n, \alpha+1)$. It is isomorphic to the group $B(m, n, \alpha+1)$, presented by the generators (16) and the defining relations (17) for all elementary periods A of rank $i \leq \alpha+1$.

The group $\Gamma(m, n, \alpha+1)$ is infinite and has solvable word problem.

12.2. If we have proved for rank $\alpha + 1$ all the properties of our notions which have been used for rank $\leq \alpha$ in the induction step, then we can say that we have completely finished the induction step (see V.2.5, p. 236). Now, as the result of our induction, all our assertions have been proved for arbitrary values of the parameter α .

Thus we can consider the set

$$A = \bigcap_{i=1}^{\infty} R_i$$

and the relation on A ,

$$X \sim Y \iff X \underset{i}{\sim} Y \text{ for some } i \ge 0 .$$

The binary operation

$$\{X\} \circ \{Y\} = \{[X, Y]_i\} , \text{ where } i = \partial(X) + \partial(Y)$$

gives us a group operation "∘" on the set B of all equivalence classes $\{X\}$ of A under the relation $\sim$. We denote this group by $\Gamma(m, n)$. Clearly, it does not depend on the parameter α . One can easily prove that $\Gamma(m, n)$ is isomorphic to the group $B'(m, n)$, defined by the generators (16) and the defining relations (17) for all elementary periods A of rank $i \ge 1$.

The group $\Gamma(m, n)$ is infinite and has solvable word problem. But we must also prove that the identical relation $x^n = 1$ holds in the group $\Gamma(m, n)$, that is $B'(m, n)$ is isomorphic to the free Burnside group $B(m, n)$ which has been introduced in §2. To prove this we must prove that for an arbitrary word $B \in A$ the relation $B^n = 1$ follows from the set of relations (17) for all elementary periods A of rank $i \ge 1$. This follows easily from the main lemma:

12.3. *For an arbitrary word $B \in A$ one can find an elementary period C of rank $i \le \partial(B)$ and a word T such that the relation*

$$B = TC^r T^{-1}$$

holds in the group $B'(m, n)$ for some $r \ge 0$ (see VI.2.5).

Hence we can say that our classification of periodic words by the classes of the elementary words of finite ranks $\alpha \ge 1$ is complete.

12.4. Now we discuss the following two questions which were raised in the Bielefeld lectures in 1977:

(a) Where do we need the assumption that the exponent n is odd?

(b) Why do we need the concept of cascade of rank α ?

To prove the Lemma III.1.12 even for rank 1 we must assume that n is odd. For example, suppose that in the reversal (11) of rank 1 the words P, Q and A_1 are empty. Then the reversal (11) has the form

$$A^t \to A^{-n+t} .$$

In this case III.1.12 states that A^{n-t} is not equal in rank 0 to A^t . But if $n = 2t$, then we have $A^t \underset{\circ}{=} A^{n-t}$. The whole theory essentially depends on III.1.12.

For instance, the proofs of IV.2.36, II.5.21, III.1.5 and IV.2.12 directly use III.1.12.

We use the notion of cascade of rank α to control the local character of dependence of the notions of normalised occurrences of active occurrence and of kernel of rank α on the neighbourhood of a given occurrence (for example see II.6.14, III.3.20, IV.1.15 and IV.3.24). This is connected with the following questions.

Consider the words

$$X \; \overline{\circ} \; PAEBQ \quad \text{and} \quad Y \; \overline{\circ} \; RAEBS \; ,$$

where X, $Y \in R_\alpha$ and E is an elementary 9-power of rank α . Under what condition can one state the following six implications:

$$PA*E*BQ \in \text{Norm}(\alpha, X, 9) \iff RA*E*BS \in \text{Norm}(\alpha, Y, 9) \; ,$$

$$PA*E*BQ \in \text{Act}(\alpha, X) \iff RA*E*BS \in \text{Act}(\alpha, Y) \; ,$$

$$PA*E*BQ \in \text{Ker}(\alpha, X) \iff RA*E*BS \in \text{Ker}(\alpha, Y) \; ?$$

Using the notion of cascade in the definition of real reversal, we can prove that these implications hold, if the words A and B contain some elementary 9-powers of rank $\alpha + 1$. In general, a cascade of rank α cannot cover an elementary 9-power of rank $\alpha + 1$. Therefore the ends P, Q, R and S of given words X and Y cannot influence the properties of an occurrence E lying between A and B .

13. Applications in Group Theory

In this paragraph we shall list the results which have been proved by the method that is described in the preceding paragraphs. Most of these results are contained in Adian (1979).

13.1. The following properties of $B(m, n)$ hold true when $m \geq 2$ and n is odd, $n \geq 665$.

(1) $B(m, n)$ is infinite (see VI.1.5).

(2) $B(m, n)$ cannot be presented by a finite set of defining relations (see VI.2.13).

(3) $B(m, n)$ has exponential growth (see VI.2.15).

For a finitely generated group G let us denote by $\overline{\gamma}(s)$ the number of different elements of G which can be presented in the form of a product of s components that are the generators of G . For the free group F_2 on two generators, we have the obvious equation $\overline{\gamma}(s) = 4.3^{s-1}$. One can prove that, for an arbitrary natural number k , there exists an odd number N such that the function $\overline{\gamma}(s)$ for the group $B(2, N)$ satisfies the inequality

$$\overline{\gamma}(s) \geq 4\left(3 - \frac{1}{k}\right)^8$$

(see VI.2.16).

(4) The word problem and the conjugacy problem are solvable for $B(m, n)$
    (see VI.2.10 and VI.3.5).

(5) The centralizer of any element $x \neq 1$ of $B(m, n)$ is cyclic (see
    VI.3.2).

(6) Every commutative subgroup of $B(m, n)$ is cyclic (see VI.3.3).

(7) The centre of $B(m, n)$ is trivial (see VI.3.4).

(8) Every finite subgroup of $B(m, n)$ is cyclic (see VII.1.8).

(9) The group $B(\infty, n)$ with countably many generators can be embedded in
    $B(2, n)$.

The result (9) has been proved in Širvanian (1976).

It follows from (9) that the group $B(m, n)$ does not satisfy the minimal and the
maximal conditions for subgroups. It has been proved in Adian (1979) that the minimum
and the maximum conditions for normal subgroups also do not hold in $B(m, n)$ if
$n = kt$, where $k > 1$, $t \geq 665$.

13.2.   The next result is connected with the finite basis problem for varieties
of groups. The question here was: is every system of group identities equivalent to
a finite system? the negative answer to this question was first obtained by
Ol'šanskiĭ (1970) in an implicit form. His proof is based on some cardinality
considerations. Shortly afterwards the author (Adian (1970)) constructed an explicit
infinite system of group identities, none of which follows from the others. Vaughan-
Lee (1970) gave another such system. We discuss the results of Adian (1970).

13.3.   For every odd $n \geq 1003$, the set of all identities of the form

(23) $\left(x^{rm}y^{rm}x^{-rm}y^{-rm}\right)^n = 1$,

where the parameter r runs over all prime numbers, is an independent system of group
identities, that is, none of them follows from the others.

To prove this theorem, we start with an arbitrary odd $n \geq 1003$ and an arbitrary
set Π of prime numbers, and construct a group $\Gamma(n, \Pi)$ on 2 generators such that
the relation (23) is satisfied if and only if $r \in \Pi$ (see VII.2.15). To this end,
we define by simultaneous induction on α the new concept of a period *distinguished* in
rank $\alpha - 1$ and analogues of all the concepts that were defined in §4 of Chapter I in
Adian (1979). Then we consider only reversals of occurrences of elementary words of
rank α that have periods distinguished in rank $\alpha - 1$. As a result of this
consideration we obtain a new classification of periodic words in the given alphabet,
that is complete relative to the identical relations (23) for all primes $r \in \Pi$.

It follows from 13.3 that there exists a group given by 2 generators and a recursively enumerable set of identical relations with unsolvable word problem.

Recently, Kleĭman (1979) has constructed a group with unsolvable word problem that is presented by finite set of identical relations.

13.4. As is well-known, the additive group of rational numbers is torsion-free, and any two cyclic subgroups have non-trivial intersection. The question which naturally arises out of this concerning the existence of non-abelian groups with the same properties remained open for a long time. We have used a modification of our method to prove that such a group exists.

We denote by $A(m, n)$ the group given by the generators

$$a_1, a_2, \ldots, a_m, d$$

and the defining relations

$$a_i d = d a_i \quad (i = 1, 2, \ldots, m) ,$$

$$A^n = d \quad \left(A^n \in \bigcup_{i=1}^{\infty} E_i\right) ,$$

where E_i is a class of elementary words of the form A^n such that for every elementary 9-power D of rank i there is one and only one word $A^n \in E_i$ that is related either with D or with D^{-1}.

Clearly $B(m, n)$ is the factor group of $A(m, n)$ by the subgroup generated by d. By 13.1 (7) this subgroup is the centre of $A(m, n)$.

We prove that the group $A(m, n)$ is torsion free and every two non-identity subgroups of $A(m, n)$ have non-trivial intersection.

Notice that we have used the group $A(m, n)$ also to prove that every finite subgroup of $B(m, n)$ is cyclic.

To study the group $A(m, n)$ we use the same classification of periodic words and concepts of generalised real reversal of rank α and generalised equivalence of rank α (see VII.1.4).

One can use the groups $A(m, n)$ to construct some interesting examples, giving the answers for some old questions.

It is known that if a Sylow subgroup of a locally finite group is central, then it splits off as a direct factor. The question naturally arises: is this true for all periodic groups (see VII.1.9 and the *Kourovka Notebook* (1973), Problem 2.13).

Consider the group $\overline{A}(m, n)$ obtained from $A(m, n)$ by adding the extra relation

$d^2 = 1$. Clearly $\overline{A}(m, n)$ has exponent $2n$, the subgroup $\langle d \rangle$ lies in the centre of $\overline{A}(m, n)$, and $\langle d \rangle$ is a Sylow 2-subgroup $\overline{A}_2(m, n)$, since n is odd. As it has been shown in VII.1.9, the subgroup $\langle d \rangle$ does not split off as a direct factor, because each of the elementary words of rank 1 , a_1^n, a_2^n and $\left[a_1 a_2^{-1} \right]^n$ is equal to d or d^{-1} in $\overline{A}(m, n)$.

The next application of $A(m, n)$ gives us a solution of Markov's problem about the existence of a countable group which admits only the discrete topology (see Markov (1946), p. 25).

Consider the group $A'(m, n)$ obtained from $A(m, n)$ by adding the extra relation $d^n = 1$. Clearly $A'(m, n)$ has exponent n^2 , the subgroup $\langle d \rangle$ is the centre of $A'(m, n)$, and d has an order n . It follows easily from VI.2.5 and VII.1.2 that for every noncentral element x of $A(m, n)$ we have some equality $x^n = d^{r+kn}$, where r, k are integers and $0 < |r| < n$. Therefore every noncentral element x of $A'(m, n)$ is a solution of some equation from the following finite set of equations:

$$x^n = d^r \quad (r = 1, 2, \ldots, n-1) \, ,$$

that is only the elements of the centre of $A'(m, n)$ do not satisfy these equations. Then the unit element of $A'(m, n)$ is isolated in any topology on $A'(m, n)$, because the set of all solutions of each equation is closed in any topology of any group. Consequently the group $A'(m, n)$ admits only the discrete topology.

The author is grateful to A.Ju. Ol'šanskiĭ for pointing out this application of $A(m, n)$.

13.5. The construction of the periodic product of a given odd exponent $n \geq 665$, for groups which do not contain involutions, was defined in Adian (1976).

Let

(24) $\{ G_i \mid i \in I \}$

be a family of groups, which do not contain involutions, and

$$G = \prod_{i \in I}{}^{*} G_i$$

be the free product of the groups (24).

Suppose that the groups (24) are presented by generators and defining relations. Suppose that all nontrivial elements of G_i belong to the set of generators. The set of all generators of G is the union of the sets of generators of all groups G_i for $i \in I$. The non-trivial elements of G have presentations of the form

(25) $$b_1 b_2 b_3 \ \cdots \ b_k$$

where an arbitrary b_j is a generator of some G_{i_j} and any two neighbouring letters b_j and b_s are from different components. We denote by R_0 the set of all words of the form (25). So we consider all our concepts relative to this set R_0 of uncancellable words, excluding all the periods which belong to some component G_i of the group G . Clearly this condition changes the meaning of all our concepts. Formally we must change only the definition of coupling of rank α . As a result of of our consideration we obtain some classification of periodic words of the form A^n and a notion of an elementary period of rank α .

We shall call the factor group of the free product G by the normal subgroup generated in G by all words A^n with the elementary periods A of finite rank α the *periodic product of exponent* n (n odd, $n \geq 665$) of the given family of groups (24). We denote the periodic product F of the groups (24) by

(26) $$F = \prod_{i \in I} {}^n G_i$$

or by

$$F = G_1 \oplus G_2 \oplus \ \cdots \ \oplus G_r \ ,$$

when the set I is finite. The operation of periodic product of groups has the following interesting properties:

(a) In the group F there are subgroups $\overline{G}_i$ isomorphic to G_i ($i \in I$) , such that all the groups $\overline{G}_i$ generate F and pairwise intersect trivially.

(b) This operation is commutative and associative.

(c) It satisfies Mal'cev's postulate, that is, it has a certain hereditary property for subgroups.

(d) For any element x which is not conjugate to any element of any component G_i , the relation $x^n = 1$ holds in F . From this it follows that when all the groups (24) are periodic groups of exponent n , so is F .

(e) If the family (24) contains at least two non-trivial groups (notice that they do not contain involutions), then F is infinite.

(f) A criterion for the simplicity of periodic products of groups was given in Adian (1978).

In order for the periodic product F of exponent $n \geq 665$ of the system (24) to be a simple group, it is necessary and sufficient that every group G_i coincides with the subgroup generated by the nth powers.

As a consequence of this criterion, we list the following corollaries:

(1) *If each group G_i of (24) is generated by elements of orders relatively prime to n, then F is a simple group.*

(2) *In the variety of all periodic groups of given odd exponent of the form nk, where $n \geq 665$, $k > 1$, and n and k relatively prime, there are infinitely many different finitely generated non-abelian simple groups.*

(3) *There is a continuum of distinct countable periodic simple groups.*

(4) *We call the set of orders of the elements of a periodic group G the spectrum of G. For each set M of odd primes containing at least one prime $p > 665$, it is possible to build a countable periodic simple group H whose spectrum coincides with M. If M is finite, then H has bounded exponent.*

In conclusion we would like to mention that recently Ol'šanskiĭ (1979) has proved the existence of a group, generated by 2 generators such that any proper subgroup is infinite cyclic. He also proved the existence of a two generator group of infinite order such that any proper subgroup is cyclic of finite order.

In the proof of these results he uses small cancellation theory and a similar classification of periodic words based on a simultaneous induction.

References

С.И. Адян [S.I. Adjan] (1970), "Бесконечные неприводимые системы групповых тождеств" [Infinite irreducible systems of group identities], *Izv. Akad. Nauk SSSR Ser. Mat.* **34**, 715-734; *Math. USSR-Izv.* **4**, 721-739 (1971). MR44#4078; Zbl.221.20047; RZ [1971], 2A181.

С.И. Адян [S.I. Adjan] (1976), "Периодические произведения групп" [Periodical products of groups], *Trudy Mat. Inst. Steklov.* **142**, 3-21. RZ [1977], 3A184.

С.И. Адян [S.I. Adian] (1978), "О простоте периодических произведений групп", [On simple periodic products of groups], *Dokl. Akad. Nauk SSSR* **241**, 745-748. RZ [1978], 12A332.

S.I. Adian (1979), *The Burnside problem and identities in groups* (translated by J. Lennox, J. Wiegold. Ergebnisse der Mathematik und ihrer Grenzgebiete, **95**. Springer-Verlag, Berlin, Heidelberg, New York).

W. Burnside (1902), "On an unsettled question in the theory of discontinuous groups", *Quart. J. Pure Appl. Math.* **33**, 230-238. FdM33,149.

Ю.Г. Клейман [Ju.G. Kleĭman] (1979), "Тождества и некоторые алгоритмические проблемы в группах" [Identities and some algorithmic problems in groups], *Dokl. Akad. Nauk SSSR* **244**, 814.818; *Soviet Math. Dokl.* **20**, 115-119.

Коуровская Тетрадь. Нерешенные задачи Теории групп [*Kourovka Notebook. Unsolved problems in the Theory of Groups*] (4th Augmented edition) (Akad. Nauk SSSR Sibirsk. Otdel. Inst. Mat., Novosibirsk, 1973). MR51#10443.

А.А. Марков [A.A. Markov] (1946), "О безусловно замкнутых множествах" [On unconditionally closed sets], *Rec. Mat.* [*Mat. Sb.*] 18 (60), No. 1, 3-28. MR7,412.

П.С. Новиков, С.И. Адян [P.S. Novikov, S.I. Adjan] (1968a), "О бесконечных периодических группах. I" [Infinite periodic groups. I], *Izv. Akad. Nauk SSSR Ser. Mat.* 32, 212-244; *Math. USSR-Izv.* 2, 209-236 (1969). MR39#1532a; Zbl.194,33; RZ [1969], 4A169.

П.С. Новиков, С.И. Адян [P.S. Novikov, S.I. Adjan] (1968b), "О бесконечных периодических группах. II" [Infinite periodic groups. II], *Izv. Akad. Nauk SSSR Ser. Mat.* 32, 251-524; *Math. USSR-Izv.* 2, 241-479 (1969). MR39#1532b; Zbl.194,33; RZ [1969], 4A170.

П.С. Новиков, С.И. Адян [P.S. Novikov, S.I. Adjan] (1968c), "О бесконечных периодических группах. III" [Infinite periodic groups. III], *Izv. Akad. Nauk SSSR Ser. Mat.* 32, 709-731; *Math. USSR-Izv.* 2, 665-685 (1969). MR39#1532c; Zbl.194,33; RZ [1969], 4A171.

А.Ю. Ольшанский [A.Ju. Ol'šanskiĭ] (1970), "О проблеме конечного базиса тождеств в группах" [On the problem of a finite basis for the identities of groups], *Izv. Akad. Nauk SSSR* 34, 376-384; *Math. USSR-Izv.* 4, 381-389 (1971). MR44#4079; Zbl.215,105; RZ [1979], 10A155.

А.Ю. Ольшанский [A.Ju. Ol'šanskiĭ] (1979), "Бесконечные группы с циклическими полгруппами" [Infinite groups with cyclic subgroups], *Dokl. Akad. Nauk SSSR* 245, 785-787; *Soviet Math. Dokl.* 20, 343-346.

В.Л. Ширванян [V.L. Širvanjan] (1976), "Вложение группы $B(\infty, n)$ в группу $B(2, n)$" [Imbedding of the group $B(\infty, n)$ in the group $B(2, n)$], *Izv. Akad. Nauk SSSR Ser. Mat.* 40, 190-208, 233; *Math. USSR-Izv.* 4, 181-199 (1977). MR54#2821; Zbl.336.20027; RZ [1976], 7A262.

M.R. Vaughan-Lee (1970), "Uncountably many varieties of groups", *Bull. London Math. Soc.* 2, 280-286. MR43#2054; Zbl.216,84; RZ [1971], 7A234.

Ордена Ленина,   Steklov Institute,
Матиметический ин-тим В.А. Стеклова,   Ul. Vavilova 42,
Академии Наук СССР,   Moscow V-333,
Москва В-333,   USSR.
Ул. Вавилова Дом 42.

ON THE WORD PROBLEM FOR GROUPS
DEFINED BY PERIODIC RELATIONS

S.I. ADIAN

By using a modification of the method described in Adian (1979), we prove the following theorem*.

THEOREM 1. *Let* G *be a group defined by a number of defining relations*

(1)
$$A_i^{r_i n} = 1 \quad (i = 1, \ldots, K)$$

in the alphabet

(2)
$$a_1, a_2, \ldots, a_m, a_1^{-1}, a_2^{-1}, \ldots, a_m^{-1},$$

where $n \geq 665$ *and* r_i *are arbitrary odd numbers. If the group* G *does not contain involutions, then the word problem for* G *is solvable.*

Proof. Obviously we may suppose that all words in (1) are uncancellable, all periods A_i are simple words and that they are not cyclic shifts of one another. We must notice here, that when we rewrite the periods A_i in terms of simple periods some new *factors* can appear for r_i. Because of the assumption of the theorem, excluding involutions from G, we may cancel powers of two from the new exponents. For this reason, we may assume that the A_i are simple periods and all exponents are odd.

The system of notions described in §4 of Chapter I of Adian (1979) is based on the set of all uncancellable periodic words of the form A^n, where n is a given odd number. We shall construct a similar system of notions starting from the finite number of periodic words

(3)
$$A_i^{n_i} \quad (i = 1, 2, \ldots, K),$$

---

* We use the notations and the rule of references introduced in Adian (1979).

where $n_i = r_i n$. For rank 0 all the notions remain the same. We call the words
(3) the periodic words of rank 1 .

We may suppose that no word of the form A_j^s for $s > \dfrac{n_j+1}{2}$ and $j \neq i$ occurs
in the word $A_i^{n_i}$, because otherwise we may replace $A_i^{r_i n} = 1$ by $B_i^{r_i n} = 1$, where B_i
is a period of length less than A_i .

In fact, suppose such a word A_j^s occurs in $A_i^{n_i}$. Then by Lemma I.2.9 of Adian
(1979) this word occurs in any period A_i of $A_i^{n_i}$ or in some cyclic shift of the
period A_i . Replace A_j^s by $A_j^{-n_j+s}$ wherever A_j^s occurs as a subword in $A_i^{n_i}$.
Notice that the resulting word is equal to $A_i^{n_i}$ in G . After possible cancellations
and conjugations in a free group, we obtain the periodic word $B_i^{n_i}$ which we want.

Obviously we may replace the defining relation $A_i^{n_i} = 1$ of the group G by the
relation $B_i^{n_i} = 1$. We can do the same for all the defining relations (1) of the
group G .

All the periods A_i are the minimal periods of rank 1 . According to I.4.10
we can pick out all the elementary periods of rank 1 relative to the given set (3)
of the periodic words. The period A_i is an elementary period of rank 1 if no
subword E of the word $A_j^{n_j}$ with $\partial(E) > 8\partial\left(A_j\right)$ occurs in the word A^4 .

Then we consider reversals of rank 1 for occurrences of such elementary words
of rank 1 . These reversals have the usual form

$$(4) \qquad X \; \overline{\circ} \; PA^t A_1 Q \rightarrow P\left(A^{-1}\right)^{n_i-t-1} A_2^{-1} Q \; \overline{\circ} \; Y \; ,$$

where A or A^{-1} is a cyclic shift of an elementary period A_i of rank 1 , n_j is
the corresponding exponent in (3), $A \; \overline{\circ} \; A_1 A_2$, the words X and Y are
uncancellable and the words $A^t A_1$ and $\left(A^{-1}\right)^{n_i-t-1} A_2^{-1}$ contain at least 9
segments. Remember that any occurrence of elementary word of rank 1 is normalised.
Then in a natural way we define real reversals among the reversals of form (4).

Basing on real reversals we define the notion of kernel of rank 1 for words $X \in N_1$, where N_1 was defined in I.4.21. Notice that we only consider reversals of form (4). Then we define sets

$$R_1, \; K_1, \; L_1, \; M_1 \;,$$

as in I.4.26, an equivalence relation in rank 1 denoted by $X \underset{\sim}{\downarrow} Y$, and all other notions for rank 1 .

One can easily verify that the proofs of all the necessary properties for the notions of rank 1 hold true without change. The only new condition is that we have restricted the class of elementary words to the elementary words obtained from (3), under the new definitions.

The condition that all exponents n_i in the reversals (4) are odd is necessary for the validity of Lemma IV.2.36 and all propositions depending on it.

In the proof of Theorem 1 we use all the notions and notations of Adian (1979) with the restrictions mentioned above without further reference.

Denote by G_1 the group with generators (2) and defining relations (1), where A_i is an elementary period of rank 1 in the new sense.

Similar to Lemma VI.2.8 we can prove that for any $X, Y \in R_1$ the following relation holds:

(5) $$X = Y \;\; \text{in} \;\; G_1 \Longleftrightarrow X \underset{\sim}{\downarrow} Y \;.$$

As in VI.2.4, for any word X in the alphabet (2) one can effectively find a word $Y \in K_1$ such that $X = Y$ in G_1 . By IV.3.12 and (5) such a word Y can even be found in $\overline{M}_1$.

If all the periods A_i in (3) are elementary words of rank 1 , then G_1 is equal to G and we can solve the word problem for G by using the relation (5).

Let some of the periods A_i in (3) be non-elementary periods of rank 1 .

Obviously we can suppose that an arbitrary word $A_i^{n_i}$ in (3) with non-elementary A_i belongs to R_1 . By II.3.7 all such periods A_i are periodic words of rank 2 .

For such a period A_i we first transform the word $A_i^{n_i}$ to a word Z , such that

$$A_i^{n_i} \underset{\sim}{\downarrow} Z \;\; \text{and} \;\; Z \in \overline{M}_1 \;.$$

Then by II.7.16 for an arbitrary period A_i we can find a word $T_i \in R_1$ and a minimal period B_i of rank 2 such that

$$A_i \underset{1}{\sim} \left[T_i, \; B_i^{k_i}, \; T_i^{-1} \right]_1$$

for some $k_i > 0$. We have $A_i = T_i B_i^{k_i} T_i^{-1}$ in G_1 . Consequently, without changing

the defining relations of the group G_1 we can rewrite all other relations (1) of the

group G obtaining the new relations

$$B_i^{k_i r_i n} = 1 \; ,$$

where B_i is a minimal period of rank 2 . All numbers k_i must be odd because

there are no involutions in the group G .

We can suppose that already the periods A_i are the minimal words of rank 2 .

By II.3.7 there are some elementary words among these minimal periods of rank 2 . We

consider all the elementary periods A_i of rank 2 . Let G_2 be the group obtained

by adding all defining relations (1) for the elementary periods A_i of rank 2 to

the group G_1 . Considering reversals of rank 2 of the form (4) for such periods

A_i we define real reversals of rank 2 , the notion of kernel of rank 2 , the sets

R_2, K_2, L_2, M_2, $\overline{M}_2$ and the relation $X \underset{2}{\sim} Y$.

Similarly to relation (5) we can prove for arbitrary $X, \; Y \in R_2$,

$$X = Y \text{ in } G_2 \Longleftrightarrow X \underset{2}{\sim} Y \; .$$

We have also an algorithm which for an arbitrary word X in the alphabet (2) finds a

word $Y \in R_2$ such that $X = Y$ in G_2 . That is the word problem in G_2 is

solvable.

Suppose that some of the periods A_i of rank 2 are not elementary periods of

rank 2 . We may suppose that for these A_i , the words $A_i^{n_i}$ belong to R_2 . By

II.3.7 all such periods A_i are periodic words of rank 3 . Then we select the

elementary periods of rank 3 from them. The corresponding words $A_i^{n_i}$ give the new

defining relations for G_3 , and so on.

Obviously, after a finite number of steps, we will have exhausted the set (1),

obtaining a group G_j which coincides with the original group G . So the solvability of the word problem of the group G is established.

Notice that one can also solve the conjugacy problem for the group G ; like in VI.3.5.

The following theorem shows that in Theorem 1 one cannot delete the condition of finiteness of the set of defining relations (1) of the group G .

THEOREM 2. *For arbitrary odd* $n > 665$ *there exists a recursive sequence of words in the alphabet*

(2) $$B_1, B_2, \ldots, B_i, \ldots$$

such that the group H *generated by* (2) *and defined by the set of relations*

(6) $$B_i^n = 1 \quad (i = 1, 2, \ldots)$$

does not contain involutions and has unsolvable word problem.

Proof. Let n be an odd number > 665 . In VI.2.1 we defined a class E of some elementary words of the form A^n . By VI.2.9 the relations

(7) $$A^n = 1 \quad (A^n \in E)$$

form a system of defining relations for the free periodic group $B(m, n)$. In Širvanian (1976) it was shown that this system of relations is independent, that is none of these relations is a consequence of the remaining ones. Let

(8) $$A_1^n, A_2^n, \ldots, A_i^n, \ldots$$

be a recursively enumerable but not recursive set $M \subset E$. Denote by H the group generated by (2) and defined by relations

(9) $$A_i^n = 1 \quad \left(A_i^n \in M\right) .$$

By the independence of the relations (9) the group H has unsolvable word problem. Obviously the group H does not contain involutions because all relations (9) hold in $B(m, n)$. We need only prove that H has a presentation by a recursive set of defining relations. We define the words B_i^n of such a set as follows:

$$B_1 = A_1$$

(10) $$B_2 = A_1^n A_2$$

$$\ldots \ldots \ldots \ldots \ldots \ldots$$

$$B_{i+1} = A_1^n A_2^n \cdots A_i^n A_{i+1}$$

$$\ldots \ldots \ldots \ldots \ldots \ldots \ldots .$$

Obviously, the group H can be defined by the relations (6) corresponding to
(10). On the other hand, by the recursive enumerability of the set M there exists
an algorithm for deciding whether or not a given word X belongs to the set M. For
this it is sufficient to use an algorithm enumerating the set M, finding the first
$\partial(X)$ words A_i^n of the sequence (8). Then establish the corresponding words B_i,
and check whether or not the word X occurs in the finite set of the B_i^n. Theorem 2
is proved.

References

S.I. Adian (1979), *The Burnside Problem and Identities in Groups* (translated by J.
     Lennox, J. Wiegold. Ergebnisse der Mathematik und ihrer Grenzgebiete, **95**.
     Springer-Verlag, Berlin, Heidelberg, New York).

В.Л. Ширванян [V.L. Širvanian] (1976), "Независимые системы определяющих соотношений
     свободной периодической группы нечетного показателя" [Independent systems of
     defining relations of the free periodic groups of odd exponent], *Mat. Sb.*
     100 (**142**), 132-136.

Ордена Ленина,   Steklov Institute,
Матиматический ин-тим В.А. Стеклова,   Ul. Vavilova 42,
Академии Наук СССР,   Moscow V-333,
Москва В-333,   USSR.
Ул. Вавилова Дом 42.

AN APPLICATION OF THE
NILPOTENT QUOTIENT PROGRAM

W.A. Alford and Bodo Pietsch

This note reports briefly on some computer calculations related to the group

$$C = \langle x, y, z \mid x^2 = y^2 = z^2 = (xy)^2 = 1, \text{ exponent 8} \rangle \; .$$

The significance of this group is explained in the paper by Hermanns in these proceedings. The nilpotent quotient program is described in the paper by Havas and Newman.

We denote the initial generators of a preimage of C by 1, 2 and 3 and define additional generators 4 to 18 by

4 = [3, 1] ,   5 = [3, 2] ,   6 = [4, 1] = [3, 1, 1] ,   7 = [4, 2] = [3, 1, 2] ,
8 = [5, 2] = [3, 2, 2] ,   9 = [6, 2] = [3, 1, 1, 2] ,   10 = [7, 3] = [3, 1, 2, 3] ,
11 = [10, 1] = [3, 1, 2, 3, 1] ,   12 = [10, 2] = [3, 1, 2, 3, 2] ,
13 = [11, 3] = [3, 1, 2, 3, 1, 3] ,   14 = [12, 3] = [3, 1, 2, 3, 2, 3] ,
15 = [13, 1] = [3, 1, 2, 3, 1, 3, 1] ,   16 = [13, 2] = [3, 1, 2, 3, 1, 3, 2] ,
17 = [14, 1] = [3, 1, 2, 3, 2, 3, 1] ,   18 = [14, 2] = [3, 1, 2, 3, 2, 3, 2] .

Then the group on this generating set which has as relators: 1^2, 2^2, 3^2, $(1.2)^2$ and the eighth powers of the following 120 words:

1.3, 2.3, 1.2.3, 1.5, 2.4, 1.2.4, 1.7, 1.8, 2.6, 3.7, 4.5, 1.3.7, 1.3.8, 1.4.5, 1.10,
2.3.6, 2.10, 3.4.5, 3.9, 4.8, 5.6, 1.2.3.6, 1.2.3.8, 1.2.10, 1.3.9, 1.5.7, 2.3.10,

2.4.8, 2.5.7, 3.4.8, 1.2.3.10, 1.2.4.8, 1.2.5.7, 1.3.12, 1.5.9, 1.5.10, 2.3.4.8,
2.3.5.7, 2.3.11, 2.4.10, 3.7.8, 1.2.3.5.7, 1.2.3.11, 1.2.3.12, 1.2.4.10, 1.2.5.10,
1.3.5.10, 1.3.6.8, 1.3.7.8, 2.3.4.10, 2.4.5.6, 2.4.5.7, 1.2.3.4.10, 1.2.3.5.9,
1.2.3.5.10, 1.2.3.6.8, 1.2.3.7.8, 1.2.5.12, 1.3.4.5.7, 1.3.4.12, 1.3.7.9, 1.3.8.9,
1.3.8.10, 1.3.16, 1.3.18, 1.5.7.8, 2.3.4.5.6, 2.3.4.11, 2.3.6.10, 2.3.7.9, 2.3.15,
2.3.17, 1.2.3.4.5.6, 1.2.3.4.5.7, 1.2.3.4.11, 1.2.3.5.12, 1.2.3.6.10, 1.2.3.8.10,
1.2.3.15, 1.3.4.5.10, 2.3.4.5.10, 1.3.5.6.8, 1.3.8.12, 1.3.9.10, 2.3.4.13, 2.3.5.6.7,
2.3.9.10, 2.3.4.6.10, 1.2.5.9, 1.4.5.7, 1.5.12, 2.4.11, 2.4.12, 1.2.4.11, 1.4.5.10,
2.4.13, 3.4.5.10, 3.7.11, 1.2.3.16, 1.2.3.17, 1.3.4.7.8, 2.3.4.14, 1.3.10, 1.3.5.6,
1.7.8, 2.3.12, 1.5.11, 2.3.5.9, 1.4.5.9, 1.5.14, 2.4.5.9, 1.2.3.18, 1.2.4.13,
1.3.5.7.8, 1.3.7.12, 2.3.6.11, 2.3.7.11, 2.3.7.12, 1.2.3.4.5.10, 2.4.14

has a largest finite quotient group of order 2^{313} and class 22 ; the ranks of the
quotients of the lower exponent 2-central series are 3, 2, 3, 2, 2, 2, 4, 5, 7, 10,
10, 13, 19, 22, 28, 35, 33, 39, 27, 29, 16, 2 .

Department of Mathematics,   Fakultät für Mathematik,
Institute of Advanced Studies,   Universität Bielefeld,
Australian National University,   Bielefeld, Germany.
Canberra, ACT 2600, Australia.

PROCEEDINGS OF THE BURNSIDE WORKSHOP
BIELEFELD, June–July 1977, 49–188.

20F50

(20-04)

GROUPS OF EXPONENT EIGHT

Fritz J. Grunewald
George Havas
J.L. Mennicke
M.F. Newman

CONTENTS

This is a report (finalized at this meeting) of the first steps in a study of groups of exponent 8 aimed at answering the special case of Burnside's question: is every 2-generator group of exponent 8 finite? It contains the bare details of the work. More informal accounts were given in lectures to the workshop of which written versions appear elsewhere in these proceedings.

INTRODUCTION

One of the driving forces behind the study of the structure of groups is a number of questions posed by Burnside (1902). In particular he asked whether every m-generator group of exponent n (every element has order dividing n) is finite. It is known that the answer is 'yes' for exponents 2, 3, 4, 6 and all (finite) numbers of generators (see, for example, Sections 5.12 and 5.13 of Magnus, Karrass, Solitar (1966)) and 'no' for all *odd* exponents greater than or equal to 665 (see Adian (1974), (1975)).

These results focus particular attention on the exponents which are powers of two. This paper is a report on some first steps in a study of groups of exponent eight.

We denote the free group of rank m and exponent n by $B(m, n)$.

Little seems to be known about groups of exponent 8 . In 1947 Sanov showed that there is a 2-generator group of exponent 8 with order 2^{136} by making use of the Schreier formula for the rank of a subgroup of finite index in a free group of finite rank. It is part of the folk-lore that the same argument applied to the subgroup F^4 generated by fourth powers of elements in a free group F of rank 2 shows that $F/(F^4)^2$ is a 2-generator group of exponent 8 and order 2^{4109} (because F/F^4 has order 2^{12} - see §6.8 of Coxeter and Moser (1957)). Recently Shield (1977) has shown that the nilpotency class of $F/(F^4)^2$ is 39 . In 1951 Sanov began another area of investigation by proving that all groups of exponent 8 satisfy the 23rd Engel congruence. The best result in this direction is due to Krause (1964); he showed they satisfy the 14th Engel congruence. Macdonald ((1973), p. 437), using a computer implementation of an algorithm for calculating nilpotent quotient (or factor) groups, showed that the largest 2-generator group of exponent 8 and nilpotency class 6 has order 2^{35} . This has since been extended to class 12 and order 2^{722} using the Canberra implementation of a related algorithm called the

nilpotent quotient algorithm (see Newman (1976)). Hermanns (1977) has recently
studied the largest metabelian quotient group of $B(2, 8)$ and has shown that it has
order 2^{63} and nilpotency class 12 (this has been confirmed by the nilpotent
quotient algorithm).

The present study was begun by two of us (Grunewald, Mennicke). It has its
origin in the possibility that the nature of the subgroups of $B(2, 8)$ might give a
hint as to whether to expect that $B(2, 8)$ is finite or not. Adian in ((1971),
Corollary 2) proved that $B(m, n)$ has all its finite subgroups cyclic for n odd and
at least 4381 (the bound has since been reduced by him to 665 ; (1975), Theorem
VII.1.8). This might suggest that a pointer to finiteness of $B(2, 8)$ would be
obtained if it were possible to show that $B(2, 8)$ has comparatively large finite
subgroups. However the situation is not quite so simple because every infinite
2-group has an infinite abelian subgroup (see Theorem 3.41 of Robinson (1972)).
Therefore infinite 2-groups have abelian subgroups of every 2-power order. In the
context of groups of exponent 8 these large finite abelian subgroups require many
generators. This suggests that a pointer to finiteness of $B(2, 8)$ would be obtained
if one could show that $B(2, 8)$ has comparatively large finite 2-generator
subgroups. Whatever one feels about such motivation, it raises finiteness problems
which are more approachable than that for $B(2, 8)$ itself. Some of these are
discussed below. For those with affirmative solution examples (described later) give
a lower bound for the order of the subgroup of $B(2, 8)$ in question. As Burnside
himself pointed out, once one knows that a group is finite the next problem is to
determine its order. This is discussed for some of the groups which arise.

Let $\{a, b\}$ be a free generating set for $B(2, 8)$ - from now on simply B .
Before turning attention to 2-generator subgroups of $B(2, 8)$ we look briefly at
1-generator or cyclic subgroups, in part to prepare for some questions which arise in
the 2-generator context. Here, of course, there is no finiteness problem and the
order of every cyclic subgroup or element divides 8 . An element which has odd
exponent sum in either a or b has order 8 for it has order 8 in a cyclic
homomorphic image. Equally clearly squares of elements have order at most 4 . For
some other elements it is comparatively easy to determine the order by constructing
suitable examples. One sees that $a^2 b^2$ and the commutator $[b, a]$ have order 8 by
considering the group presented by

$$\langle k, h \mid k^8 = h^4 = 1, k^{h^2} = k^{-1}, kk^h = k^h k \rangle .$$

This group is a splitting extension of a direct product of two cyclic groups of order
8 (the normal closure of k) by a cyclic group of order 4 (generated by h). It
is easily seen to have exponent 8 , so the mapping $a \mapsto hk$, $b \mapsto h$ can be
extended to a homomorphism from B . Under this homomorphism $\left(a^2 b^2\right)^4$ and $[b, a]^4$

both map to $(kk^h)^4$ which is not the identity. Considering the group K presented by

$$\langle k, h \mid k^8 = h^8 = 1, k^{h^4} = k^{-1}, kk^h = k^h k, kk^{h^2} = k^{h^2} k, kk^{h^3} = k^{h^3} k \rangle ,$$

which is a splitting extension of a direct product of four cyclic groups of order 8 by a cyclic group of order 8 and has exponent 8 , settles the order problem for more complicated elements; for instance, $a^4 b^4$ and $[b, a, a, a]$ both have order 8 . For commutators of higher weight we record a few coarser results. Let B_c denote the cth term of the lower central series of B and B'' the second derived group of B . The work of Hermanns (1977) shows that $(B_5)^4 \leq B''$ and $(B_9)^2 \leq B''$. Calculating in B/B_{12} gives $(B_6)^2 \leq B_7$ (cf. p. 437 of Macdonald (1973)), $(B_5)^4 \leq B_{12}$ and $(B_7)^2 \leq B_{11}$. Hence, for example, $(B_c)^2 \leq B_{c+1}$ for all c greater than or equal to 6 . Finally we mention, for later use, that $(a^4 b^4)^4$ lies in B_{12} so that $[(a^4 b^4)^4, b^2, (a^4 b^4)^4]$ is in B_{26} .

We now turn to 2-generator subgroups of B . Observe first that if the two generators have order 2 , then the subgroup they generate is dihedral of order 4, 8 or 16 . The first case of interest is $\langle a^4, b^4 \rangle$ (here $\langle \ \rangle$ denotes subgroup generated by). This has order 16 ; consider $\langle (hk)^4, h^4 \rangle$ in the group K above. If either generator does not have order 2 finiteness problems immediately arise. Even the simplest seems difficult.

PROBLEM 1. *Is the subgroup of* $B(2, 8)$ *generated by* $\{a^4, b^2\}$ *finite?*

Since the paper was submitted it has been shown (by Havas and Newman) that this subgroup has a largest finite quotient - of order at most 2^{205} and class at most 26 . The subgroup $\langle a^4, b^2 \rangle$ may be regarded as a product of its finite subgroups $\langle a^4, b^4 \rangle$ and $\langle b^2 \rangle$; these meet in $\langle b^4 \rangle$ (in other words $\langle a^4, b^2 \rangle$ is generated by an amalgam of finite groups). In $\langle a^4, b^4 \rangle$ there are two proper subgroups $\langle (a^4 b^4)^4, b^4 \rangle$ and $\langle (a^4 b^4)^2, b^4 \rangle$ containing b^4 . This suggests two intermediate finiteness problems. Is $L = \langle (a^4 b^4)^4, b^2 \rangle$ finite? Is $M = \langle (a^4 b^4)^2, b^2 \rangle$ finite? The first of these is easy to answer for b^4 is central in L and $L/\langle b^4 \rangle$ is generated by two elements of order 2 , so L is finite of order at most 32 . In more detail L is a homomorphic image of

$$\langle T, Y \mid T^2 = Y^4 = [T, Y^2] = (TY)^8 = 1 \rangle$$

under the mapping $T \mapsto (a^4 b^4)^4$, $Y \mapsto b^2$. Thus L has order 32 if and only if $[(a^4 b^4)^4, b^2, (a^4 b^4)^4]$ is non-trivial; so the earlier observation that this element lies in B_{26} implies that, if L has order 32 , an example of nilpotency class at least 26 will be involved in exhibiting this order. The second question is harder. Its solution is the main result of the investigation so far.

The subgroup of $B(2, 8)$ *generated by* $\{(a^4 b^4)^2, b^2\}$ *is finite of order at most* 2^{31} .

The best lower bound we have for the order of this subgroup M is 2^5 which comes by considering $\langle ((hk)^4 h^4)^2, h^2 \rangle$ in the group K .

PROBLEM 2. *What are the orders of the subgroups of* $B(2, 8)$ *generated by* $\{(a^4 b^4)^4, b^2\}$ *and* $\{(a^4 b^4)^2, b^2\}$?

The proof of the finiteness of M , as can be seen by glancing ahead, is highly computational and combinatorial. In the later stages it has been significantly machine assisted. Below we give an outline of the proof and its development. The proof has evolved in a way which is hard, and not always illuminating, to describe. The account here contains some traces of the evolution. They have been kept partly because their removal would not, we believe, significantly aid understanding (and could introduce errors which hinder it) and partly because their retention may shed some light on what was involved in developing the proof.

Observe to begin with that M is a homomorphic image of the group Γ^* presented by

$$\langle Y, W \mid Y^4 = W^4 = (Y^2 W)^2 = 1, \text{ exponent } 8 \rangle$$

under the mapping $Y \mapsto b^2$, $W \mapsto (a^4 b^4)^2$. We prove that Γ^* is finite of order 2^{31} . Another presentation for Γ^* was actually used, namely

$$\langle Y, Z \mid Y^4 = (YZ)^4 = (YZ^{-1})^2 = 1, \text{ exponent } 8 \rangle .$$

A finiteness proof need only use finitely many eighth powers, so it can be set in the context of some suitable finitely presented preimage of Γ^* . The one we ended up with is the group Γ with the presentation given at the beginning of Chapter 2. This group has a subgroup Γ_1 of index 8 generated by

$$\{(ZY)^2, YZ^2 Y^{-1} = (YZ)^2 Z^{-2}, ZYZ^2 Y^{-1} Z^{-1} = (ZY)^2 Z^{-2}\}$$

(Theorem 1 of Chapter 2). A major step in the proof is to show $\langle (ZY)^2 Z^{-2}, (YZ)^2 Z^{-2} \rangle$ is a finite 2-group. This is done by considering a preimage G of the group G^* presented by

$$\langle p, q \mid p^4 = q^4 = (pq)^4 = \left(pq^{-1}\right)^4 = \left(p^2 q^2\right)^4 =$$
$$= \left(pqp^{-1}q^{-1}\right)^4 = \left(p^2 qpq^2 pq\right)^2 = 1, \text{ exponent } 8 \rangle .$$

Theorem 3 of Chapter 2 shows that the group G is a preimage of the subgroup via the mapping $p \mapsto (ZY)^2 Z^{-2}$, $q \mapsto (YZ)^2 Z^{-2}$. (The relation $\left(qp^{-1}qpq^{-1}p\right)^4 = 1$ - p. 126 was found after the proof in Chapter 1 was essentially complete.) The proof of the finiteness of G was the subject of Grunewald's dissertation (1973) and of a lecture (by Mennicke) at Oberwolfach in May 1974. The collaboration reported here arose out of a discussion (between Mennicke and Newman) following that lecture. It seemed that a judicious mixture of hand and machine calculations should result in better progress. This has been the case, firstly because great care and considerable time are needed to try to ensure that hand calculations are accurate, and secondly because the nilpotent quotient algorithm is able to give quite detailed information about the groups involved. A presentation for G is given at the beginning of Chapter 1. In that chapter G is shown to be finite of order at most 2^{21} (Theorem 4). Moreover G has exponent 8 (Theorem 7), so G and G^* coincide. The nilpotent quotient algorithm was applied to the presentation for G. It terminated, showing that G has a largest finite 2-quotient of order 2^{21} and nilpotency class 9. Combined with the finiteness proof this gives that G^* has order 2^{21}. A consistent power commutator presentation for G^* is given in Chapter 3. (This presentation was used to help check the calculations in Chapter 1.) The nilpotent quotient algorithm also yields that the rank of the multiplicator of G^* is 7. This suggests the possibility that G^* has a presentation involving only two eighth powers in addition to the explicitly given relations. This has been confirmed by a technique based on coset enumeration. This is described in Chapter 3 and provides an alternative and independent proof for the finiteness of G^*. Chapter 2 contains the rest of the finiteness proof for Γ^* (via one for Γ). This proof involves less detailed calculations than those in Chapter 1 because in the final stages of completing the proof it was known that the nilpotent quotient algorithm applied to Γ^* terminates and shows that Γ^* has a largest finite 2-quotient of order 2^{31} and nilpotency class 19. The resulting consistent power commutator presentation can be deduced from that for Γ given in Chapter 4. This presentation was used to help guide and check the proof in Chapter 2. From the presentation can be "read off" properties of Γ^* such as $\left((\Gamma^*)^4\right)^2$ is non-trivial and $\left(\left((\Gamma^*)^4\right)^2\right)^2$ is trivial. The group Γ which is shown to be a finite 2-group in Chapter 2, is a proper preimage of Γ^*; the nilpotent quotient algorithm shows Γ to have order 2^{33} and nilpotency class 21.

The image of G^* in Γ has order 2^{14}. The kernel is generated as a normal subgroup of G^* by

$$\{[[q, p], (q^2p)^4], [q, p^2]^2[q^2, p]^2, (qpq^{-1}p)^4\} .$$

The elements of the kernel when mapped into B via Γ^* yield words in a, b which may be viewed as non-obvious consequences of the exponent 8 condition. The images of the generators of the kernel are rather unpleasant to write down, some other examples of non-obvious consequences are

$$[(a^4b^4a^4b^2)^2, b^4(a^4b^4a^4b^2)^4b^4]^4 ,$$

$$((a^4b^4a^4b^2)^3b^4(a^4b^4a^4b^2)^3b^2(a^4b^4a^4b^2)^2b^2)^4 .$$

We are indebted to Mr W.A. Alford of the Australian National University for doing some of the computing and to Mr R.I. Yager, a vacation scholar at the Australian National University during the summer of 1976-1977, for his careful checking of the proofs in Chapter 2 which enabled us to eliminate a number of errors and misprints from a draft.

CHAPTER 1

A FINITENESS PROOF FOR G

We shall prove that the group

$$G = \langle p, q \mid$$

(i) $p^4 = 1$

(ii) $q^4 = 1$

(iii) $(pq)^4 = 1$

(iv) $\left(pq^{-1}\right)^4 = 1$

(v) $\left(p^2 q^2\right)^4 = 1$

(vi) $\left(pqp^{-1}q^{-1}\right)^4 = 1$

(vii) $\left(p^2 qpq^2 pq\right)^2 = 1$

(viii) $\left(q^2 pq^2 p^{-1}\right)^8 = \left(p^2 qp^2 q^{-1}\right)^8 = 1$

(ix) $\left(p^2 q\right)^8 = \left(q^2 p\right)^8 = 1$

(x) $\left(pqpq^{-1}\right)^8 = \left(qpqp^{-1}\right)^8 = 1$

(xi) $\left(p^2 qpq^{-1}p^2 qp^{-1}q^{-1}\right)^8 = \left(q^2 pqp^{-1}q^2 pq^{-1}p^{-1}\right)^8 = 1$

(xii) $\left(qpq^{-1}p^2\right)^8 = \left(pqp^{-1}q^2\right)^8 = 1$

(xiii) $\left(p^2 q^2 p^2 q^{-1}\right)^8 = \left(q^2 p^2 q^2 p^{-1}\right)^8 = 1$

(xiv) $\left(pqp^2 q^2\right)^8 = \left(qpq^2 p^2\right)^8 = 1$

(xv) $\left(pqp^2 q^{-1}\right)^8 = \left(qpq^2 p^{-1}\right)^8 = 1\rangle$

is finite. The elementary computations which follow will also show the structure of this group.

A. The Main Theorems

THEOREM 1. *The commutator subgroup* $G' \leq G$ *having the index*[*]

$$|G : G'| = 2^4$$

---

[*] If $H_1 \leq H_2$ are two groups we write
$$|H_2 : H_1|$$
for the index of H_1 in H_2 .

*is generated by:**

(1) $a = qpq^{-1}p^{-1}$

(4) $d = q^2pq^{-2}p^{-1}$

(2) $b = p\cdot qpq^{-1}p^{-1}\cdot p^{-1}$

(5) $e = p\cdot q^2pq^2p^{-1}\cdot p^{-1}$.

(3) $c = p^2\cdot qpq^{-1}p^{-1}\cdot p^{-2}$

This theorem is proved by the transformation formulae in Lemma 1.

DEFINITION. $H := \langle a, b, c \rangle$.

Let R_H be a defining system of relations for H . Let $R_{G'}$ be a defining system of relations for G' . Let H_g be the subgroup of H consisting of the words with an even number of letters.

$G'_d := \mathrm{Ker}\big(G' \to \langle a, b, c, d, e \mid R_{G'}; a = b = c = 1; d^2 = e^2 = (ed)^2 = 1\rangle\big)$.

$D := d^2 = \big(q^2pq^{-2}p^{-1}\big)^2$.

$F := ebe^{-1}a^{-1}ba^{-1} = \big(pq^2\big)^4$.

H_g is generated by

$$a^2, b^2, c^2, ab, ac, bc$$

and has in H the index

$$|H : H_g| \le 2 .$$

G'_d is generated by

$$a, dad^{-1}, eae^{-1}, deae^{-1}d^{-1}, b, dbd^{-1}, ebe^{-1}, debe^{-1}d^{-1}, c, dcd^{-1}, ece^{-1}, dece^{-1}d^{-1},$$
$$d^2, ede^{-1}d^{-1}, dede^{-1}, e^2, de^2d^{-1}$$

and has in G' the index

$$|G' : G'_d| \le 2^2 .$$

Obviously, the systems of generators stated are reductions of the systems of generators obtained by the Reidemeister-Schreier method.

THEOREM 2. (1) D *normalizes* $\langle a, b, c, F \rangle$.

(2) $D^2 \in \langle a, b, c \rangle$. *Moreover* D^2 *lies in the centre of* G .

(3) $G'_d = \langle a, b, c, F, D \rangle$.

---

* We use here a, b , which were already used for generators of $B(2, 8)$ in the introduction. This should cause no confusion.

(4) $|G' : \langle a, b, c, F, D \rangle| \leq 2^2$.

(5) $|\langle a, b, c, F, D \rangle : \langle a, b, c, F \rangle| \leq 2$.

(6) $|\langle a, b, c, F \rangle : \langle a, b, c \rangle| \leq 2^3$.

Proof. (1), (2) follow by Lemma 16.

(3) follows by Lemma 2 (36)-(48).

(6) Lemma 16 and the transformation formulae derived from Lemmas 13, 15 show that F normalizes the normal subgroup N generated by $a^2, b^2, c^2, ac, (ab)^2$ in $\langle a, b, c \rangle$. Since the elements $FaFa^{-1}$ and $FbFb^{-1}$ lie in the centre of $\langle a, b, c, F \rangle$ by Lemma 16 and since both are of the order 2 , the factor group

$$\langle a, b, c, F \rangle / N$$

is of an order smaller than 2^5 . Since it is obvious that N in $\langle a, b, c \rangle$ has index 4 , the proof of (6) is complete.

THEOREM 3. $\langle (ac)^2, (c^{-1}a)^2, (abcb^{-1})^2 \rangle$ *is an elementary abelian subgroup of exponent* 2 *in the centre of* H . *The following relations hold in the factor group*

(1) $a^4 = b^4 = c^4 = 1$,

(2) $(ac)^2 = (abcb^{-1})^2 = (c^{-1}a)^2 = (abcb)^2 = 1$,

(3) $(ab)^4 = (bc)^4 = 1$,

(4) $bcab^{-1} = ac$,

(5) $bacb^{-1} = ca$,

(6) $(bcb^{-1}a^{-1})^2 = 1$.

Proof. The first relation follows by Theorem 10, Lemma 11 (28), Lemma 13 (19), Lemma 11 (13), Lemma 16.

(2) follows by Lemma 16.

(3) Lemma 11 (26), (22).

(4) Lemma 11 (24).

(5) Lemma 11 (24), Lemma 4 (1).

(6) Lemma 11 (7).

Obviously, in the group $H / \langle (ac)^2, (c^{-1}a)^2, (abcb^{-1})^2 \rangle$, $\langle ac, ca \rangle$ is an elementary abelian normal subgroup of order 2^2 . The factor group is a factor group of the group studied by Coxeter (1940):

$$\langle P, Q \mid P^4 = Q^4 = (PQ)^4 = (P^{-1}Q)^4 = (P^{-1}Q^{-1}PQ)^2 = (PQ^{-1}PQ)^2 = (P^{-1}QPQ)^2 = 1 \rangle .$$

At most, it is of order 2^6. So the proof of the following theorem is complete.

THEOREM 4. *G is finite and*

$$|G| \le 2^{21} .$$

We shall now describe the normal subgroup which is generated in G by the 4th powers.

DEFINITION.

$$v_1 := b^{-1}a^{-1}Dba = (p^2q)^4 ,$$

$$v_2 := ab^{-1}a^{-1}Db ,$$

$$v_3 := c^{-1}b^{-1}cb^{-1}D^{-1}cb^{-1}Fac^{-1}b^{-1}a^{-1}b^{-1}Fb ,$$

$$v_4 := (ac)^2 = (qpq^{-1}p)^4 ,$$

$$v_5 := (abcb^{-1})^2 ,$$

$$v_6 := (a^{-1}c)^2 ,$$

$$V := \langle v_1, v_2, v_3, v_4, v_5, v_6 \rangle .$$

Let α be the automorphism of G which interchanges p and q.

$$u_i := \alpha(v_i) , \quad i = 1, 2, 3, 4, 5, 6 ,$$

$$U := \langle u_1, u_2, u_3, u_4, u_5, u_6 \rangle .$$

THEOREM 5. *V is the normal subgroup generated by $(p^2q)^4$ in G. U is the normal subgroup generated by $(q^2p)^4$ in G. $V \cdot U$ is elementary abelian of exponent 2. $V \cap U \ge \langle v_4v_5, v_6 \rangle$. $|V \cdot U| \le 2^{10}$.*

Proof. The above statements follow by Lemma 13, Lemma 15, Lemma 16.

$$B(2, 4) := \langle x, y \mid w^4 = 1 \rangle \simeq \langle x, y \mid x^4 = y^4 = (xy)^4 = (x^{-1}y)^4 = (x^2y)^4$$
$$= (xy^2)^4 = (x^2y^2)^4 = (x^{-1}y^{-1}xy)^4 = (x^{-1}yxy)^4 = 1 \rangle .$$

This presentation is found in Coxeter-Moxer (1957, p. 81). Now we shall show that

$$((xy)^2(yx)^2)^2 \in Z(B(2, 4)) .$$

It is sufficient to show the identity

$$\left((xy)^2(yx)^2\right)^2 = \left(x^2yxy^2xy\right)^2 .$$

this is equivalent to $x \not\updownarrow \left((xy)^2(yx)^2\right)^2$; where $\updownarrow$ denotes commutes with. To obtain the commutator relation which remains to be shown, it suffices to exchange x and y in the above identity:

$$\left((xy)^2(yx)^2\right)^2 \cdot \left(x^2yxy^2xy\right)^2 = xyxy \cdot yxyx \cdot xyxy \cdot yxyx^{-1} \cdot yxyx^{-1} \cdot xyxy \cdot x \cdot xyxy \cdot yxy$$

$$= xyxy \cdot x^{-1}y^{-1} \cdot x^{-1}y^2x^{-1}y^2 \cdot x^{-1}y^{-1}xy^{-1} \cdot x^{-1}y^2x^{-1}y^2 \cdot x^{-1}y^{-1}x^{-1}yxy$$

$$= (xy)^4 = 1 .$$

It is well-known that $\left((xy)^2(yx)^2\right)^2 \neq 1$ in $B(2, 4)$.

A proof of this fact is also given by Theorem 8.

THEOREM 6. $G/V \cdot U$ *is of exponent* 4 *and*

$$|G/V \cdot U| = 2^{11} .$$

Proof. Obviously, it is sufficient to show that the relation

$$\left(pqpq^{-1}\right)^4 = 1$$

holds in $G/V \cdot U$. This follows by the definition of v_4 . When determining the order, remember that $B(2, 4)$ is exactly of order 2^{12} . Thus we have shown that $V \cdot U$ is exactly the normal subgroup of G which is generated by the 4th powers.

Next follows:

THEOREM 7. G *is of exponent* 8 ; G^4 *is elementary abelian of exponent* 2 . *For* G *we have the following diagram:*

where $\langle\langle x \rangle\rangle_G$ denotes the normal subgroup generated by x in G .

The following computations yield also a description of the commutator subgroup of

$$G/V \cdot U = G/G^4 .$$

For the generators of $(G/V \cdot U)'$ we shall use the same notation as for the generators of G' .

THEOREM 8. *In* $(G/G^4)'$ *we have the following relations:*

(1) $a^2 = b^2 = c^2$;

(2) $(ab)^2 = (bc)^2 = d^2 = e^2$;

(3) $ac = ca$;

(4) $a^2, e^2 \in Z\big((G/G^4)'\big)$;

(5) $dad^{-1} = e^2 a$,

$dbd^{-1} = e^2 b$,

$dcd^{-1} = e^2 c$;

(6) $eae^{-1} = e^2 a$,

$ebe^{-1} = e^2 b$,

$ece^{-1} = e^2 c$.

Proof. These relations are all stated in Lemma 16. Dividing $(G/G^4)'$ by the normal subgroup $\langle a^2, e^2 \rangle$ one obviously obtains an elementary abelian group of order 2^5 . This implies also that G/G^4 is of order 2^{11} at most.

THEOREM 9. (1) *The solubility length of* G *is at most* 3 .

(2) *The nilpotency class of* G *is at most* 9 .

Proof. (1) Lemma 16 yields that

$$[G', G'] \leq \langle V \cdot U, D \rangle$$

where $\langle V \cdot U, D \rangle$ is an abelian group.

(2) Let $\zeta_i(G)$ be the ith term of the upper central series of our group G , that is, $\zeta_0(G) = \langle 1 \rangle$ and $\zeta_{i+1}(G)/\zeta_i(G) = Z\big(G/\zeta_i(G)\big)$.

Considering the formulas in Lemmas 13, 15, 16 one finds that $\langle U \cdot V, D \rangle \subseteq \zeta_5(G)$. The formulas of Theorem 8 and Lemma 1 then imply:

$$\langle de, a^2, ac, U \cdot V, D \rangle \subseteq \zeta_6(G) ;$$

$$\langle d, ab, de, a^2, ac, U \cdot V, D \rangle \subseteq \zeta_7(G) ;$$

$$[G, G] \subseteq \zeta_8(G) .$$

This proves (2).

B. The Proofs

LEMMA 1. *We have the following rules of conjugation:*

(1) $pap^{-1} = b$,   (6) $qaq^{-1} = da^{-1}$,

(2) $pbp^{-1} = c$,   (7) $qbq^{-1} = aeb^{-1}a^{-1}$,

(3) $pcp^{-1} = c^{-1}b^{-1}{}^{-1}$,   (8) $qcq^{-1} = adbc^{-1}b^{-1}a^{-1}$,

(4) $pdp^{-1} = e$,   (9) $qdq^{-1} = c^{-1}e^{-1}a^{-1}$,

(5) $pep^{-1} = b^{-1}db$,   (10) $qeq^{-1} = ad^{-1}abcb^{-1}a^{-1}$.

Proof. (1) to (4) follow by the definition of $a - e$ using (i), (ii).

(5) follows by (i), (ii), (iii), (iv).

(6) and (7) follow by the definition of $a - e$.

(8) By the definition of $a - e$ and (i), (ii) we have

$$qcq^{-1}abcb^{-1}d^{-1}a^{-1} = qp^2qpq\,(qp^{-1})^4(pq)^4q^{-1}p^{-1}q^{-1}p^{-2}q^{-1} .$$

Then (iii), (iv) imply (8).

(9) follows immediately by (ii), (iv).

(10) follows by (iii).

LEMMA 2. *In* $\langle a, b, c, D, F \rangle$ *we have the relations:*

(1) $a^4 = b^4 = c^4 = D^4 = F^2 = 1$;

(2) $(ac)^4 = (abc)^4 = (abcb^{-1})^4 = (ca^{-1})^8 = 1$;

(3) $(Dbc^{-1}bc)^2 = 1$;

(4) $(Dbc^{-1})^2 = (ab)^2$;

(5) $d \not\equiv abcb^{-1}$, $D \not\equiv abcb^{-1}$;

(6) $c^2 \not\equiv b^{-1}a^{-1}Dba$;

(7) $(D^{-1}ab)^2 = (bc^{-1})^2$;

(8) $(Dbab^{-1}a^{-1})^2 = 1$;

(9) $(Dbaba^{-1})^2 = 1$;

(10) $Dbacb^{-1}D^{-1}aba^{-1}c^{-1}b^{-1}a^{-1} = 1$;

(11) $Dbc^{-1}bDb \neq abca^{-1}$;

(12) $abc^{-1}b^{-1}a^{-1}Dbc^{-1}bca^{-1}b^{-1}D^{-1}b^{-1}ca = 1$;

(13) $\left(D^{-1}abca^{-1}Db^{-1}c^{-1}\right)^2 = 1$;

(14) $\left(Dba^2b\right)^4 = 1$;

(15) $\left(Dba^2bc^{-1}bc^{-1}b\right)^2 = 1$;

(16) $\left(c^{-1}b^{-1}a^{-1}Dca^{-1}\right)^4 = 1$;

(17) $b^2 \neq \left(c^{-1}b^{-1}a^{-1}Dca^{-1}\right)^2$;

(18) $\left(DbaDbc^{-1}bDbac^{-1}\right)^2 = 1$;

(19) $Fa^2F = a^2$;

(20) $Fbc^2b^{-1}F = bc^2b^{-1}$;

(21) $Fab^2c^2aF = a^{-1}c^2b^2a^{-1}$;

(22) $F\left(ab^{-1}\right)^2F = a^{-1}c^2b^{-1}a^{-1}c^2b^{-1}$;

(23) $Fab^{-1}abca^{-1}F = a^{-1}c^{-1}abcacb^{-1}$;

(24) $cb^{-1}Fac^{-1}b^{-1}a^{-1} \neq b^2$;

(25) $c^{-1}bcbcFac^{-1}b^{-1}FDbac^2b^{-1}a^{-1}c^{-1}Fac^{-1}b^{-1}FDbac^2b^{-1}a^{-1} = 1$;

(26) $caFac^{-1}b^{-1}FDbac^2b^{-1}Fac^{-1}b^{-1}FDba^2bc = 1$;

(27) $aca^{-1}b^{-1}D^{-1}Fbca^{-1}Fa^{-1}c^{-1}Fac^{-1}b^{-1}FDba = 1$;

(28) $bc^2a^{-1}b^{-1}D^{-1}Fbca^{-1}Fc^{-1}b^{-1}cabcFac^{-1}b^{-1}FDbac^2b^{-1}a^{-1}c^{-1} = 1$;

(29) $Dbc^{-1}bDbac^{-1}b^{-1}a^{-1}b^{-1}D^{-1}ca^{-1}b^{-1}D^{-1}Fbcb^{-1}aF = 1$;

(30) $aca^{-1}b^{-1}D^{-1}Fbcb^{-1}aFa^{-1}b^{-1}D^{-1}b^{-1}cac^{-1}Dbac^{-1}b^{-1}cb^{-1}D^{-1} = 1$;

(31) $Dbac^{-1}b^{-1}a^{-1}b^{-1}D^{-1}c^{-1}b^{-1}Fc^{-1}b^{-1}cb^{-1}Fac^{-1}b^{-1}a^{-1}c^{-1}b = 1$;

(32) $D^{-1}aD^{-1}ab^{-1}aFac^{-1}abFac^{-1} = 1$;

(33) $\left(Fac^{-1}b^{-1}a^{-1}cb^{-1}\right)^2 = 1$;

(34) $\left(D^{-1}Fac^{-1}abcb^{-1}\right)^2 = 1$;

(35) $\left(caFab^{-1}a\right)^4 = 1$;

(36) $dad^{-1} = Dbabca^{-1}Fbc^{-1}a^{-1}b^{-1}D^{-1}$;

(37) $eae^{-1} = babca^{-1}Fbc^{-1}Dba$;

(38) $de \cdot a \cdot e^{-1}d^{-1} = b^{-1}a^{-1}b^{-1}D^{-1}$;

(39) $dbd^{-1} = cb^{-1}Fac^{-1}b^{-1}a^{-1}b^{-1}D^{-1}$;

(40) $ebe^{-1} = Fab^{-1}a$;

(41) $de \cdot b \cdot e^{-1}d^{-1} = c^{-1}b^{-1}Fc^{-1}b^{-1}cb^{-1}Fac^{-1}b^{-1}a^{-1}b^{-1}D^{-1}$;

(42) $dcd^{-1} = DFac^{-1}b^{-1}a^{-1}b^{-1}D^{-1}$;

(43) $ece^{-1} = cb^{-1}D^{-1}cb^{-1}Fac^{-1}b^{-1}a^{-1}b^{-1}$;

(44) $de \cdot c \cdot e^{-1}d^{-1} = Dbc^{-1}b$;

(45) $ede^{-1}d^{-1} = babca^{-1}Fbabca^{-1}Fbc^{-1}b^{-1}cabca^{-1}Fbc^{-1}bcFbc$;

(46) $dede^{-1} = cFbabca^{-1}Fbc^{-1}Dbc^{-1}$;

(47) $e^2 = cb^{-1}D^{-1}cb^{-1}Fac^{-1}b^{-1}a^{-1}b^{-1}Fc^{-1}$;

(48) de^2d^{-1}

$\quad = c^{-1}b^{-1}Fc^{-1}b^{-1}cb^{-1}Fac^{-1}b^{-1}a^{-1}c^{-1}bcb^{-1}Fac^{-1}b^{-1}a^{-1}b^{-1}Fac^{-1}b^{-1}ab^{-1}D^{-1}$;

(49) $\left(dbc^{-1}\right)^2 = (db)^4 = (dba)^2 = 1$.

Proof. (1) follows by (vi), (viii), (ix).

(2) follows by (x), (vi), the transformation formulae in Lemma 1 and (xi).

(3)

$$\left(Dbc^{-1}bc\right)^2 = \left(q^2pq^{-1} \cdot q^{-1}p^{-1}q^{-1}\left(q^{-1}p\right)^4 p^{-1}q^{-1} \cdot q^{-1}p^{-1}q^{-1}p^{-1} \cdot qp^2q^{-1}p\right)^2$$
$$= \left(q^2 \cdot pqp^{-1}q^{-1} \cdot pqp^{-1}\left(pq^{-1}\right)^4 \cdot q^{-1} \cdot q^2\right)^2 \quad \text{by (iv), (iii), (vii)}$$
$$= 1 \quad \text{by (iv), (vi).}$$

(4)

$$Dbc^{-1}Dbc^{-1}b^{-1}a^{-1}b^{-1}a^{-1} = q^2pq^2 \cdot p^{-1}q^{-1}p^{-1} \cdot \left(pq^{-1}\right)^4 q^2p^{-1}q^{-1}p^2 \cdot q^2pq^2\; p^{-1}q^{-1}p^{-1}\left(pq^{-1}\right)^4 \; \cdot$$

$$q^{-1} \cdot q^{-1}pq^{-1}p \cdot pqp^2q^{-1} = q \cdot qpq^2qpq^2 \cdot p^{-1}q^2 \cdot pq^{-1}pq^{-1} \cdot pq^2p^{-1} \cdot qpqp^2 \cdot q^{-1}$$

$$\text{by (iv), (iii), (vii)}$$

$$= 1 \quad \text{by (iv), (iii), (vii).}$$

(5) Lemma 1 yields

$$p^4 d p^{-4} = abcb^{-1} \cdot d \cdot bc^{-1} b^{-1} a^{-1} .$$

Then (i) yields Lemma 2 (5).

(6)

$$a^{-1} b^{-1} D^{-1} abc^2 b^{-1} a^{-1} Dbac^2 = p \cdot q^2 pqp \cdot (p^{-1} q^{-1})^4 \cdot (qp^{-1})^4 pqpq^2 \cdot p^{-1} q^{-1} p^{-1} q^{-1} \cdot p^{-1} q$$

$$\cdot pq^{-1} p^{-1} qp^2 qpq^2 p^{-1} q^{-1} p^{-1} (pq^{-1})^4 \cdot qpqp \cdot q^{-1} pqpq^{-1} p^{-1} \cdot qpq^{-1} p$$

$$= p \cdot q^2 pqp^2 qp \cdot q^{-1} pq^2 pq^{-1} p^{-1} qp^2 qpq^2 \cdot p^{-1} q^{-1} p^2 q^{-1} p^{-1} q^2$$

$$\cdot pqpq^{-1} p^{-1} qpq^{-1} p \text{ by (iii), (iv)}$$

$$= q^{-1} p^2 q^{-1} p^{-1} qpq^2 \cdot pq^{-1} pq^{-1} \cdot p^2 q^{-1} (qp^2)^8 \cdot p^2 \cdot q^{-1} p^{-1} q^{-1} p^{-1} \cdot qpq^{-1} p$$

$$\text{by (vii)}$$

$$= q^{-1} pq^{-1} \cdot (qpq^{-1} p^{-1})^4 \cdot pq^{-1} pq^{-1} p \text{ by (ix), (iv), (iii)}$$

$$= 1 \text{ by (vi), (iv).}$$

(7)

$$D^{-1} abD^{-1} abcb^{-1} cb^{-1} = pq^2 p^{-1} q^2 p \cdot q^2 p^{-1} q^{-1} p^2 q^{-1} p^{-1} \cdot q^2 p^{-1} q^2 pq^2 \cdot p^{-1} q^{-1} p^{-1} q^{-1} \cdot p^{-1} qp^{-1} q^{-1}$$

$$\cdot pqpq^{-1} p^{-1} qp^{-1} q^{-1} p^{-1}$$

$$= pq^2 p^{-1} q^2 \cdot p^2 qpq \cdot (q^{-1} p)^4 \cdot qpqp \cdot (p^{-1} q^{-1})^4 \cdot pqpq^{-1} p^{-1} \cdot qp^{-1} q^{-1} p^{-1}$$

$$\text{by (iii), (vii)}$$

$$= pq \cdot (qp^{-1})^4 q^{-1} p^{-1} \text{ by (iii), (iv), (vii)}$$

$$= 1 \text{ by (iv).}$$

(8)

$$(Dbab^{-1} a^{-1})^2 = (q^2 pq^{-1} \cdot pqpq \cdot (q^{-1} p^{-1})^4 \cdot (pq^{-1})^4 \cdot qpqp \cdot q^{-1} pqp^2 q^{-1})^2$$

$$= (q^2 p \cdot q^{-1} p^{-1} q^{-1} p^{-1} \cdot q^2 \cdot p^{-1} q^{-1} p^{-1} q^{-1} \cdot pqp^2 q^{-1})^2 \text{ by (iv), (iii), (vii)}$$

$$= q^2 (p^2 q)^8 q^{-2} \text{ by (iii)}$$

$$= 1 \text{ by (ix).}$$

(9)

$$(Dbaba^{-1})^2 = (q^2 pq^{-1} \cdot p \cdot qpq \cdot (q^{-1} p^{-1})^4 \cdot (pq^{-1})^4 \cdot qpqp^2 \cdot q^{-1} p^{-1} qp^{-1} q^{-1})^2$$

$$= (q^2 p^2 \cdot qp \cdot (p^{-1} q^{-1})^4 \cdot q^{-1} p^{-1} \cdot qp \cdot (p^{-1} q)^4 \cdot q^{-1} p^{-1} \cdot p^2 q^2)^2 \text{ by (iii), (iv), (vii)}$$

$$= 1 \text{ by (iii), (iv), (vi).}$$

(10)

$$Dbacb^{-1}D^{-1}aba^{-1}c^{-1}b^{-1}a^{-1} = q^2pq^{-1}\cdot pqpq\cdot\left(q^{-1}p^{-1}\right)^4\cdot\left(pq^{-1}\right)^4\cdot qpqp\cdot q^{-1}\cdot pqpq^2\cdot\left(qp^{-1}\right)^4$$
$$\cdot pqp\cdot q^2p^{-1}q^{-1}p^2q^{-1}p^{-1}\cdot q^2\cdot q^{-1}p^{-1}q^{-1}p^{-1}\cdot qpq^{-1}$$
$$= q^2pq^{-1}p\cdot qpq^2pqp^2\cdot qpq^{-1}\cdot pqp^2qpq^2pq^{-1} \quad\text{by (iii), (vii)}$$
$$= q^2p\left(q^{-1}p^{-1}\right)^4p^{-1}q^2 \quad\text{by (vii)}$$
$$= 1 \quad\text{by (iii).}$$

(11)

$$Dbc^{-1}bDbabca^{-1}b^{-1}D^{-1}b^{-1}cb^{-1}D^{-1}ac^{-1}b^{-1}a^{-1}$$
$$= q^2pq^2\cdot p^{-1}q^{-1}p^{-1}\left(pq^{-1}\right)^4\cdot q^2p^{-1}q^{-1}p^{-1}\cdot qpq^{-1}p^2q^2pq^{-1}\cdot q^{-1}p^{-1}q^{-1}p^{-1}\cdot\left(pq^{-1}\right)^4\cdot qpqp^{-1}q^{-1}p^2q$$
$$\cdot p^{-1}q^{-1}p^{-1}q^{-1}\cdot\left(qp^{-1}\right)^4\cdot pqpq^2\cdot p^{-1}q^2p^2\cdot qp^{-1}q^{-1}\cdot pqpq^2\cdot p\cdot\left(p^{-1}q\right)^4\cdot qp\cdot q^2p^{-1}q^{-1}pq^{-1}p^2qpq^{-1}$$
$$= q^2pqp^{-1}q^{-1}\cdot p^2q^2p^2q^2\cdot pq^{-1}pqpq^2pqp^{-1}q^{-1}p^2q^2\cdot pqp^2qpq^2\cdot p^{-1}q^2p^2q\cdot p^{-1}q^{-1}p^{-1}q^{-1}\cdot p^{-1}q^2$$
$$\cdot p^{-1}q^{-1}p^{-1}q^{-1}\cdot q^{-1}p^{-1}q^{-1}p^{-1}pq^{-1}p^2qpq^{-1} \quad\text{by (iv), (iii), (vii)}$$
$$= q^2p\cdot qp^{-1}qp^{-1}\cdot p^{-1}q^2p^{-1}q^2p^{-1}q^{-1}p^{-1}\cdot(pq)^4\cdot qpqp^{-1}\cdot q^{-1}pq^{-1}p\cdot pqp$$
$$\cdot p\left(p^2q^2\right)^4\cdot q^{-1}pq^{-1}\cdot p^{-1}\cdot q^{-1}pq^{-1}p\cdot pqpq^{-1} \quad\text{by (v), (vii), (iii)}$$
$$= q^2p^2q^{-1}pq^{-1}p^{-1}q^{-1}p^2q^{-1}\cdot q^2p^{-1}q^{-1}p^2q^{-1}p^{-1}\cdot q^{-1}pqp^2q^{-1}p\cdot pqpq\cdot p^2q^{-1}p^{-1}q^2$$
$$\text{by (v), (iii), (iv)}$$
$$= q^2p^2q^{-1}pq^{-1}p^{-1}q^{-1}\cdot pq^{-1}\cdot(qp)^4\cdot pq^{-1}pq^{-1}\cdot p^{-1}q^{-1}pq^{-1}p^{-1}q^2 \quad\text{by (iii), (vii)}$$
$$= q^2p\left(pq^{-1}\right)^4p^{-1}q^2 \quad\text{by (iii), (iv)}$$
$$= 1 \quad\text{by (iv).}$$

(12)

$$abc^{-1}b^{-1}a^{-1}Dbc^{-1}bca^{-1}b^{-1}D^{-1}b^{-1}ca$$
$$= qp^2q^2p^{-1}\cdot(pq)^4\cdot q\cdot p^{-1}q^{-1}p^{-1}\left(pq^{-1}\right)^4q^2\cdot p^{-1}q^{-1}p^{-1}\cdot qp^2q^{-1}p^2q$$
$$\cdot p^{-1}q^{-1}p^{-1}q^{-1}\left(qp^{-1}\right)^4\cdot pqpq^2p^{-1}q^2p^2qp^{-1}q^{-1}pqpq^{-1}pqpq^{-1}p^{-1}$$
$$= qp^2q^2p^{-1}\cdot qpqp\cdot q^2\cdot pqpq\cdot p^2q^{-1}p^2\cdot q^2pqp^2qp\cdot q^2p^{-1}q^2p^2qp^{-1}\cdot q^{-1}pqpq^{-1}\cdot pqpq^{-1}p^{-1}$$
$$\text{by (iii), (iv), (vii)}$$
$$= q^{-1}\cdot q^2p^2q^2p^2q^2\cdot qpq\cdot\left(q^{-1}p\right)^4\cdot p\cdot q\cdot q^2p^2q^2p^2q^2p^2q^2p^2\cdot p^{-1}\cdot p^{-1}q^{-1}p^{-1}q^{-1}$$
$$\cdot pqpq^{-1}\cdot pqpq^{-1}p^{-1} \quad\text{by (iii), (iv), (vii)}$$
$$= q^{-1}p^2q^2\cdot pq^{-1}pq^{-1}\cdot q^{-1}p^{-1}\cdot qpqp^2qpq\cdot q^2pqpq^{-1}p^{-1} \quad\text{by (iii), (iv), (v)}$$
$$= q^{-1}p^2\cdot q^{-1}p^{-1}q^{-1}q^2\left(q^{-1}p^{-1}\right)^4p^{-1}q^{-1}p^2q^{-1}p^{-1}(pq)^4\cdot q^2p^{-1} \quad\text{by (iv), (vii)}$$
$$= 1 \quad\text{by (iii), (vii).}$$

(13)

$$D^{-1}abca^{-1}Db^{-1}c^{-1} = pq^2p^{-1}q^{-1}\cdot q^{-1}pq^{-1}p\cdot(p^{-1}q^{-1})^4\cdot(qp^{-1})^4\cdot pq^{-1}p^2q^2p^{-1}q^2pq^2$$
$$qpp^{-1}q^{-1}p\cdot pq\cdot p^{-1}q^{-1}p^{-1}q^2p^{-1}q^{-1}p^{-1}\cdot pqpq^2p^{-1}$$
$$= pq^2p^{-1}q^{-1}p^{-1}qp\cdot p^2qpq^{-1}\cdot p^2qpq^{-1}\cdot(qp^{-1})^4\cdot pq^{-1}\cdot pqpq^2pqp$$
$$p^2q^{-1}pq^{-1}p^{-1}(pq)^4\cdot qpqpqp\cdot p^{-1}q^{-1}pqpq^2p^{-1} \quad\text{by (iii), (iv), (vii)}$$
$$= pq^2p^{-1}q^{-1}p^{-1}qp\cdot(p^2qpq^{-1})^4\cdot p^{-1}q^{-1}pqpq^2p^{-1} \quad\text{by (iii), (iv), (vii)}.$$

Then (xii) yields Lemma 2 (13).

(14)

$$Dba^2b = q^2p\cdot q^2p^{-1}q^{-1}p^2\cdot p(pq^{-1})^4\cdot qpqp\cdot q^{-1}p^{-1}qp^2q^{-1}p^2$$
$$= q^2p^2qp^2q^2p^2q^{-1}p^2 \quad\text{by (iii), (iv), (vii)}$$
$$= q^2p^2q^{-1}p^2\cdot q^2p^2qp^2 \quad\text{by (v)}.$$

By (v), (xiii) follows Lemma 2 (14).

(15)

$$Dba^2bc^{-1}bc^{-1}b = q^2p\cdot q^2p^{-1}q^{-1}p^2q^{-1}p^{-1}q^{-1}\cdot(qp)^2\cdot(pq^{-1})^2\cdot(qp)^2\cdot q^{-1}p^{-1}qp^2q^{-1}pq$$
$$p^{-1}q^{-1}p^{-1}qpq^{-1}pqp^{-1}q^{-1}p^{-1}qpq^{-1}p^2$$
$$= q^2p^2\cdot q^{-1}\cdot q^2p^2q^2p^2q^2p^2\cdot p\cdot(pq)^4\cdot q^{-1}p^{-1}q^{-1}p^2q^{-1}p^{-1}q^{-1}\cdot q^2pq^{-1}$$
$$pqp^{-1}q^{-1}p^{-1}qpq^{-1}p^2 \quad\text{by (iii), (iv), (vii)}$$
$$= q^2p^2q^{-1}p^2q^2\cdot pqpq\cdot p^2qp\cdot q^{-1}pq^{-1}p\cdot qp^{-1}q^{-1}p^{-1}qpq^{-1}p^2 \quad\text{by (iii), (v), (vii)}$$
$$= q^2p^2q^{-1}p^2qp^{-1}\cdot q^{-1}pq^{-1}p^{-1}p\cdot p^{-1}q^{-1}p^{-1}q^2p^{-1}q^{-1}p^{-1}\cdot qpq^{-1}p^2 \quad\text{by (iii), (iv)}$$
$$= q^2p^2q^{-1}\cdot p^2qp^2qp^{-1}\cdot qpq\cdot p\cdot q^2\cdot pqpq\cdot pq^{-1}p^2 \quad\text{by (iv), (vii)}$$
$$= q^2p^2q^{-1}\cdot p^2qp^2qp^2q^{-1}\cdot p^2q^2p^2q^2p^2q^{-1}\cdot qp^2q^2 \quad\text{by (iii)}$$
$$= q^2p^2q^{-1}\cdot(p^2q)^4qp^2q^2 \quad\text{by (v)}$$
$$= 1 \quad\text{by (ix)}.$$

(16) and (17)

$$(c^{-1}b^{-1}a^{-1}Dca^{-1})^2 = (p^{-1}\cdot p^{-1}q^{-1}\cdot p^{-1}(pq)^4\cdot qp^{-1}qp^{-1}\cdot pqpq^2pqp\cdot q^{-1}p^2qp^{-1}q^{-1})^2$$
$$= (p^2q^2p\cdot q^{-1}p^{-1}q^{-1}p^{-1}\cdot q^2\cdot qp(p^{-1}q^{-1})^4\cdot qp^{-1}qp^{-1}\cdot q^{-1})^2$$
$$\text{by (iii), (iv), (vii)}$$
$$= p^2q^2p^2qp^{-1}q^{-1}p\cdot q^2p^2q^2p^2\cdot qp^{-1}q^{-1}pq^2 \quad\text{by (iii), (iv)}$$
$$= p^2q^2p^2q^2\cdot q^{-1}p^{-1}q^{-1}p^{-1}q^{-1}\cdot q^{-1}p^{-1}\cdot p^{-1}q^{-1}p^{-1}q^{-1}\cdot pq^2 \quad\text{by (v)}$$
$$= q^2p^2q^{-1}\cdot q^{-1}p^{-1}qp\cdot q^{-1}p^{-1}qp\ qp^2q^2 \quad\text{by (iii), (v)}.$$

Then follows Lemma 2 (16) by (vi) and moreover

$$\left(c^{-1}b^{-1}a^{-1}Dca^{-1}\right)^2 \cdot b^2$$

$$= q^2p^2q^{-1}\cdot q^{-1}p^{-1}qpq^{-1}p^{-1}qp\cdot p^{-1}\cdot q^2p^2q^2p^2q^2p^2p^{-1}\cdot p^{-1}q^{-1}p^{-1}q^{-1}\cdot pqp^{-1}q^{-1}p^{-1}$$

$$= q^2p^{-1}\cdot p^{-1}q^{-1}p^{-1}q^{-1}\cdot pq\cdot p^{-1}q^{-1}p^{-1}q^{-1}\cdot q^{-1}p^{-1}\cdot qpqp^2qpqq^{-1}p^2q^{-1}p^{-1} \quad \text{by (iii), (v), (vi)}$$

$$= q^2p^{-1}q\cdot pqp^2q^2\cdot pqp^2q\cdot p^{-1}q^{-1}p^{-1}q^{-1}\cdot q^{-1}p^2q^{-1}p^{-1} \quad \text{by (iii), (vii)}$$

$$= q^2p^{-1}q\cdot pqp^2q^2\cdot pqp^2q^2\cdot pqp^2q^2\cdot q^2p^{-1}q^{-1}p^2q^{-1}p^{-1} \quad \text{by (iii)}$$

$$= q^2p^{-1}q\cdot\left(pqp^2q^2\right)^4\cdot q^{-1}pq^2 \quad \text{by (vii).}$$

Then follows Lemma 2 (17) by (xiv).

(18)

$$DbaDbc^{-1}bDbac^{-1} = q^2p\cdot q^2p^{-1}q^{-1}\cdot\left(q^{-1}p\right)^4p^{-1}\cdot qpqp\cdot q^{-1}p^{-1}q^2pq^2\cdot p^{-1}q^{-1}p^{-1}\cdot\left(pq^{-1}\right)^4$$

$$q^2p^{-1}q^{-1}p^{-1}\cdot qp\cdot q^{-1}p^2q^2p\cdot q^2p^{-1}q^{-1}p^{-1}\cdot\left(pq^{-1}\right)^4\cdot qpqp\cdot q^{-1}p^2qp^{-1}q^{-1}p^2$$

$$= q^2p^2qp^2q^{-1}pqp^{-1}q^{-1}\cdot p^2q^2p^2q^2\cdot p^2qp^2\cdot qp^{-1}qp^{-1}\cdot q^{-1}p^2 \quad \text{by (iii), (iv), (vii)}$$

$$= q^2p^2q^2\cdot q^{-1}p^2\cdot q^{-1}p^2\cdot\left(p^{-1}q\right)^4\cdot q^{-1}pq^{-1}\cdot qp\left(p^{-1}q^{-1}\right)^4\cdot q\cdot p^2q^2p^2q^2\cdot q^2$$

$$= q^2p^2q^2\cdot\left(q^{-1}p^2\right)^4\cdot q^2p^2q^2 \quad \text{by (iii), (iv), (v).}$$

Then follows Lemma 2 (18) by (ix).

(19)

$$Fa^2 = pq^2pq\cdot qpq^{-1}p^{-1}\left(pq^{-1}\right)^4qpq^{-1}p^{-1}(pq)^4\cdot q^{-1}p^{-1}q^2p^{-1} \ .$$

Then (iii), (iv), (vi) obviously yield Lemma 2 (19).

(20)

$$b^{-1}Fbc^2b^{-1}Fbc^2 = p^2qp^{-1}\cdot qpq^2pqp^2\cdot p^2\cdot qpq^2pqp^2\cdot q^{-1}p^{-1}qpq^{-1}\cdot p^{-1}qp^{-1}q$$

$$pq^2pq^2pq^2pqp^2q^{-1}p^{-1}qpq^{-1}p$$

$$= p^2\cdot qpqp\cdot\left(p^{-1}q^2\right)^8\cdot q^2pq^2\cdot pq^{-1}pq^{-1}\cdot q^{-1}pq^{-1}p\cdot pq^2pq^2pq^2$$

$$pqp^2q^{-1}p^{-1}qpq^{-1}p \quad \text{by (vii), (iv)}$$

$$= p\cdot q^{-1}p^{-1}qpq^{-1}p^{-1}qp\cdot pqp^{-1}q\cdot pq^2pq^2pq^2p^2\cdot q^{-1}p^2q^{-1}$$

$$p^{-1}qpq^{-1}p \quad \text{by (ix), (iv), (iii)}$$

$$= q^{-1}pqp^{-1}q^{-1}\cdot pqpq\cdot p\cdot q^{-1}p^{-1}q^2p^{-1}q^2p^{-1}\cdot q^2p^{-1}q^{-1}p^2q^{-1}p^{-1}\cdot qpq^{-1}p$$

$$\text{by (ix), (vi)}$$

$$= q^{-1}pqp^{-1}q^2p^{-1}q^{-1}\cdot p^2q^{-1}p^{-1}q^2p^{-1}q^{-1}\cdot p^2q\cdot pq^{-1}pq^{-1}\cdot p \quad \text{by (vii), (iii)}$$

$$= 1 \quad \text{by (vii), (iv).}$$

(21)

$$a^{-1}Fa^{-1}b^2c^2 = p^{-1}\cdot p^{-1}q^{-1}\left(qp^{-1}\right)^4 pq\cdot p^{-1}q^{-1}pq\left(q^{-1}p^{-1}\right)^4 p^2 qpq^{-1}p^{-1}qp^2 q^{-1}p^{-1}qpq^{-1}p$$

$$= p^{-1}q^{-1}p^2\cdot pqpq^2 p\cdot \left(p^{-1}q\right)^4\cdot qp\cdot p^{-1}q^2 p^2 q^{-1}p^{-1}qp^2 q^{-1}p^{-1}qpq^{-1}p \quad \text{by (iv), (iii)}$$

$$= p^{-1}q^{-1}pq^{-1}p^{-1}q\cdot qp^{-1}qp^{-1}\cdot p\cdot q^2 p^2 q^2 p^2 q^2 p\cdot p^{-1}qp^{-1}q\cdot p^2 q^{-1}\cdot p^{-1}qpq^{-1}p \quad \text{by (iv), (vii)}$$

$$= p^{-1}q^{-1}pq^{-1}p^{-1}\cdot qpq^{-1}pq^{-1}p^{-1}q^2 p^{-1}q^{-1}p^2\cdot p^{-1}q^{-1}p^{-1}q^{-1}p^{-1}\cdot qpq^{-1}p \quad \text{by (v), (iv)}$$

$$= p^{-1}q^{-1}pq^{-1}p^{-1}qp\cdot q^{-1}p^{-1}q^{-1}p^{-1}\left(pq^2\right)^8\cdot q^2 p^{-1}q^2 p^{-1}q^2 p^{-1}qp \quad \text{by (vii), (iii)}$$

$$= p^{-1}q^{-1}pq^{-1}p^{-1}qp^2\cdot qpq^{-1}p^{-1}\cdot qp\cdot \left(p^{-1}q\right)^4\cdot q^{-1}p^{-1}\cdot p^2 q^{-1}pqp^{-1}qp \quad \text{by (ix), (iii).}$$

Then by (iv), (vi) we have $\left(a^{-1}Fa^{-1}b^2c^2\right)^2 = 1$. Hence Lemma 2 (19) yields Lemma 2 (21).

(22)

$$Fab^{-1}ab^{-1}Fbc^2abc^2a = pq^2 pq^2\cdot p^2\cdot p^{-1}q^{-1}p^{-1}\cdot \left(pq^{-1}\right)^4\cdot q^2 p^{-1}q^{-1}p^{-1}\cdot qpq^{-1}pqp^{-1}\cdot qpq^2 pqp^2\cdot p^2$$
$$\cdot qpq^2 pqp^2\cdot q^{-1}p^{-1}qpq^{-1}pqp^{-1}q^{-1}p^{-1}qpq^{-1}pqpq^{-1}p^{-1}$$

$$= pq^2 pq^2 p^{-1}qpq\cdot (qp)^4\cdot p^{-1}q\cdot (qp)^4\cdot p^{-1}q^{-1}\cdot p^{-1}q^{-2}p^{-1}q^2 p^{-1}q^2 p^{-1}q^2 p^{-1}q^2$$
$$\cdot p^{-1}qpq^{-1}pqp^{-1}q^{-1}p^{-1}qpq^{-1}pqpq^{-1}p^{-1} \quad \text{by (iv), (vii)}$$

$$= pq^2 pq^2 p^{-1}qpq\cdot p^{-1}qp^{-1}q\cdot pq^{-1}\cdot q^{-1}pq^{-1}p\cdot q^{-1}pqp^{-1}q^{-1}p^{-1}qpq^{-1}pqpq^{-1}p^{-1} \quad \text{by (iii), (ix)}$$

$$= pq^2 pq\cdot qp^{-1}qp^{-1}\cdot p^{-1}q^{-1}\cdot p^2 q^{-1}p^{-1}q^2 p^{-1}q^{-1}\cdot p^2 q^{-1}p^{-1}\cdot pqpqp$$
$$\cdot q^{-1}pqpq^{-1}p^{-1} \quad \text{by (iv)}$$

$$= pq(pq)^4 q^{-1}p^{-1} \quad \text{by (vii), (iv), (iii)}$$
$$= 1 \quad \text{by (iii).}$$

(23)

$$Fab^{-1}abca^{-1}Fbc^{-1}a^{-1}c^{-1}b^{-1}a^{-1}ca$$
$$= pq^2 pq^2 pq^{-1}p^{-1}\cdot \left(pq^{-1}\right)^4\cdot q^2 p^{-1}q^{-1}p^2\cdot pqp^{-1}q^{-1}\cdot p^2\cdot qp^{-1}qp^{-1}q^2 p^{-1}q^2 p^{-1}\cdot q^{-1}pq^{-1}p$$
$$\cdot qp^{-1}q^{-1}p^{-1}qp^{-1}q^{-1}p^{-1}qpq^{-1}p^2 qpq^{-1}pqpq^{-1}p^{-1}$$

$$= pq^2 pq^2 p^{-1}qpq^{-1}pqp^2\cdot qpqpqp^{-1}\cdot q^2 p^2 q^2 p^2\cdot qpq^2\cdot pqpq$$
$$p^{-1}q^{-1}p^{-1}qpq^{-1}p^2 qpq^{-1}pqpq^{-1}p^{-1} \quad \text{by (iv), (vii), (iii)}$$

$$= pq^2 pq^2 p^{-1}qpq^{-1}\cdot pqpq\cdot p^2 q^2\cdot p^2 q^{-1}p^{-1}q^2 p^{-1}q^{-1}\cdot p^{-1}qp^{-1}q^{-1}p^{-1}$$
$$qpq^{-1}p^2 qpq^{-1}pqpq^{-1}p^{-1} \quad \text{by (v), (iii)}$$

$$= pq^2pq^2p^{-1}qpq^2p^2qp^{-1}\left(pq^{-1}\right)^4 \cdot q^{-1}\cdot q^{-1}p^{-1}q^{-1}(qp)^4 \cdot p^2q^{-1}p^{-1}$$

$$qpq^{-1}p^2qpq^{-1}pqpq^{-1}p^{-1} \quad \text{by (vii), (iii)}$$

$$= pq^2pq\cdot qp^{-1}qp^{-1}\cdot p^2q^2p^2q^2p^2q^2\left(q^{-1}p\right)^4p^{-1}\cdot qpqp\cdot q^{-1}pqpq^{-1}p^{-1}$$

$$\text{by (iv), (iii), (vii)}$$

$$= pq(qp)^4q^{-1}p^{-1} = 1 \quad \text{by (v), (iv), (iii).}$$

(24)

$$abca^{-1}Fbc^{-1} = q\cdot p^{-1}q^{-1}p^{-1}q^{-1}\cdot\left(qp^{-1}\right)^4pqp^{-1}q^2p^2q\cdot\left(q^{-1}p\right)^4q\cdot q^{-1}p^{-1}q^2p^{-1}q^{-1}p^2$$

$$= q^2pqp^2qp\cdot p^2q^2p^2q^2\cdot p^2qpq^2pq \quad \text{by (iv), (iii), (vii)}$$

$$= p^{-1}q^{-1}pq\cdot\left(q^{-1}p\right)^4p^{-1}q^{-1}pq \quad \text{by (v), (vii).}$$

By (vi), (iv) we have then $\left(abca^{-1}Fbc^{-1}\right)^2 = 1$. Moreover

$$abca^{-1}Fbc^{-1}\cdot b^2\cdot abca^{-1}Fbc^{-1}b^2 = p^{-1}q^{-1}pqp^{-1}q^{-1}\cdot pqpq\cdot pq^{-1}p^{-1}qp\cdot q^{-1}pq^{-1}p\cdot qp^{-1}q^{-1}\cdot pqpq$$

$$\cdot pq^{-1}p^{-1}qpq^{-1}p^2$$

$$= p^{-1}\cdot q^{-1}pq^{-1}p\cdot\left(p^{-1}q^2\right)^8\cdot q^{-1}pq^{-1}p\cdot p \quad \text{by (iii), (iv)}$$

$$= 1 \quad \text{by (ix), (iv).}$$

(25)

$$c^{-1}bcbcFac^{-1}b^{-1}FDbac^2b^{-1}a^{-1}c^{-1}Fac^{-1}b^{-1}FDbac^2b^{-1}a^{-1}$$

$$= p^{-1}qp^{-1}q^{-1}p^{-1}qp^2q^{-1}p^2qp^2q^{-1}p^2q^2pq^2pq^2\cdot pq^{-1}pq^{-1}\cdot p\cdot pqp^2qpq^2\cdot pq^2p^2q^2p^{-1}q^2\cdot pq^{-1}pq^{-1}$$

$$\cdot p^2qpq^{-1}p\cdot qpq^{-1}p^{-1}qpq^{-1}p^{-1}\cdot qp^2q^{-1}p^{-1}q\cdot p^{-1}q^{-1}p^{-1}q^{-1}\cdot q^{-1}pq^2pq^2\cdot pq^{-1}pq^{-1}\cdot p\cdot pqp^2qpq^2$$

$$\cdot pq^2p^2q^2p^{-1}q^2\cdot pq^{-1}pq^{-1}\cdot p^2qpq^{-1}p\cdot qpq^{-1}p^{-1}\cdot qpq^{-1}p^{-1}\cdot qp^2q^{-1}$$

$$= p^{-1}qp^{-1}q^{-1}p^{-1}qp^2q^{-1}p^2qp^2q^{-1}p^2q^2pq^2p\cdot q^{-1}p^{-1}q^{-1}p^{-1}\cdot q^{-1}p^2qp^2q^{-1}\cdot q^{-1}p^{-1}q^{-1}p^{-1}\cdot qpqp$$

$$\cdot q^{-1}p^2qp^{-1}q^{-1}pqpq^{-1}p^{-1}qp^{-1}q^{-1}p^{-1}q^2pq^2p\cdot q^{-1}p^{-1}q^{-1}p^{-1}q^{-1}p^2q\cdot p^2q^{-1}\cdot q^{-1}p^{-1}q^{-1}p^{-1}\cdot qpqp$$

$$\cdot q^{-1}p^2qp^{-1}q^{-1}pqpq^{-1} \quad \text{by (iii), (iv), (vi), (vii)}$$

$$= p^{-1}qp^{-1}q^{-1}p^{-1}qp^2q^{-1}p^2qp^2q^{-1}p^2q^2pq^2p^{-1}q^{-1}\cdot p^{-1}qp^{-1}q\cdot p^{-1}q^{-1}pqpq^{-1}$$

$$\cdot p^{-1}qp^{-1}q^{-1}p^{-1}q^2pq^2p^2q\cdot p^{-1}qp^{-1}q\cdot p^2\cdot qp^{-1}qp^{-1}\cdot q^{-1}pqpq^{-1} \quad \text{by (iii), (iv), (vii)}$$

$$= p^{-1}qp^{-1}q^{-1}p^{-1}qp^2q^{-1}p^2qp^2q^{-1}p^2q^2pq\cdot qp^{-1}qp^{-1}\cdot pqpq^2pqp\cdot q^{-1}p^{-1}\cdot q\cdot p^{-1}q^{-1}p^{-1}q^{-1}\cdot q^{-1}p$$

$$\cdot q^2p^{-1}q^2pq\cdot pqpq^{-1} \quad \text{by (iii), (iv), (vii)}$$

$$= p^{-1}qp^{-1}q^{-1}p^{-1}qp^2q^{-1}p^2q\cdot p^2q^{-1}p^2q^2pqpq^{-1}p^2q\cdot p^2q^{-1}pqpq^{-1}pqpq^{-1}\cdot p^2qp^2qp \quad \text{by (iii), (iv), (vii)}$$

$$= p^{-1}qp^{-1}q^{-1}p^{-1}q\cdot p^2q^{-1}p^2qpqp^{-1}qpq^{-1}p^{-1}q^{-1}pqp^{-1}qp^2\cdot q^{-1}pqpq^{-1}pqp\cdot q^{-1}p^2qp^2qp \quad \text{by (iii), (iv), (vii)}$$

$= p^{-1}qp^{-1}q^{-1}p^{-1}qpqp^{-1}qpqp^{-1}q^2p\cdot q^2p^2q^{-1}p^{-1}q^{-1}pqpq^{-1}\cdot pqpq^{-1}p^2qp^2qp$ by (iii), (vii)

$= p^{-1}qp^{-1}q^{-1}p^{-1}qpqp^{-1}qpqp^{-1}\cdot q^2pq^{-1}pqp^{-1}q^{-1}p^{-1}qpqpq^{-1}p^2qp^2\cdot qp$ by (iii), (vii)

$= p^{-1}qp^{-1}q^{-1}p^{-1}qpq\cdot p^{-1}qpqp^{-1}q^{-1}p^{-1}q^{-1}q^{-1}p^{-1}q^{-1}p\cdot q^{-1}p^{-1}qp$ by (iii), (vii)

$= p^{-1}qp^{-1}q^{-1}p^{-1}\cdot qpqp^2qpq\cdot p^{-1}q^{-1}p^{-1}qp$ by (iii), (vii)

$= p^{-1}q^2\left(q^{-1}p^{-1}\right)^4\left(p^{-1}q^{-1}\right)^4q^2p$ by (iii), (vii)

$= 1$ by (iii).

$\quad$ (26)

$caFac^{-1}b^{-1}FDbac^2b^{-1}Fac^{-1}b^{-1}FDba^2bc$

$= p^2qpq^{-1}\cdot pqpq\cdot pq^2pq^2\cdot pq^{-1}pq^{-1}\cdot p\cdot pqp^2qpq^2\cdot pq^2p^2q^2p^{-1}q^2pq^{-1}pq^{-1}\cdot p^2qpq^{-1}p\cdot qpq^{-1}p^{-1}$

$\quad\cdot qpq^{-1}p^{-1}\cdot qp^{-1}qpq^2pq^2\cdot pq^{-1}pq^{-1}\cdot p\cdot pqp^2qpq^2\cdot pq^2p^2q^2p^{-1}q^2\cdot pq^{-1}pq^{-1}\cdot p^2qpq^{-1}p^{-1}qp^{-1}q^{-1}p$

$= p^2qpq^2p^{-1}qp\cdot q\cdot^{-1}p^{-1}q^{-1}p^{-1}\cdot q^{-1}p^2qp^2q^{-1}\cdot q^{-1}p^{-1}q^{-1}p^{-1}\cdot qpqp\cdot q^{-1}p^2\cdot qp^{-1}q^{-1}\cdot pqp^2qpq^2$

$\qquad\qquad\cdot p\cdot q^{-1}p^{-1}q^{-1}p^{-1}\cdot q^{-1}p^2qp^2q^{-1}\cdot q^{-1}p^{-1}q^{-1}p^{-1}\cdot qpqp\cdot q^{-1}p^{-1}qp^{-1}q^{-1}p$

$\qquad\qquad\qquad$ by (iii), (iv), (vi), (vii)

$= p^2qpq^2p^{-1}qp^{-1}\cdot p^{-1}qp^{-1}q\cdot p^2\cdot q^{-1}p^{-1}q^{-1}p^{-1}\cdot q^2\cdot p^{-1}q^{-1}p^{-1}q^{-1}\cdot p^2\cdot qp^{-1}qp^{-1}\cdot q^{-1}p^2q^{-1}p$

$\qquad\cdot qp^{-1}qp^{-1}\cdot p^{-1}\cdot q^{-1}p^{-1}q^{-1}p^{-1}\cdot q^2\cdot p^{-1}q^{-1}p^{-1}q^{-1}p^{-1}qp^{-1}q^{-1}p$ by (iii), (iv), (vii)

$= p^2qpq^2\cdot p^{-1}qp^{-1}q\cdot p^2q^{-1}\cdot p^2q^{-1}p^{-1}q^2p^{-1}q^{-1}\cdot p$ by (iii), (iv)

$= 1$ by (iii), (iv), (vii).

$\quad$ (27)

$aca^{-1}b^{-1}D^{-1}Fbca^{-1}Fa^{-1}c^{-1}Fac^{-1}b^{-1}FDba$

$= qpq^{-1}pqpq^{-1}p^2qp^{-1}q^{-1}p^2\cdot qp^{-1}qp^{-1}\cdot q^2pq^2p^2q^{-1}\cdot q^{-1}p^{-1}q^2p^{-1}q^{-1}p^2\cdot q^{-1}p^2\cdot qp^{-1}qp^{-1}\cdot q^2p^{-1}q^2$

$\qquad\cdot p^{-1}q^{-1}p^{-1}q^{-1}\cdot p^{-1}q\cdot p^{-1}q^{-1}p^{-1}q^{-1}\cdot q^{-1}pq^2pq^2\cdot pq^{-1}pq^{-1}\cdot p\cdot pqp^2qpq^2\cdot pq^2p^2q^2p^{-1}q^2$

$\qquad\qquad\qquad\cdot pq^{-1}pq^{-1}\cdot p^2qpq^{-1}p^{-1}$

$= qpq^{-1}pqpq^{-1}p^2q\cdot p^{-1}q^{-1}p^{-1}q^{-1}\cdot pqpq^2p^2q^{-1}p^2\cdot qpqp^{-1}\cdot q^{-1}q^{-1}pq^{-1}p\cdot qpq^{-1}p^2pq^{-1}p^{-1}q^{-1}p^{-1}$

$\qquad\qquad\cdot q^{-1}p^2qp^2q^{-1}\cdot q^{-1}p^{-1}q^{-1}p^{-1}\cdot qpqp\cdot q^{-1}p^{-1}$ by (iii), (iv), (vii)

$= qpq^{-1}pqpq^{-1}p^2\cdot q^2pqp^2qp\cdot q^2p^2q^{-1}pq^{-1}\cdot p\cdot qp^{-1}q^{-1}q\cdot q^{-1}pq^{-1}p\cdot q^2p^{-1}p\cdot p^{-1}qp^{-1}q\cdot p^2q^{-1}p^{-1}q^{-1}p^{-1}$

$\qquad\qquad\qquad\cdot q^2\cdot p^{-1}q^{-1}p^{-1}q^{-1}\cdot p^{-1}$ by (iii), (iv), (vii)

$= qpq^{-1}pq\cdot pq^{-1}pq^{-1}pq^{-1}\cdot q^{-1}p^{-1}q^{-1}p^{-1}\cdot q\cdot p^{-1}q^{-1}p^{-1}q^{-1}\cdot p^{-1}q$ by (iii), (iv), (vii)

$= q\left(pq^{-1}\right)^4q^{-1} = 1$ by (iii), (iv).

(28)

$bc^2a^{-1}b^{-1}D^{-1}Fbca^{-1}Fc^{-1}b^{-1}cabcFac^{-1}b^{-1}FDbac^2b^{-1}a^{-1}c^{-1}$

$= pqp^2q^{-1}p^{-1}qpq^{-1}p^2q \cdot p^{-1}q^{-1}p^{-1}q^{-1} \cdot (qp^{-1})^4 \cdot pqpq^2p^2q^{-1} \cdot q^{-1}p^{-1}q^2p^{-1}q^{-1}p^2 \cdot q^{-1}p^2 \cdot qp^{-1}qp^{-1}$

$\qquad \cdot q^2p^{-1}q^2p^{-1}q^2p^2qp^2q^{-1} \cdot pqpq^{-1}pqp^{-1}q^{-1}p^2q^2pq^2pq^2 \cdot pq^{-1}pq^{-1} \cdot p \cdot pqp^2qpq^2 \cdot pq^2p^2q^2p^{-1}q^2$

$\qquad \cdot pq^{-1}pq^{-1} \cdot p^2qpq^{-1}p \cdot qpq^{-1}p^{-1}qpq^{-1}p^{-1} \cdot qp^2q^{-1}p^{-1}qp^{-1}q^{-1}p^2$

$= pqp^2q^{-1}p^{-1}qpq^{-1}p^2q \cdot p^{-1}q^{-1}p^{-1}q^{-1} \cdot pqpq^2p^2q^{-1}p^2 \cdot qpqp \cdot qp^{-1}q^2p^{-1}q^2p^2qp^2q^{-1}pqpq^{-1}pqp^{-1}q^{-1}$

$\qquad p^2q^2pq^2p \cdot q^{-1}p^{-1}q^{-1}p^{-1} \cdot q^{-1}p^2qp^2q^2p^{-1}q^{-1}p^{-1} \cdot qpqp \cdot q^{-1}p^2qp^{-1}q^{-1}pqpq^{-1}p^{-1}qp^{-1}q^{-1}p^2$

$\qquad\qquad\qquad\qquad\qquad\qquad\qquad\qquad\qquad \text{by (iv), (vi), (vii)}$

$= pqp^2q^{-1}p^{-1}qpq^{-1}p^2 \cdot q^2pqp^2qp \cdot q^2p \cdot pq^{-1}pq^{-1} \cdot p^2q^2p^{-1}q^2p^2qp^2q^{-1}pqpq^{-1}pqp^{-1}q^{-1}p^2q^2pq^2p^{-1}$

$\qquad \cdot p^{-1}qp^{-1}q \cdot p^2 \cdot q^2p^{-1}q^{-1}p^2q^{-1}p^{-1} \cdot q^2p^2qp^{-1}q^{-1}pqpq^{-1}p^{-1}qp^{-1}q^{-1}p^2 \quad \text{by (iii)}$

$= p \cdot qp^2q^{-1}p^{-1}q^2p^{-1} \cdot (pq^{-1})^4 \cdot q^{-1} \cdot pq^2p^{-1}q^2p^2qp^2q^{-1}pqpq^{-1}pqp^{-1}q^{-1}p^2q^2p \cdot q^2p^{-1}q^2 \cdot (qp^{-1})^4$

$\qquad\qquad\qquad\qquad\qquad \cdot pq^2pqpq^{-1}p^{-1}q \cdot p^{-1}q^{-1}p^2 \quad \text{by (iv), (vii)}$

$= pq^2pq \cdot qpqp \cdot pq^{-1}pq^{-1} \cdot q^{-1}p^{-1}q^2p^2qp^2q^{-1}pqpq^{-1}pqp^{-1}q^{-1} \cdot p^2q^2pqpq^{-1}p^2qp^2q^2p^{-1}q^{-1}p^2$

$\qquad\qquad\qquad\qquad\qquad\qquad\qquad\qquad \text{by (iii), (iv), (vi)}$

$= pq^2pqp^{-1}q^{-1}p^2qp^{-1}q^{-1}p^{-1}q^2p^2qp^2q^{-1} \cdot pqpq^{-1}pqp^{-1} \cdot q^{-1}pq^{-1}p^{-1}qpqp^{-1}q^{-1}p^{-1}qpq^2pq$

$\qquad\qquad\qquad\qquad\qquad\qquad\qquad\qquad \text{by (iii), (iv), (vii)}$

$= q^{-1}p^{-1}q \cdot pq^2p^{-1}q^2pq^2p^2q^{-1}p^{-1}q^{-1} \cdot pqpq^{-1}pqp^{-1} \cdot q^{-1}pq^{-1}p^{-1}qpqp^{-1}q^{-1}p^{-1} \cdot qpq^{-1} \cdot q^{-1}pq$

$\qquad\qquad\qquad\qquad\qquad\qquad\qquad\qquad \text{by (iii), (iv), (vii)}$

$= q^{-1}p^{-1}q \cdot pq^{-1}pq \cdot pq^{-1}pq^{-1} \cdot p^{-1}q^{-1} \cdot pq^{-1}p \cdot q^{-1}p^{-1}qpqp^{-1}q^{-1}p^{-1}qpqq^{-1}pq$

$\qquad\qquad\qquad\qquad\qquad\qquad\qquad\qquad \text{by (iii), (iv), (vii)}$

$= q^{-1}p^{-1}q \cdot pq^{-1}pq^{-1} \cdot q^{-1}p^{-1} \cdot qpqp^2qpq \cdot p^{-1}q^{-1}p^{-1}qpq^{-1} \cdot q^{-1}pq \quad \text{by (iv)}$

$= q^{-1}p^{-1}q \cdot (pq^{-1})^4 \cdot q^{-1}pq = 1 \quad \text{by (iii), (iv), (vii)}.$

(29)

$Dbc^{-1}bDbac^{-1}b^{-1}a^{-1}b^{-1}D^{-1}ca^{-1}b^{-1}D^{-1}Fbcb^{-1}aF$

$= q^2pq^2 \cdot p^{-1}q^2 \cdot pq^{-1} \cdot pq^{-1} \cdot pqp^{-1}q^{-1}p^{-1} \cdot qpq^{-1}p^2q^2pq^2p^{-1}q^2 \cdot pq^{-1}pq^{-1} \cdot p^2qpq^{-1}p^2qpq^{-1}p^2$

$\qquad \cdot qp^{-1}qp^{-1} \cdot q^2pq^2p^{-1}q^2p^2qpq^{-1}p^2qp^{-1}q^{-1}p^2 \cdot qp^{-1}qp^{-1} \cdot q^2pq^2p^2q^2p^{-1} \cdot q^2p^{-1}q^{-1}p^2q^{-1}p^{-1}$

$\qquad\qquad\qquad\qquad\qquad\qquad\qquad\qquad \cdot qp^{-1}q^{-1}p^{-1} \cdot qpqp \cdot q^2pq^2pq^2$

$= q^2pq^2 \cdot pqpq \cdot qpqp \cdot qpq^{-1}p^2q^2pq^2p^{-1}q^{-1}p^{-1} \cdot qpqp \cdot q^{-1}p^2qpq^{-1}p^{-1}q^{-1} \cdot pqpq \cdot qp^{-1}q^2p^2qpq^{-1}p^2q$

$\qquad \cdot p^{-1}q^{-1}p^{-1}q^{-1} \cdot pqpq^2p^2q^{-1}p^2qp \cdot q^{-1}p^{-1}q^{-1}p^2q^{-1}p^{-1}q^{-1} \cdot q^2pq^2pq^2 \quad \text{by (iii), (iv), (vii)}$

$= q^2pqp^{-1}q^{-1} \cdot p^2q^2p^2q^2 \cdot p \cdot q \cdot p^2q^{-1}p^{-1}q^2p^{-1}q^{-1} \cdot q^2p^2qp^2 \cdot p^{-1}q^{-1}p^{-1}q^2p^{-1}q^{-1}p^{-1} \cdot qp^{-1}q^2p^2qpq^{-1}p^2$

$\qquad\qquad \cdot q^2pqp^2qp \cdot q^2p^2q^{-1}p^2 \cdot qpqp \cdot qp^2qp \cdot q^{-1}pq^2pq^2 \quad \text{by (iii), (iv), (vii)}$

$= q^2pqp^{-1}qp^2q^{-1}p^2qp^{-1}qp^{-1}\cdot qpq^2\cdot pqpq\cdot p^{-1}q^2p^2q\cdot pq^{-1}pq^{-1}\cdot p^2\cdot q^{-1}pq^{-1}p\cdot q^{-1}pqpq^{-1}pq^2pq^2$
by (v), (iii), (vii), (iv)

$= q^2p\cdot qp^{-1}qp^{-1}\cdot p^{-1}q^{-1}p^{-1}q^{-1}\cdot p^2qp^{-1}q^{-1}\cdot p^2q^2p^2q^2\cdot p^{-1}\cdot q^{-1}pq^{-1}p\cdot q^2pq^2$ by (iii), (iv)

$= q^2p^2q^{-1}p^2\cdot qp^{-1}qp^{-1}\cdot qp^{-1}\cdot p^{-1}q^{-1}p^{-1}q^{-1}\cdot pq^2 = 1$ by (v), (iii), (iv).

(30)

$aca^{-1}b^{-1}D^{-1}Fbcb^{-1}aFa^{-1}b^{-1}D^{-1}b^{-1}cac^{-1}Dbac^{-1}b^{-1}cb^{-1}D^{-1}$

$= qpq^{-1}pqpq^{-1}p^2qp^{-1}q^{-1}p^2\cdot qp^{-1}qp^{-1}\cdot q^2pq^2p^2q^2\cdot p^{-1}q^2p^{-1}q^{-1}p^2q^{-1}\cdot p^{-1}qp^{-1}q^{-1}p^{-1}\cdot qpqp$
$\cdot q^2pq^2\cdot pq^2pqp^{-1}q^{-1}p^2\cdot qp^{-1}qp^{-1}\cdot q^2pq^2p^{-1}q^2p^2qp^{-1}q^{-1}pqpq^{-1}pqpq^{-1}p^2qp^{-1}q^{-1}p^2q^2pq^2p^{-1}q^2$
$\cdot pq^{-1}pq^{-1}\cdot p^2qpq^{-1}p^2qp^2q^{-1}pqpq^{-1}\cdot p^{-1}qp^{-1}q\cdot p^{-1}q^2pq^2p^{-1}q^2$

$= qpq^{-1}pqpq^{-1}p^2q\cdot p^{-1}q^{-1}p^{-1}q^{-1}\cdot pqpq^2p^2q^{-1}p^2qp\cdot q^{-1}p^{-1}q^{-1}p^2q^{-1}p^{-1}q^{-1}\cdot q^2pq^2pq^2pq$
$\cdot p^{-1}q^{-1}p^{-1}q^{-1}\cdot pqpq^2p^{-1}q^2p^2qp^{-1}q^{-1}pqpq^{-1}pqpq^{-1}\cdot p^2qp^{-1}q^{-1}p^2q^2pq^2p^{-1}q^{-1}p^{-1}$
$\cdot qpqp\cdot q^{-1}p^2qp^2q^{-1}\cdot pqpq^2pqp\cdot q^2p^{-1}q^2$

$= qpq^{-1}pqpq^{-1}p^2\cdot q^2pqp^2qp\cdot q^2p^2q^{-1}p^2\cdot qpqp\cdot qp^2qpq^{-1}pq^2pq^2p\cdot q^2pqp^2qp\cdot q^2p^{-1}q^2p^2qp^{-1}q^{-1}$
$pqpq^{-1}pqpq^{-1}p^2qp^{-1}q^{-1}p^2q^2p\cdot q^2p^{-1}q^{-1}p^2q^{-1}p^{-1}\cdot q^2p^2qp^2\cdot q^{-1}p^{-1}q^{-1}p^{-1}\cdot q^2\cdot p^{-1}q^{-1}p^{-1}q^{-1}$
$\cdot q^{-1}p^{-1}q^2$ by (iii), (vi), (vii)

$= qpq^{-1}pq\cdot pq^{-1}pq^{-1}\cdot p^2\cdot q^{-1}pq^{-1}p\cdot q^{-1}pqpq^{-1}pq^2pqp^2\cdot q^{-1}p^2q^2p^2q\cdot p^{-1}q^{-1}pqpq^{-1}pqpq^{-1}p^2qp^{-1}$
$\cdot q^{-1}p^2q^2p^2q\cdot p^2\cdot qp^{-1}qp^{-1}\cdot qp^2qpq^{-1}p^{-1}q^2$ by (iii), (vii)

$= qpq^{-1}pq^2p^{-1}\cdot q^{-1}pq^{-1}p\cdot q^2pqp^2qp\cdot pq^2p^{-1}\cdot p^{-1}q^{-1}p^{-1}q^{-1}\cdot pqpq^{-1}pqpq^{-1}p^2qp^{-1}qp^2q^2p^2$
$\cdot q^{-1}p^{-1}q^{-1}p^{-1}\cdot qpq^{-1}p^{-1}q^2$ by (v), (iv)

$= qpq^{-1}pq^2p^{-1}q^{-1}pq^2p^2q^{-1}p^{-1}q^2pq^2\cdot p^{-1}q^{-1}p^{-1}q^{-1}\cdot p^2q^{-1}p^{-1}\cdot qpqpq^{-1}p^2qp^{-1}qp^2q^2p^{-1}qpq^2$
$pq^{-1}p^{-1}q^2$ by (iii), (iv), (vii)

$= qpq^{-1}pq^2p^{-1}q^{-1}pq^2p^2q^{-1}p^{-1}q^{-1}\cdot q^{-1}pq^{-1}p\cdot q\cdot p^{-1}q^{-1}p^2q^{-1}p^{-1}q^2\cdot p^{-1}\cdot p^{-1}qp^{-1}qp^2q^2p^{-1}qpq^2$
$pq^{-1}p^{-1}q^2$ by (iv), (vii)

$= qpq^{-1}pq^2p^{-1}q^{-1}pq^2p^2\cdot q^{-1}p^{-1}q^{-1}p^{-1}\cdot q^2\cdot p^{-1}q^{-1}p^{-1}q^{-1}\cdot q^{-1}p^{-1}qpq^2pq^{-1}p^{-1}q^2$ by (iv), (vii)

$= qpq^{-1}pq^2p^{-1}q^{-1}pqpq^{-1}p\cdot q^{-1}p^{-1}qpq^{-1}p^{-1}qp\cdot q^2pq^{-1}p^{-1}q^2$ by (iii)

$= qpq^{-1}pq^2p^{-1}q^{-1}\cdot pqpq^2pqp\cdot p^2q^{-1}\cdot pq^{-1}pq^{-1}\cdot p^{-1}q^2$ by (vi)

$= qpq^{-1}pq^2\cdot p^{-1}q^{-1}p^{-1}q^{-1}\cdot p^{-1}q^2pq^2$ by (vii), (iv)

$= 1$ by (iii), (iv).

(31)

$Dbac^{-1}b^{-1}a^{-1}b^{-1}D^{-1}c^{-1}b^{-1}Fc^{-1}b^{-1}cb^{-1}Fac^{-1}b^{-1}a^{-1}c^{-1}b$

$= q^2pq^2p^{-1}q^2\cdot pq^{-1}pq^{-1}\cdot p^2qpq^{-1}p^2qpq^{-1}p^2\cdot qp^{-1}qp^{-1}\cdot q^2pq^2p^{-1}q^2p^{-1}\cdot qp^2qpq^2p\cdot q^2pq^2p^{-1}$

$$qp^2q^{-1}pqpq^{-1}p^{-1}qp^{-1}qpq^2pq^2 \cdot pq^{-1}pq^{-1} \cdot p^2qpq^{-1}p^{-1}qp^{-1}q^{-1}p^{-1}qpq^{-1}p^2$$

$$= q^2pq^2p^{-1}q^{-1}p^{-1} \cdot qpqp \cdot q^{-1}p^2qpq^{-1}p^{-1}q^{-1} \cdot pqpq \cdot qpq^2p^2q^{-1} \cdot q^{-1}pq^{-1}p \cdot pqpq \cdot qp^{-1}qp^2q^{-1}pqpq^{-1}$$

$$\cdot p^{-1}qp^{-1}q \cdot pq^2pq^{-1}p^{-1} \cdot qpqp \cdot q^{-1}p^{-1}qp^{-1}q^{-1}p^{-1}qpq^{-1}p^2 \quad \text{by (iv), (vii)}$$

$$= q^2p^2qp^2qp^{-1}q \cdot pq^{-1}p^{-1}q^2p^{-1}q^{-1}p^{-1} \cdot qpq^2p^2q^{-1}p^{-1}qp^2q^{-1} \cdot p^{-1}qp^{-1}q \cdot p^2q^{-1}pqpq^2pq^{-1}p^2q^2p$$

$$\cdot q^{-1}p^2q^{-1}p^{-1}q^2p^{-1} \cdot qp^{-1}q^{-1}p^{-1}qpq^{-1}p^2 \quad \text{by (iii), (iv), (vii)}$$

$$= q^2p^2qp^2 \cdot qp^{-1}qp^{-1} \cdot qpq^2 \cdot pqpq \cdot pq^2p^2q^{-1}p^{-1}qp^2q^2pq^{-1}p^{-1}q^{-1} \cdot pqpq \cdot qpq^{-1} \cdot p^2q^2p^2q^2$$

$$\cdot pqp^2q^2q^{-1}p^{-1}q^{-1}p^{-1}qpq^{-1}p^2 \quad \text{by (iii), (iv), (vii)}$$

$$= q^2p^2qp^{-1}q^{-1}p^2 \cdot qp^{-1}qp^{-1} \cdot p^{-1}q^{-1}p^{-1}qp^2q^2 \cdot pq^{-1}p^{-1}q^2p^{-1}q^{-1}p^{-1}$$

$$\cdot qpqp^2q^2p^{-1}qp^2q^2p^{-1}q^{-1}p^{-1}qpq^{-1}p^2 \quad \text{by (v), (iv), (iii), (vii)}$$

$$= q^2p^2q \cdot p^{-1}q^{-1}p^{-1}q^{-1} \cdot p \cdot q^{-1}p^{-1}q^{-1}p^{-1} \cdot qp^2q^2p^{-1}qpq^2 \cdot pqpq$$

$$\cdot pqp^2q^2p^{-1}qp^2q^2p^{-1}q^{-1}p^{-1}qpq^{-1}p^2 \quad \text{by (iv), (vii), (iii)}$$

$$= q^2p^2q^2p \cdot qp^{-1}qp^{-1} \cdot q^2p^2q^2 \cdot pqpq \cdot pq^2p^{-1}qp^2q^{-1}pqpq^2pq^{-1}p^2 \quad \text{by (iii)}$$

$$= q^2p^2q^2p^2q^{-1}pqp^2 \cdot qp^{-1}qp^{-1} \cdot qp^2q^{-1}pqpq^2pq^{-1}p^2 \quad \text{by (iii), (iv)}$$

$$= q^2p^2q^2p^2q^{-1}pq \cdot p^{-1}q^{-1}p^{-1}q^{-1} \cdot pqpq^2pq^{-1}p^2 \quad \text{by (iv)}$$

$$= q^2p^2q^2p^2 \cdot q^{-1}pq^2pqp^2q \cdot pq^2pq^{-1}p^2 = 1 \quad \text{by (vii), (v).}$$

(32)

$$D^{-1}aD^{-1}ab^{-1}aFac^{-1}abFac^{-1}$$

$$= pq^2p^{-1}q^2pq^2p^{-1}q^{-1}pqp^{-1}q^2pq^2p^{-1}q^{-1}pq^{-1}pqp^{-1} \cdot q^{-1}p^2 \cdot q^{-1}p^{-1}(pq)^4 \cdot qpq^{-1}p^{-1} \cdot (pq^{-1})^4$$

$$\cdot qp \cdot qp^{-1}q^{-1}p^2qp^2q^{-1}p^{-1}q^2pq^2pq^2 \cdot pq^{-1}pq^{-1} \cdot p^2qp^{-1}q^{-1}p^2$$

$$= pq^2p^{-1}q^2pq^2p^{-1}q^{-1}pqp^{-1}q^2pq^2p^{-1}q^{-1}pq \cdot q^2pqp^2qp \cdot p^{-1} \cdot (q^{-1}p)^4p^{-1}q^2p^{-1} \cdot q^{-1}pq^{-1}p \cdot (p^{-1}q^{-1})^4$$

$$\cdot qp \cdot qp^{-1}qp^{-1} \cdot qpqp(p^{-1}q^{-1})^4 \cdot q^{-1}pq^2pq^{-1}p^{-1}qpqp^{-1}q^{-1}p^2 \quad \text{by (iii), (iv), (vi)}$$

$$= pq^2p^{-1}q^2pq^2p^{-1}q^{-1}pqp^{-1}q^2pq \cdot qp^{-1}q^{-1}pqp^{-1}q^{-1}pqp^{-1} \cdot pq^{-1}pq^{-1}p^{-1} \cdot q^2p^2q^2p^2 \cdot q^2p^2q$$

$$\cdot q^{-1}p^2q^{-1}p^{-1}q^2p^{-1} \cdot p^{-1}q^{-1}p^2qpq^{-1}pq^2pq^{-1}p^{-1}qpqp^{-1}q^{-1}p^2 \quad \text{by (iii), (iv), (vii)}$$

$$= pq^2p^{-1}q^2pq^2p^{-1}q^{-1}pqp^{-1} \cdot q^2p^2(p^{-1}q)^4q^2p^2 \cdot q \cdot pq^{-1}pq^{-1} \cdot q^{-1}pq^{-1}p^{-1}qpqp^{-1}q^{-1}p^2$$

$$\text{by (iii), (iv), (vii)}$$

$$= pq^2p^{-1}q^2pq^2 \cdot p^{-1}q^{-1}p^{-1}q^{-1} \cdot qp \cdot pqpq^2pqp \cdot p \cdot pq^{-1}pq^{-1} \cdot p^{-1}qpqp^{-1}q^{-1}p^2 \quad \text{by (iii), (v), (iv)}$$

$$= pq^2p^{-1}q^{-1}p^{-1}(pq^{-1})^4 \cdot qpqp^2qpq \cdot p^{-1}q^{-1} \cdot p^2 \quad \text{by (iv), (vii), (iii)}$$

$$= 1 \quad \text{by (iv), (vii), (iii).}$$

(33)

$$Fac^{-1}b^{-1}a^{-1}cb^{-1} = pq^2pq^2pq^2 \cdot pq^{-1}pq^{-1} \cdot pq^{-1} \cdot qpqp \cdot q^{-1}p^2qpq^{-1}p^{-1}qp^{-1}q^{-1}p^{-1}$$

$$= pq^2pq^2p \cdot q^{-1}p^2q^{-1}p^{-1}q^2p^{-1} \cdot p^{-1}qpq^{-1}p^{-1}qp^{-1}q^{-1}p^{-1} \quad \text{by (iii), (iv)}$$

$$= pq^2 \cdot pq^2p^2q^2p \cdot qp^2qp^{-1}qpq^{-1}p^{-1} \cdot qp^{-1}q^{-1}p^{-1} \quad \text{by (vii)}$$

$$= pq^2p^{-1}q^2p^2q \cdot qp^{-1}qp^{-1} \cdot p^{-1}qp^{-1}q \cdot qp^2qpq^2p \cdot q^{-1}pq^2p^{-1}$$

$$\text{by (v), (iv), (vii)}$$

$$= pq^2p^{-1}q^2p^2qpq^2 \cdot (qpq^{-1}p^{-1})^2 \cdot q^2p^{-1}q^{-1}p^{-1}q^2q^2pq^2p^{-1} \quad \text{by (iv), (vii).}$$

Then Lemma 2 (33) follows by (vi).

(34)

$$D^{-1}Fac^{-1}abcb^{-1} = pq^2p^{-1}q^2pq^2p^2q^2p^{-1}q^2p^{-1}q^2p^{-1}qpq^{-1}p^2q \cdot p^{-1}q^{-1}p^{-1}q^{-1} \cdot qp^{-1}qp^{-1} \cdot q^2$$
$$\cdot qp^{-1}qp^{-1}q^{-1}p^{-1}$$

$$= pq^2p^{-1}q^2pq^2p^2q^2p^{-1}q^2p^{-1}q^2p^{-1}q \cdot pq^{-1}pq^{-1} \cdot q \cdot pq^2pqp^2q \cdot q \cdot p^{-1}q^{-1}p^{-1}q^2pq^2p^{-1}$$

$$\text{by (iii), (iv)}$$

$$= pq^2p^{-1}q^2pq^2p^2q^2p^{-1}q^2p^{-1}q^2 \cdot p^{-1}q^{-1}p^{-1}q^{-1} \cdot qpq^{-1} \cdot p^{-1}qpq^{-1}p^{-1}qpq^{-1}$$
$$\cdot pq^2p^{-1}q^2pq^2p^{-1} \quad \text{by (iv), (vii)}$$

$$= pq^2p^{-1}q^2pq^2p^2q^2p^{-1}qp^{-1} \cdot pqp^{-1}q^{-1}pqp^{-1}q^{-1}(qp^{-1})^4 \cdot pq^{-1}p$$
$$\cdot q^2p^2q^2p^{-1}q^2pq^2p^{-1} \quad \text{by (iii), (vi).}$$

Obviously we have Lemma 2 (34) by (iv), (vi).

(35)

$$caFab^{-1}a = p^2qpq^{-1} \cdot pqpq \cdot pq^2pq^2 \cdot pq^{-1}pq^{-1} \cdot pqp^{-1}q^{-1}p^{-1} \cdot qpq^{-1}p^{-1}$$

$$= p \cdot pqpq \cdot qp^{-1}qp^{-1} \cdot qp^{-1} \cdot p^{-1} \cdot p^{-1}q(qp)^4p^{-1}q \cdot qp^{-1} \quad \text{by (iii), (iv), (vii)}$$

$$= pq^{-1}p^{-1}q \cdot qp^{-1}q^{-1}p \cdot q^{-1}pqp^{-1} \quad \text{by (iii), (iv).}$$

Then Lemma 2 (35) follows by (vi).

(36)

$$dad^{-1}Dbacb^{-1}Fac^{-1}b^{-1}a^{-1}b^{-1}D^{-1}$$
$$= q^2pq^2p^{-1}qp \cdot q^{-1}p^{-1}q^{-1}p^{-1} \cdot (pq^{-1})^4 \cdot qpqpq^{-1} \cdot pqpq^2pqpp^{-1}q^{-1} \cdot (p^{-1}q)^4q^{-1}p^2q^2pq^2 \cdot pq^{-1}pq^{-1}$$
$$\cdot pq^{-1} \cdot qpqp \cdot q^{-1}p^2 \cdot qp^{-1}qp^{-1} \cdot q^2pq^2p^{-1}q^2$$

$$= q^2pq^2p^{-1}q \cdot p^2qpq^2pq \cdot p \cdot q^{-1}p^{-1}q^{-1}p^{-1} \cdot q^2p^{-1}q^{-1} \cdot p^2q^2p^2q^2 \cdot pq^{-1} \cdot p^2q^{-1}p^{-1}q^2p^{-1}q^{-1}$$
$$\cdot pqpq^2p^{-1}q^2 \quad \text{by (iv), (vii), (iii)}$$

$$= q^2pq^2p^{-1} \cdot (p^{-1}q)^4pq^2p^{-1}q^2 = 1 \quad \text{by (iv), (iii), (vii).}$$

(37)

$eae^{-1}a^{-1}b^{-1}D^{-1}cb^{-1}Fac^{-1}b^{-1}a^{-1}b^{-1}$

$= pq^2qp^2p^2q \cdot pq^{-1}pq^{-1} \cdot q^{-1} \cdot p^{-1}q^{-1}p^{-1}q^{-1} \cdot p^2 \cdot qp^{-1}qp^{-1} \cdot q^2pq^2p^{-1}q^2p^2 \cdot qpq^{-1} \cdot p^{-1}qp^{-1}q \cdot pq^2pq^2$

$\qquad\qquad\qquad\qquad\qquad\qquad\qquad \cdot pq^{-1}pq^{-1} \cdot p^2qpq^{-1}p^2qp^{-1}q^{-1}p^{-1}$

$= pq^2 \cdot pq^2p^2q^2p^{-1} \cdot qpqp \cdot q^2p^{-1}q \cdot qp^2qpq^2p \cdot q^{-1}p^2q^2pq^{-1}p^{-1} \cdot qpqp \cdot q^{-1}p^2qp^{-1}q^{-1}p^{-1}$

$\qquad\qquad\qquad\qquad\qquad\qquad\qquad\qquad\qquad\qquad\qquad\qquad\qquad\text{by (iv), (iii)}$

$= pq^2p^{-1}q^2p^2 \cdot qp^{-1}qp^{-1} \cdot qp^{-1}q^2p^{-1}q^{-1} \cdot p^2q^2p^2q^2 \cdot p \cdot q^{-1}p^2q^{-1}p^{-1}q^2p^{-1} \cdot p^{-1}qp^{-1}q^{-1}p^{-1}$

$\qquad\qquad\qquad\qquad\qquad\qquad\qquad\qquad\qquad\qquad\qquad \text{by (iv), (iii), (vii), (v)}$

$= pq^2p^{-1}q^2 \cdot p^{-1}qp^{-1}q \cdot p^{-1}qp^2 \cdot qp^{-1}qp^{-1} \cdot q^{-1}p^{-1} = 1 \quad \text{by (v), (iv), (vii).}$

(38)

$de \cdot a \cdot e^{-1}d^{-1}Dbab = q^2p^2q^2p^2 \cdot q \cdot pq^{-1}pq^{-1} \cdot p \cdot p^{-1}q^{-1}p^{-1}q^2p^{-1}q^{-1}p^{-1} \cdot pq^{-1}pq^{-1}pq^{-1} \cdot p^2qp^2q^{-1}p^2$

$\qquad\qquad\qquad = p^2q^2 \cdot pqpq \cdot pq^2 \cdot pqpq \cdot pqp^2q^{-1}p^2 = 1 \quad \text{by (iii), (iv), (vii), (v).}$

(39)

$dbd^{-1}Dbabca^{-1}Fbc^{-1} = q^2 \cdot pq^{-1}pq^{-1} \cdot p^2q^2 \cdot pq^{-1}pq^{-1} \cdot p^2q \cdot p^{-1}q^{-1}p^{-1}q^{-1} \cdot (qp^{-1})^4 \cdot pqp^{-1}q^2p^2$

$\qquad\qquad\qquad\qquad\qquad\qquad \cdot (pq^{-1})^4q^{-1} \cdot q^{-1}p^{-1}q^2p^{-1}q^{-1}p^2$

$\qquad\qquad = q^{-1}p^{-1}qpq^{-1}p^{-1} \cdot qpq^2pqp^2 \cdot qp^{-1} \cdot q^2p^2q^2p^2 \cdot qpq^2pq$

$\qquad\qquad = q^{-1}p^{-1}q \cdot (pq^{-1})^4 \cdot q^{-1}pq = 1 \quad \text{by (iii), (iv), (v), (vii).}$

(40)

$\qquad ebe^{-1}a^{-1}ba^{-1}F = pq^2pq^2p^{-1} \cdot qpq^2pq \cdot (q^{-1}p^{-1})^4p^2qpq^2p(p^{-1}q)^4 \cdot qp^2 \cdot pq^2p^{-1}q^2p^{-1}$

$\qquad\qquad\qquad = 1 \quad \text{by (iii), (iv), (vii).}$

(41)

$de \cdot b \cdot e^{-1}d^{-1}Dbabca^{-1}Fbc^{-1}bcFbc$

$= q^2p^2q^2p^{-1}qpqp^{-1}q^2p^{-1}q^2 \cdot pq^{-1}pq^{-1} \cdot pq^{-1} \cdot qpq \cdot p^{-1}q^{-1}p^{-1}q^{-1} \cdot (qp^{-1})^4 \cdot pqp^{-1}q^2p^2q^2p$

$\qquad\qquad \cdot p^{-1}q^{-1}(q^{-1}p)^4 \cdot p^{-1}q^2p^{-1}q^{-1}p^{-1} \cdot qp^2q^{-1}pq^2p^{-1}q^2p^{-1}q^2p^{-1}q^{-1}p^2q^{-1} \cdot p$

$= q^2p^2q^2p^{-1}qpqp^{-1}q^2p^{-1}q^{-1}pq^{-1}p^{-1} \cdot q^2p^2q^2p^2 \cdot q^2p^2qpq^2 \cdot pqpq \cdot p^2q^{-1}pq^2p^{-1}q^{-1}p^2qpq^2p^2$

$\qquad\qquad\qquad\qquad\qquad\qquad\qquad\qquad\qquad\qquad\qquad\qquad\text{by (iii), (iv), (vii)}$

$= q^2p^2q^2p^{-1}qpqp^{-1}q^2p^{-1} \cdot q^{-1}pq^{-1}p \cdot q^{-1}pqp^{-1} \cdot q^{-1}pq^{-1}p \cdot q^2p^{-1}q^{-1} \cdot p^2qpq^2p^2 \quad \text{by (v), (iii)}$

$= q^2p^2q^2p^{-1}qpqp^{-1} \cdot q^2p^2q^2p^2 \cdot q \cdot p^{-1}q^{-1}p^{-1}q^{-1} \cdot p^2qpq^2p^2$

$= q^2p^2q^2p^{-1} \cdot qpqp \cdot q^2p^{-1}qp^{-1}q \cdot pq^2p^2 = 1 \quad \text{by (iii), (iv), (v).}$

(42)

$$dcd^{-1}Dbabca^{-1}FD^{-1} = q^2pq^2pq\ pq^{-1}pq^{-1}\cdot q^{-1}\cdot pq^{-1}pq^{-1}\cdot p^2q\cdot p^{-1}q^{-1}p^{-1}q^{-1}(qp^{-1})^4\cdot pq$$
$$\cdot p^{-1}q^2p^2q^2p\cdot q^2p^{-1}q^2$$
$$= q^2pq^2pq^2p^{-1}q\cdot (p^2qpq^2pq)^2\cdot q^{-1}pq^2p^{-1}q^2p^{-1}q^2 = 1$$
$$\text{by (iii), (iv), (vi).}$$

(43)

$$ece^{-1}babca^{-1}Fbc^{-1}Dbc^{-1}$$
$$= qp^2\cdot pq^{-1}pq^{-1}\cdot p^{-1}q^2p^{-1}\cdot q^{-1}pq^{-1}\cdot pq\cdot p^{-1}q^{-1}p^{-1}q^{-1}(qp^{-1})^4\cdot pqp^{-1}q^2p^2q(q^{-1}p)^4$$
$$\cdot p^{-1}q^2p^{-1}q^{-1}p^2q^{-1}\cdot q^{-1}pq^{-1}\cdot q^{-1}p^{-1}q^{-1}p^{-1}\cdot (pq^{-1})^4\cdot q^2p^{-1}q^{-1}p^2$$
$$= pq^{-1}p^{-1}q^{-1}p^2q^2p^2q^{-1}p^{-1}\cdot qpq^2pqp^2\cdot qp^{-1}\cdot q^2p^2q^2p^2\cdot qpq^2\cdot pq^{-1}pq^{-1}\cdot pqpq^{-1}p^{-1}q^{-1}p^2$$
$$\text{by (v), (iii), (iv), (vii)}$$
$$= pq^{-1}p^{-1}q^{-1}p^2q^2p\cdot (pq^{-1})^4\cdot p^{-1}q^2pq^{-1}p^{-1}q^{-1}p^2 = 1 \quad \text{by (vii), (iii), (iv).}$$

(44)

$$dece^{-1}d^{-1}\cdot b^{-1}cb^{-1}D^{-1} = q^2p\cdot qp^{-1}qp^{-1}\cdot p^{-1}\cdot q^2p^2q^2p^2\cdot qp^{-1}q^{-1}pqpq^{-1}p^{-1}\cdot qp^{-1}qp^{-1}\cdot q^2pq^2p^{-1}q^2$$
$$= q^2pqp^{-1}q^{-1}p^2q^{-1}p^{-1}q^{-1}\cdot pqpq^2pqp\cdot q^2p^{-1}q^2$$
$$= q^2pqp^{-1}q^{-1}p^2\cdot q^{-1}p^{-1}q^{-1}p^{-1}\cdot q^{-1}p^{-1}q^{-1}\cdot q^{-1}p^{-1}q^{-1}p^{-1}q^{-1}\cdot q^2p^{-1}q^2 = 1$$
$$\text{by (iii), (iv), (vii).}$$

(47)

$$e^2cFbabca^{-1}Fbc^{-1}Dbc^{-1}$$
$$= pq^2pq^2p^{-1}q^{-1}p^{-1}\cdot (pq^{-1})^4\cdot q^{-1}p^{-1}q^2p^{-1}q^2p^{-1}q^{-1}pq^{-1}p^2q\cdot p^{-1}q^{-1}p^{-1}q^{-1}\cdot (pq^{-1})^4pqp^{-1}q^2p^{-1}$$
$$\cdot (q^{-1}p)^4\cdot p^{-1}qp^{-1}q^2p^{-1}q^{-1}p^2q^2pq^2p^{-1}\cdot q^{-1}\cdot (q^{-1}p)^4p^{-1}\cdot q^2p^{-1}q^{-1}p^2$$
$$= pq^2\cdot pq^{-1}pq^{-1}\cdot p^{-1}q^2p^{-1}\cdot q^{-1}pq^{-1}p\cdot pq^2pqp^2q\cdot p^{-1}\cdot q^2p^2q^2p^2\cdot qpq^2\cdot pq^{-1}pqp^{-1}q^{-1}p^{-1}qpq$$
$$\text{by (iii), (iv), (vii)}$$
$$= pq^{-1}p^{-1}q^{-1}p^2q^2p^2\cdot q^{-1}pq^{-1}p\cdot q^{-1}pq^2pq^{-1}pqp^{-1}q^{-1}p^{-1}qpq$$
$$= pq^{-1}p^{-1}q^{-1}p^2q^2\cdot pq^{-1}pq^{-1}\cdot pqp^{-1}q^{-1}p^{-1}qpq$$
$$= pq^{-1}\cdot p^{-1}q^{-1}p^2q^{-1}p^{-1}q^2\cdot p^{-1}q^{-1}p^{-1}\cdot qpq = 1 \quad \text{by (iii), (iv), (vii).}$$

(46)

$$dede^{-1} = (de)^2\cdot e^2 .$$

But by (v) we have $dede = (q^2p^2)^4 = 1$. Hence Lemma 2 (46) follows by Lemma 2 (47).

(48)

$$de^{-2}d^{-1} \cdot debe^{-1}d^{-1} \cdot dece^{-1}d^{-1} \cdot dbd^{-1} \cdot dcd^{-1} = de^{-1}bce^{-1}bcd^{-1}$$
$$= q^2pq^2p(q^2p^{-1}q^{-1}p^2q^{-1}p^{-1})^2 \cdot p^{-1}q^2p^{-1}q^2 = 1$$

Hence Lemma 2 (48) follows by Lemma 2 (41), (44), (39), (42).

(45)

$$d^2 \cdot ede^{-1}d^{-1} \cdot de^2d^{-1} = d \cdot (de)^2 \cdot d^{-1} = 1 .$$

Lemma 2 (48) yields Lemma 2 (45).

(49)

$$(dbc^{-1})^2 = (q^2 \cdot pq^{-1}pq^{-1} \cdot pq \cdot p^{-1}q^{-1}p^2)^2 = (q^{-1}p^{-1}q^2p^{-1}q^{-1}p^2)^2 = 1 \text{ by (iv), (vii),}$$

$$(db)^4 = (q^2 \cdot pq^{-1}pq^{-1} \cdot p^2)^4 = (q^{-1}p^{-1}qp)^4 = 1 \text{ by (iv), (vi),}$$

$$(dba)^2 = (q^2 \cdot pq^{-1}pq^{-1}pq^{-1} \cdot qpqp \cdot q^{-1}p^{-1})^2$$
$$= (q^{-1}p^2q^{-1}p^{-1}q^2p^{-1})^2 = 1 \text{ by (iii), (iv), (vii).}$$

LEMMA 3. *D operates on the generators of* H_g *as follows:*

(1) $D^{-1}a^2D = bab^2a^{-1}c^{-1}bca^{-1}b^2a^{-1}c^{-1}b^{-1}cb^{-1}$;

(2) $D^{-1}b^2D = bc^{-1}bcab^2ac^2b^2c^2a^{-1}b^2a^{-1}c^{-1}b^{-1}cb^{-1}$;

(3) $D^{-1}c^2D = bc^{-1}bcab^2ac^2a^{-1}b^2a^{-1}c^{-1}b^{-1}cb^{-1}$;

(4) $D^{-1}abD = bac^2a^{-1}b^2a^{-1}c^{-1}b^{-1}cb^{-1}$;

(5) $D^{-1}baD = bc^{-1}bcab^2ac^2a^{-1}c^{-1}bca^{-1}b^2a^{-1}c^{-1}b^{-1}cb^{-1}$;

(6) $D^{-1}acD = bab^2a^{-1}c^{-1}bc^{-1}ab^2cab^2ac^2a^{-1}b^2a^{-1}c^{-1}b^{-1}cb^{-1}$;

(7) $D^{-1}caD = bc^{-1}b^{-1}a^{-1}c^2a^{-1}b^2a^{-1}c^{-1}b^{-1}cb^{-1}$;

(8) $D^{-1}bcD = bc^{-1}b^{-1}a^{-1}b^{-1}a^{-1}c^{-1}b^{-1}cb^{-1}$;

(9) $D^{-1}cbD = bc^{-1}bcab^2ac^2a^{-1}b^{-1}ab^2cab^2ac^2b^2c^2a^{-1}b^2a^{-1}c^{-1}b^{-1}cb^{-1}$;

(10) $D^2 = bc^{-1}bcab^2ac^2a^{-1}c^{-1}bc^{-1}b^{-1}a^{-1}$.

Proof. (8) follows by Lemma 2 (3), (4).

(4) By Lemma 2 (6) we have $D^{-1}abc^{-1}b^{-1}bc^{-1}b^{-1}a^{-1}D = bac^2a^{-1}b^{-1}$. So Lemma 2 (5) and Lemma 3 (8) yield Lemma 3 (4).

(3) follows by Lemma 2 (6) and Lemma 3 (4).

(7) By Lemma 2 (5), (7), (8) we have $D^{-1}cab^{-1}a^{-1}D = bc^{-1}b^{-1}a^2b^{-1}$. Lemma 3 (4) yields Lemma 3 (7).

(2) By Lemma 2 (8) we have $D^{-1}ab^2a^{-1}D = bab^2a^{-1}b^{-1}$. Lemma 3 (4) yields Lemma 3 (2).

(10) By Lemma 2 (1), (7) we have $D^{-1}abDD^2ab = bc^{-1}bc^{-1}$. Lemma 3 (4) yields Lemma 3 (10).

(5) By Lemma 2 (8) we have $D^2D^{-1}baDD^{-1}b^{-1}a^{-1}D = aba^{-1}b^{-1}$. Lemma 3 (4), (10) yield Lemma 3 (5).

(1) By Lemma 2 (8), (9) we have $D^{-1}ab^2a^{-1}D = bab^2a^{-1}b^{-1}$. Lemma 3 (4), (5) yield Lemma 3 (1).

(9) By Lemma 2 (5) we have $D^{-1}abDD^{-1}cbDD^{-1}b^2D = abcb^{-1}$. Lemma 3 (4), (2) yield Lemma 3 (9).

(6) We have the identity $D^{-1}acD = D^{-1}a^2DD^{-1}a^{-1}c^{-1}DD^{-1}c^2D$. Lemma 3 (1), (3), (7) yield Lemma 3 (6).

LEMMA 4. *In $H = \langle a, b, c \rangle$ we have*

(1) $b^2cab^{-2} = a^2ca^{-1}$;

(2) $b(c^{-1}a)^2b^{-1} = a^{-1}(c^{-1}a)^2a$;

(3) $b(ac^{-1})^2b^{-1} = (c^{-1}a)^2$;

(4) $b(ac)^2b^{-1} = (ca)^2$;

(5) $b(ca)^2b^{-1} = a^{-1}(ca)^2a$;

(6) $(bca^2c^{-1}a^2b^{-1}a^2)^2 = 1$;

(7) $b^{-1}a(ac)^2a^{-1}(ac^{-1})^2b = (a^{-1}c)^2(ac)^2$;

(8) $ba(ac)^2a^{-1}(ac^{-1})^2b^{-1} = (c^{-1}a)^2(ac)^2$;

(9) $(ca)^2 \ne (a^{-1}c)^2$, $(ca)^2 \ne (ac^{-1})^2$, $(ac)^2 \ne (a^{-1}c)^2$, $(ac)^2 \ne (ac^{-1})^2$;

(10) $b^{-1}(ac^{-1})^2b = a(ac^{-1})^2a^{-1}$;

(11) $c^{-1}a \ne (ac^{-1})^4$;

(12) $b^{-1}(a^{-1}c)^2 c^{-1}(ac)^2 cb = c(ca)^2 c^{-1}(ca^{-1})^2$;

(13) $b^{-1}c(ac^{-1})^2 c^{-1}b = ac(ac^{-1})^2 c^{-1}a^{-1}$;

(14) $(aca^{-1}cac^{-1})^4 = 1$;

(15) $aca^{-1}c^{-1}ac \neq (ca)^2$;

(16) $(ac)^2 \neq (ca)^2$;

(17) $bc^{-1}(ac)^2 c(c^{-1}a)^2 b^{-1} = a^{-1}c^{-1}(ac)^2 c(c^{-1}a)^2 a$;

(18) $bc^{-1}(ac)^2 cb^{-1} = a^{-1}c^{-1}(ac)^2 ca$;

(19) $ba^{-1}(a^{-1}c)^2 ab^{-1} = a(ca^{-1})^2 a^{-1}$;

(20) $ba^{-1}c^{-1}(ac)^2 cab^{-1} = c^{-1}a^{-1}(ca)^2 ac$;

(21) $b^{-1}c(ca)^2 c^{-1}b = c^{-1}a^{-1}(ca)^2 ac$.

Proof. (1) By Lemma 2 (10) we have

$$p^2 Dbacb^{-1}D^{-1}aba^{-1}c^{-1}b^{-1}a^{-1}p^{-2} = 1 \ .$$

Lemma 1 (1)-(5) and Lemma 2 (5) yield $D^{-1}caD = bc^{-1}b^{-1}a^{-1} \cdot ca \cdot abcb^{-1}$. A comparison with Lemma 3 (7) yields Lemma 4 (1).

(2) and (3). By Lemma 4 (1) we have

$$p^2 b^2 cab^2 p^{-2} = p^2 a^2 ca^{-1}p^{-2} \ \Rightarrow \ b^{-1}a^{-1}c^{-1}b^2 a^{-1}c^{-1}b^{-1} = c^{-1}ac^{-1}a$$

by Lemma (1)-(5). Use Lemma 4 (1) in both possible ways obtaining Lemma 4 (2) and Lemma 4 (3).

(4) By Lemma 3 (1) we have

$$D^{-2}a^2 D^2 = D^{-1}bab^2 a^{-1}c^{-1}bca^{-1}bca^{-1}b^2 a^{-1}c^{-1}b^{-1}cb^{-1}D \ .$$

By Lemma 3 (5), (2), (7), (8), (4), (9) follows

$$a^{-1}c^{-1}bc^{-1}ab^2 cab^2 ac^2 a^{-1}b^{-1}ab^2 cab^2 ac^2 = 1 \ .$$

Lemma 4 (1) and Lemma 2 (2) yield Lemma 4 (4).

(5) By Lemma 4 (4) we have

$$p^2 b(ac)^2 b^{-1}p^{-2} = p^2(ca)^2 p^{-2} \ .$$

Lemma 1 (1)-(5) yield Lemma 4 (5).

(6) follows by Lemma 3 (1), Lemma 4 (2) and $a^4 = 1$.

(7) By Lemma 4 (2), (4) we have

$$b(a^{-1}c)^2b^{-1}b(ac)^2b^{-1} = a^{-1} \cdot a^{-1}ca^{-1}c \cdot acac \cdot a \ .$$

Lemma 2 (2) yields Lemma 4 (7).

(8) By Lemma 4 (1) we have

$$D^{-1}b^2cab^2ac^{-1}a^2D = 1 \ .$$

Using Lemma 3 (2), (7), (6), (3), (1) we have

$$c^{-1}b^2c^2a^{-1}b^2a^{-1}c^{-1}b^2a^{-1}b^2c^2a^{-1}b^2a^{-1}c^{-1}b^{-1}cab^2a^{-1}c^{-1}bc^{-1}abcab^2a^{-1}c^{-1}b = 1 \ .$$

By Lemma 4 (1), (7) follows

$$ca^{-1}ca^2bca^2c^{-1}a^2b^{-1}a^{-1}caca^{-1} = 1 \ .$$

Lemma 4 (6) and Lemma 2 (2) yield Lemma 4 (8).

(9) By Lemma 4 (8), (3), (4), (1) we have

$$a^2ca^2c^{-1} = b^{-1}(c^{-1}a)^2bb^2b(ac)^2b^{-1}b^2 = (ac^{-1})^2a^2caca^{-1} \ .$$

Hence we have

$$(ac)^2 \not\simeq (c^{-1}a)^2 \ \Rightarrow \ p^2(ac)^2p^{-2} \not\simeq p^2(c^{-1}a)^2p^{-2} \ \Rightarrow \ (ca)^2 \not\simeq (a^{-1}c)^2$$

by Lemma 1 (1)-(5).

Lemma 4 (7) and Lemma 2 (2) yield

$$b^{-1}a^2ca^2c^{-1}bb^{-1}a^2ca^2c^{-1}b = (a^{-1}c)^4 \ .$$

By Lemma 4 (2) follows

$$(ca)^2 \not\simeq (ac^{-1})^2 \ .$$

Applying p^2 , we obtain the missing relation.

(10) By Lemma 4 (2) we have

$$b^2c^{-1}ac^{-1}ab^2 = ba^{-1}c^{-1}a^{-1}c^{-1}b^{-1}bca^2c^{-1}a^2b^{-1} \ .$$

Lemma 4 (5), (8), (9) and Lemma 2 (2) yield

$$b^2c^{-1}ac^{-1}ab^2 = a^{-1}cacaca^{-1}cacac \ .$$

Applying Lemma 2 (2) once again, we obtain

$$b^2(c^{-1}a)^2b = a(ac^{-1})a^{-1} \ .$$

By Lemma 4 (3) we obtain Lemma 4 (10).

(11) By $a^4 = 1$ we have

$$(a^{-1}a^{-1}c^{-1}a^{-1}a^{-1}a^{-1}ca^{-1}a^{-1})^2 = 1 \ .$$

Lemma 4 (10), Lemma 2 (2) yield

$$\left(ca^{-1}ca^{-1}ca^{-1}ca\right)^2 = 1 \ .$$

Applying p^2 we obtain Lemma 4 (11).

(12) By Lemma 4 (8) we have

$$p^2ba^2ca^2c^{-1}b^{-1}p^{-2} = p^2\left(c^{-1}a\right)^2(ac)^2p^{-2} \ .$$

Lemma 1 (1)-(5), Lemma 4 (9) yield Lemma 4 (12).

(13) By Lemma 4 (4) and Lemma 4 (12) follows

$$b^{-1}cacaa^{-1}c^2ac^2b = acacc^2ac^2a^{-1} \ .$$

(14) By Lemma 2 (2), Lemma 3 (6), Lemma 4 (1) we have

$$\left(a^2c^{-1}a^2b^{-1}c^{-1}a^{-1}c^2a^2bc\right)^4 = 1 \ .$$

Lemma 4 (8) yields Lemma 4 (14).

(15) By Lemma 4 (4) and Lemma 2 (2) we have

$$D^{-1}acacb^{-1}cacabD = 1 \ .$$

Use Lemma 3 (6), (2), (8), (4) and Lemma 4 (1) to conclude

$$bca^2c^{-1}a^2b^{-1}c^{-1}a^{-1}c^2a^2b^{-1}a^2ca^2c^{-1}bc^{-1}a^{-1}cb^2ca$$
$$\cdot b^{-1}a^{-1}c^{-1}b^2 \cdot b^{-1}a^2ca^2c^{-1}bc^{-1}a^{-1}c^2a^2bcaca^2 =$$

By Lemma 4 (8), (7), (1) we have

$$b^{-1}\left(ac^{-1}\right)^2c\left(ca^{-1}\right)^2c^{-1}ca^2c^{-1}a^2bcaca^{-1}caca^{-1} \cdot ca^2c^2aca^{-1}cac^2a^2ca^{-1} = 1 \ .$$

By Lemma 4 (10), (7), (13) follows

$$a^2c^{-1}aca^{-1}cac^{-1}acaca^2acaca^{-1}ca^{-1}cc^{-1}a^{-1}c^2ac \cdot a^{-1}cac^2a^2ca^{-1} = 1 \ .$$

By Lemma 2 (2) and Lemma 4 (9) we have

$$a^2c^{-1}aca^{-1}cac^2a^2cac^{-1}aca^{-1}cac^2a^2ca^{-1} = 1 \ .$$

Apply the automorphism induced by p^2 obtaining

$$\left(c^{-1}aca^{-1}cac^{-1}aca^{-1}caa^{-1}c^{-1}a^{-1}c^{-1}\right)^2 = 1 \ .$$

By Lemma 4 (14) and Lemma 2 (2) we have

$$\left(c^{-1}aca^{-1}ca\right)^2 \overset{.}{\not=} (ca)^2 \ .$$

Then Lemma 4 (9) implies Lemma 4 (15).

(16) By Lemma 4 (15) and Lemma 2 (2) we have

$$D^{-1}aca^{-1}c^{-1}accacac^{-1}a^{-1}cac^{-1}a^{-1}cacaD = 1 .$$

By Lemma 3 (6), (7) and Lemma 4 (1) we have

$$a^2b^{-1}a^2ca^2c^{-1}bc^{-1}a^{-1}cab^{-1}a^2ca^2c^{-1}bc^{-1}a^{-1}c^2acac^{-1}ac$$
$$ba^2ca^2c^{-1}b^{-1}a^{-1}c^{-1}acba^2ca^2c^{-1}b^{-1}a^{-1}cac = 1 .$$

Lemma 4 (7), (8) imply

$$cacaca^{-1}cac^{-1}a^{-1}cac^{-1}acac^{-1}a^2c^{-1}a^2caca^{-1}c^{-1}ac(c^{-1}a)^2(ac)^2a^{-1} = 1 .$$

Lemma 4 (9) and Lemma 2 (2) yield

$$a^{-1}c^{-1}a^{-1}cac^{-1}acac^{-1}ac(c^{-1}a)^2(ac)^2a^{-1}c^2a^{-1}c^{-1}ac^{-1}\cdot(ca)^2(ca^{-1})^2a^{-1} = 1 .$$

Lemma 4 (9) and Lemma 2 (2) imply

$$(ca)^2c^{-1}a^{-1}(ac^{-1})^2(ac)^2(ca^{-1})^2ac(ca)^2(ac)^2 = 1 .$$

Hence Lemma 4 (9) and Lemma 2 (2) imply Lemma 4 (16).

(17) By Lemma 4 (4), (10) we have

$$p^2b^{-1}ac^{-1}ac^{-1}acacbp^{-2} = p^2a^2c^{-1}a^2cp^{-2} .$$

By Lemma 1 (1)-(5) follows Lemma 4 (17).

(18) By Lemma 4 (17) we have

$$bc^2a^{-1}c^{-1}a^{-1}cc^{-1}ac^{-1}ab^{-1} = a^{-1}c^2a^{-1}c^2a^2 .$$

Lemma 4 (2) and Lemma 2 (2) yield Lemma 4 (18).

(19) By Lemma 4 (8), (6) we have

$$ba^2ca^2c^{-1}cacab^{-1} = c^{-1}ac^{-1}aacaca^{-1}(ca)^2a .$$

Lemma 4 (9) implies Lemma 4 (19).

(20) By Lemma 2 (2) we have

$$bcabcabcabca = 1 .$$

Lemma 4 (1) yields

$$b^{-1}a^2ca^{-1}ba^{-1}(a^{-1}c)^2aa^{-1}c^{-1}b^{-1}cabca = 1 .$$

Lemma 4 (19) implies

$$ba^{-1}c^{-1}b^{-1}cabcab^{-1} = a^2c^{-1}ac^2a^2 .$$

Raise this relation to the second power obtaining

$$ba^{-1}c^{-1}b^{-1}(ca)^2bcab^{-1} = a^2c^{-1}\cdot acac\cdot ca^2 .$$

Use Lemma 4 (4), (9), (16) and Lemma 2 (2) to conclude Lemma 4 (20).

(21) By Lemma 4 (20) we have

$$b^{-1}b^2a^{-1}c^{-1}(ac)^2cab^2b = c^{-1}a^{-1}(ca)^2ac \ .$$

Lemma 4 (1), (5), (4), (9) yield Lemma 4 (21).

LEMMA 5. *The following rules of conjugation apply:*

(1) $a(ac)^2a^{-1} = a(ac)^2a^{-1}$;

(2) $b(ac)^2b^{-1} = (ca)^2$ *by Lemma* 4 (4);

(3) $c(ac)^2c^{-1} = (ca)^2$;

(4) $a^2(ac)^2a^2 = a^{-1}(ca)^2a$;

(5) $b^2(ac)^2b^2 = a^{-1}(ca)^2a$ *by Lemma* 4 (4), (5);

(6) $c^2(ac)^2c^2 = c(ca)^2c^{-1}$;

(7) $a^{-1}(ac)^2a = (ca)^2$;

(8) $b^{-1}(ac)^2b = a(ac)^2a^{-1}$ *by Lemma* 4 (4), (1) ;

(9) $c^{-1}(ac)^2c = c^{-1}(ac)^2c$;

(10) $a(ca)^2a^{-1} = (ac)^2$;

(11) $b(ca)^2b^{-1} = a^{-1}(ca)^2a$ *by Lemma* 4 (5);

(12) $c(ca)^2c^{-1} = c(ca)^2c^{-1}$;

(13) $a^2(ca)^2a^2 = a(ac)^2a^{-1}$;

(14) $b^2(ca)^2b^2 = a(ac)^2a^{-1}$ *by Lemma* 4 (1);

(15) $c^2(ca)^2c^2 \doteq c^{-1}(ac)^2c$;

(16) $a^{-1}(ca)^2a = a^{-1}(ca)^2a$;

(17) $b^{-1}(ca)^2b = (ac)^2$ *by Lemma* 4 (4);

(18) $c^{-1}(ca)^2c = (ac)^2$;

(19) $aa(ac)^2a^{-1}a^{-1} = a^{-1}(ca)^2a$;

(20) $ba(ac)^2a^{-1}b^{-1} = (ac)^2$ *by Lemma* 4 (4), (1);

(21) $ca(ac)^2a^{-1}c^{-1} = a^{-1}c^{-1}(ac)^2ca$ *by Lemma* 4 (16);

(22) $a^2a(ac)^2a^{-1}a^2 = (ca)^2$;

(23) $b^2a(ac)^2a^{-1}b^2 = (ca)^2$ *by Lemma 4 (1);*

(24) $c^2a(ac)^2a^{-1}c^2 = c^{-1}a^{-1}(ca)^2ac$ *by Lemma 4 (9), Lemma 2 (2);*

(25) $a^{-1}a(ac)^2a^{-1}a = (ac)^2$;

(26) $b^{-1}a(ac)^2a^{-1}b = a^{-1}(ca)^2a$ *by Lemma 4 (7), (10), Lemma 2 (2);*

(27) $c^{-1}a(ac)^2a^{-1}c = a^{-1}(ca)^2a$ *by Lemma 4 (9);*

(28) $aa^{-1}(ca)^2aa^{-1} = (ca)^2$;

(29) $ba^{-1}(ca)^2ab^{-1} = a(ac)^2a^{-1}$ *by Lemma 4 (7), (10), Lemma 2 (2);*

(30) $ca^{-1}(ca)^2ac^{-1} = a(ac)^2a^{-1}$ *by Lemma 4 (9);*

(31) $a^2a^{-1}(ca)^2aa^2 = (ac)^2$;

(32) $b^2a^{-1}(ca)^2ab^2 = (ac)^2$ *by Lemma 4 (4), (5), (7), (9);*

(33) $c^2a^{-1}(ca)^2ac^2 = a^{-1}c^{-1}(ac)^2ca$ *by Lemma 4 (9), (16);*

(34) $a^{-1}a^{-1}(ca)^2aa = a(ac)^2a^{-1}$;

(35) $b^{-1}a^{-1}(ca)^2ab = (ca)^2$ *by Lemma 4 (5);*

(36) $c^{-1}a^{-1}(ca)^2ac = c^{-1}a^{-1}(ca)^2ac$;

(37) $ac(ca)^2c^{-1}a^{-1} = c^{-1}a^{-1}(ca)^2ac$ *by Lemma 4 (16);*

(38) $bc(ca)^2c^{-1}b^{-1} = c^{-1}(ac)^2c$ *by Lemma 4 (3), (9), (12);*

(39) $cc(ca)^2c^{-1}c^{-1} = c^{-1}(ac)^2c$;

(40) $a^2c(ca)^2c^{-1}a^2 = a^{-1}c^{-1}(ac)^2ca$ *by Lemma 4 (9);*

(41) $b^2c(ca)^2c^{-1}b^2 = a^{-1}c^{-1}(ac)^2ca$ *by Lemma 4 (1), (4), (5), (9);*

(42) $c^2c(ca)^2c^{-1}c^2 = (ac)^2$;

(43) $a^{-1}c(ca)^2c^{-1}a = c^{-1}(ac)^2c$ *by Lemma 4 (9);*

(44) $b^{-1}c(ca)^2c^{-1}b = c^{-1}a^{-1}(ca)^2ac$ *by Lemma 4 (21);*

(45) $c^{-1}c(ca)^2c^{-1}c = (ca)^2$;

(46) $ac^{-1}(ac)^2ca^{-1} = c(ca)^2c^{-1}$ *by Lemma 4 (9);*

(47) $bc^{-1}(ac)^2cb^{-1} = a^{-1}c^{-1}(ac)^2ca$ *by Lemma* 4 (18);

(48) $cc^{-1}(ac)^2cc^{-1} = (ac)^2$;

(49) $a^2c^{-1}(ac)^2ca^2 = c^{-1}a^{-1}(ca)^2ac$ *by Lemma* 4 (9), (16);

(50) $b^2c^{-1}(ac)^2cb^2 = c^{-1}a^{-1}(ca)^2ac$ *by Lemma* 4 (18), (20);

(51) $c^2c^{-1}(ac)^2cc^2 = (ca)^2$;

(52) $a^{-1}c^{-1}(ac)^2ca = a^{-1}c^{-1}(ac)^2ca$;

(53) $b^{-1}c^{-1}(ac)^2cb = c(ca)^2c^{-1}$ *by Lemma* 4 (12), (3), (9);

(54) $c^{-1}c^{-1}(ac)^2cc = c(ca)^2c^{-1}$;

(55) $aa^{-1}c^{-1}(ac)^2caa^{-1} = c^{-1}(ac)^2c$;

(56) $ba^{-1}c^{-1}(ac)^2cab^{-1} = c^{-1}a^{-1}(ca)^2ac$ *by Lemma* 4 (20);

(57) $ca^{-1}c^{-1}(ac)^2cac^{-1} = c^{-1}a^{-1}(ca)^2ac$ *by Lemma* 4 (16), (9);

(58) $a^2a^{-1}c^{-1}(ac)^2caa^2 = c(ca)^2c^{-1}$ *by Lemma* 4 (9);

(59) $b^2a^{-1}c^{-1}(ac)^2cab^2 = c(ca)^2c^{-1}$ *by Lemma* 4 (1), (4), (5), (9);

(60) $c^2a^{-1}c^{-1}(ac)^2cac^2 = a^{-1}(ca)^2a$ *by Lemma* 4 (16), (9);

(61) $a^{-1}a^{-1}c^{-1}(ac)^2caa = c^{-1}a^{-1}(ca)^2ac$;

(62) $b^{-1}a^{-1}c^{-1}(ac)^2cab = c^{-1}(ac)^2c$ *by Lemma* 4 (18);

(63) $c^{-1}a^{-1}c^{-1}(ac)^2cac = a(ac)^2a^{-1}$ *by Lemma* 4 (16);

(64) $ac^{-1}a^{-1}(ca)^2aca^{-1} = a^{-1}c^{-1}(ac)^2ca$ *by Lemma* 4 (16), (9);

(65) $bc^{-1}a^{-1}(ca)^2acb^{-1} = c(ca)^2c^{-1}$ *by Lemma* 4 (21);

(66) $cc^{-1}a^{-1}(ca)^2acc^{-1} = a^{-1}(ca)^2a$;

(67) $a^2c^{-1}a^{-1}(ca)^2aca^2 = c^{-1}(ac)^2c$ *by Lemma* 4 (16), (9);

(68) $b^2c^{-1}a^{-1}(ca)^2acb^2 = c^{-1}(ac)^2c$ *by Lemma* 4 (18), (20);

(69) $c^2c^{-1}a^{-1}(ca)^2acc^2 = a(ac)^2a^{-1}$ *by Lemma* 4 (9);

(70) $a^{-1}c^{-1}a^{-1}(ca)^2aca = c(ca)^2c^{-1}$ *by Lemma* 4 (16);

(71) $b^{-1}c^{-1}a^{-1}(ca)^2acb = a^{-1}c^{-1}(ac)^2ca$ *by Lemma* 4 (20);

(72) $c^{-1}c^{-1}a^{-1}(ca)^2acc = a^{-1}c^{-1}(ac)^2ca$ by Lemma 4 (9), (16).

The formulae in Lemma 5 show that the elements $(ac)^2$, $(ca)^2$, $a(ac)^2a^{-1}$, $a^{-1}(ca)^2a$, $c(ca)^2c^{-1}$, $c^{-1}(ac)^2c$, $a^{-1}c^{-1}(ac)^2ca$, $c^{-1}a^{-1}(ca)^2ac$ form a full class of conjugates in $H = \langle a, b, c \rangle$. Moreover, the formulae show that they commute pairwise. Hence it is shown that $N_1 = \langle\langle (ac)^2 \rangle\rangle_H$ is an elementary abelian normal subgroup of exponent 2 and of order 2^8 in $H = \langle a, b, c \rangle$. It is also shown that the elements $(ac^{-1})^2$ and $(c^{-1}a)^2$ commute elementwise with N.

LEMMA 6. *We have the following rules of conjugation:*

(1) $a(ac^{-1})^2a^{-1} = (a^{-1}c)^2 \cdot a^{-1}(ca)^2a \cdot (ac)^2$ by Lemma 5;

(2) $b(ac^{-1})^2b^{-1} = (c^{-1}a)^2$ by Lemma 4 (3);

(3) $c(ac^{-1})^2c^{-1} = (a^{-1}c)^2(ca)^2c^{-1}(ac)^2c$ by Lemma 5;

(4) $a^2(ac^{-1})^2a^2 = (ca^{-1})^2(ca)^2a(ac)^2a^{-1}$ by Lemma 5;

(5) $b^2(ac^{-1})^2b^2 = (ca^{-1})^2(ca)^2a(ac)^2a^{-1}$ by Lemma 4 (3), (7), Lemma 5;

(6) $c^2(ac^{-1})^2c^2 = (ca^{-1})^2(ac)^2c(ca)^2c^{-1}$ by Lemma 5;

(7) $a^{-1}(ac^{-1})^2a = (c^{-1}a)^2$;

(8) $b^{-1}(ac^{-1})^2b = (a^{-1}c)^2 \cdot a^{-1}(ca)^2a \cdot (ac)^2$ by Lemma 4 (7), Lemma 5;

(9) $c^{-1}(ac^{-1})^2c = (c^{-1}a)^2$;

(10) $a(c^{-1}a)^2a^{-1} = (ac^{-1})^2$;

(11) $b(c^{-1}a)^2b^{-1} = (ca^{-1})^2 \cdot (ca)^2 \cdot a(ac)^2a^{-1}$ by Lemma 4 (17);

(12) $c(c^{-1}a)^2c^{-1} = (ac^{-1})^2$;

(13) $a^2(c^{-1}a)^2a^2 = (a^{-1}c)^2a^{-1}(ca)^2a(ac)^2$ by Lemma 5;

(14) $b^2(c^{-1}a)^2b^2 = (a^{-1}c)^2a^{-1}(ca)^2a(ac)^2$ by Lemma 4 (3), (7), Lemma 5;

(15) $c^2(c^{-1}a)^2c^2 = (a^{-1}c)^2(ca)^2c^{-1}(ac)^2c$ by Lemma 5;

(16) $a^{-1}(c^{-1}a)^2a = (ca^{-1})^2(ca)^2a(ac)^2a^{-1}$ by Lemma 5;

(17) $b^{-1}(c^{-1}a)^2b = (ac^{-1})^2$ by Lemma 4 (3);

(18) $c^{-1}(c^{-1}a)^2c = (ca^{-1})^2c(ca)^2c^{-1}(ac)^2$ by Lemma 5.

LEMMA 7. *The following elements form a conjugacy class of involutions in* $\langle a, b, c \rangle$:

$$\left(acbcb^{-1}\right)^2, \; \left(b^{-1}abc\right)^2, \; a^{-1}\left(cb^{-1}ab\right)^2 a, \; \left(cb^{-1}ab\right)^2, \; a^{-1}\left(b^{-1}abc\right)^2 a, \; a\left(acbcb^{-1}\right)^2 a^{-1},$$
$$a^2\left(cb^{-1}ab\right)^2 a^2, \; a\left(cb^{-1}ab\right)^2 a^{-1} .$$

Rules of conjugation: (see Proof for comments)

(1) $a\left(acbcb^{-1}\right)^2 a^{-1} = a\left(acbcb^{-1}\right)^2 a^{-1}$;

(2) $b\left(acbcb^{-1}\right)^2 b^{-1} = a^{-1}\left(cb^{-1}ab\right)^2 a$ *by* p *Lemma* 5 (8), (9) p^{-1} ;

(3) $c\left(acbcb^{-1}\right)^2 c^{-1} = \left(b^{-1}abc\right)^2$ *by* p *Lemma* 5 (2) p^{-1} ;

(4) $a^2\left(acbcb^{-1}\right)^2 a^2 = a^{-1}\left(b^{-1}abc\right)^2 a$;

(5) $b^2\left(acbcb^{-1}\right)^2 b^2 = \left(cb^{-1}ab\right)^2$;

(6) $c^2\left(acbcb^{-1}\right)^2 c^2 = \left(cb^{-1}ab\right)^2$ *by* (3);

(7) $a^{-1}\left(acbcb^{-1}\right)^2 a = \left(b^{-1}abc\right)^2$;

(8) $b^{-1}\left(acbcb^{-1}\right)^2 b = \left(b^{-1}abc\right)^2$;

(9) $c^{-1}\left(acbcb^{-1}\right)^2 c = a^{-1}\left(cb^{-1}ab\right)^2 a$ *by* p *Lemma* 5 (1), (8) p^{-1} ;

(10) $a\left(b^{-1}abc\right)^2 a^{-1} = \left(acbcb^{-1}\right)^2$ *by* (*);

(11) $b\left(b^{-1}abc\right)^2 b^{-1} = \left(acbcb^{-1}\right)^2$;

(12) $c\left(b^{-1}abc\right)^2 c^{-1} = \left(cb^{-1}ab\right)^2$;

(13) $a^2\left(b^{-1}abc\right)^2 a^2 = a\left(acbcb^{-1}\right)^2 a^{-1}$ *by* (10);

(14) $b^2\left(b^{-1}abc\right)^2 b^2 = a^{-1}\left(cb^{-1}ab\right)^2 a$ *by* (2);

(15) $c^2\left(b^{-1}abc\right)^2 c^2 = a^{-1}\left(cb^{-1}ab\right)^2 a$ *by* p *Lemma* 5 (13), (14) p^{-1} ;

(16) $a^{-1}\left(b^{-1}abc\right)^2 a = a^{-1}\left(b^{-1}abc\right)^2 a$;

(17) $b^{-1}\left(b^{-1}abc\right)^2 b = \left(cb^{-1}ab\right)^2$ *by* (*);

(18) $c^{-1}\left(b^{-1}abc\right)^2 c = \left(acbcb^{-1}\right)^2$ *by* (3);

(19) $aa^{-1}\left(cb^{-1}ab\right)^2 aa^{-1} = \left(cb^{-1}ab\right)^2$;

(20) $ba^{-1}\left(cb^{-1}ab\right)^2 ab^{-1} = \left(cb^{-1}ab\right)^2$ *by* (2), (5);

(21) $ca^{-1}\left(cb^{-1}ab\right)^2 ac^{-1} = \left(acbcb^{-1}\right)^2$ *by* p *Lemma* 5 (17), (53), (43) p^{-1} ;

(22) $\quad a^2 a^{-1} (cb^{-1}ab)^2 aa^2 = a(cb^{-1}ab)^2 a^{-1}$;

(23) $\quad b^2 a^{-1} (cb^{-1}ab)^2 ab^2 = (b^{-1}abc)^2 \quad by$ (*);

(24) $\quad c^2 a^{-1} (cb^{-1}ab)^2 ac^2 = (b^{-1}abc)^2 \quad by$ (3), (9);

(25) $\quad a^{-1} a^{-1} (cb^{-1}ab)^2 aa = a^2(cb^{-1}ab)^2 a^2$;

(26) $\quad b^{-1} a^{-1} (cb^{-1}ab)^2 ab = (abcb^{-1})^2 \quad by$ (2);

(27) $\quad c^{-1} a^{-1} (cb^{-1}ab)^2 ac = (cb^{-1}ab)^2 \quad by$ (24), (12);

(28) $\quad a(cb^{-1}ab)^2 a^{-1} = a(cb^{-1}ab)^2 a^{-1}$;

(29) $\quad b(cb^{-1}ab)^2 b^{-1} = (b^{-1}abc)^2 \quad by$ (17);

(30) $\quad c(cb^{-1}ab)^2 c^{-1} = a^{-1}(cb^{-1}ab)^2 a \quad by$ (27);

(31) $\quad a^2(cb^{-1}ab)^2 a^2 = a^2(cb^{-1}ab)^2 a^2$;

(32) $\quad b^2(cb^{-1}ab)^2 b^2 = (abcb^{-1})^2 \quad by$ (5);

(33) $\quad c^2(cb^{-1}ab)^2 c^2 = (abcb^{-1})^2 \quad by$ (3);

(34) $\quad a^{-1}(cb^{-1}ab)^2 a = a^{-1}(cb^{-1}ab)^2 a$;

(35) $\quad b^{-1}(cb^{-1}ab)^2 b = a^{-1}(cb^{-1}ab)^2 a \quad by$ (20);

(36) $\quad c^{-1}(cb^{-1}ab)^2 c = (b^{-1}abc)^2 \quad by$ (12);

(37) $\quad aa^{-1}(b^{-1}abc)^2 aa^{-1} = (b^{-1}abc)^2$;

(38) $\quad ba^{-1}(b^{-1}abc)^2 ab^{-1} = a(cb^{-1}ab)^2 a^{-1} \; by \; p \; Lemma \; 5 \; (17), (47), (63), (20) \; p^{-1}$;

(39) $\quad ca^{-1}(b^{-1}abc)^2 ac^{-1} = a(abcb^{-1})^2 a^{-1} \; by \; p \; Lemma \; 5 \; (8), (27), (35) \; p^{-1}$;

(40) $\quad a^2 a^{-1}(b^{-1}abc)^2 aa^2 = (abcb^{-1})^2 \quad by$ (10);

(41) $\quad b^2 a^{-1}(b^{-1}abc)^2 ab^2 = a^2(cb^{-1}ab)^2 a^2 \; by \; p \; Lemma \; 5 \; (8), (24), (68), (2) \; p^{-1}$;

(42) $\quad c^2 a^{-1}(b^{-1}abc)^2 ac^2 = a^2(cb^{-1}ab)^2 a^2 \; by \; p \; Lemma \; 5 \; (8), (27), (34), (26),$
$$(68), (2) \; p^{-1} ;$$

(43) $\quad a^{-1} a^{-1}(b^{-1}abc)^2 aa = a(abcb^{-1})^2 a^{-1} \quad by$ (13);

(44) $\quad b^{-1} a^{-1}(b^{-1}abc)^2 ab = a(abcb^{-1})^2 a^{-1} \; by \; p \; Lemma \; 5 \; (8), (27)_{\bar{5}} \; (35) \; p^{-1}$;

(45) $\quad c^{-1} a^{-1}(b^{-1}abc)^2 ac = a(cb^{-1}ab)^2 a^{-1} \; by \; p \; Lemma \; 5 \; (8), (21), (56), (64),$
$$(62), (2) \; p^{-1} ;$$

(46) $aa\left(abcb^{-1}\right)^2a^{-1}a^{-1} = a^{-1}\left(b^{-1}abc\right)^2a$ by (4);

(47) $ba\left(abcb^{-1}\right)^2a^{-1}b^{-1} = a^{-1}\left(b^{-1}abc\right)^2a$ by (44);

(48) $ca\left(abcb^{-1}\right)^2a^{-1}c^{-1} = a^2\left(cb^{-1}ab\right)^2a^2$ by p Lemma 5 (11), (34), (26), (68),

$\qquad\qquad\qquad\qquad\qquad$ (2) p^{-1} ;

(48) $a^2a\left(abcb^{-1}\right)^2a^{-1}a^2 = \left(b^{-1}abc\right)^2$ by (7);

(50) $b^2a\left(abcb^{-1}\right)^2a^{-1}b^2 = a\left(cb^{-1}ab\right)^2a^{-1}$ by p Lemma 5 (11), (33), (62), (2) p^{-1} ;

(51) $c^2a\left(abcb^{-1}\right)^2a^{-1}c^2 = a\left(cb^{-1}ab\right)^2a^{-1}$ by p Lemma 5 (11), (30), (19), (29),

$\qquad\qquad\qquad\qquad\qquad$ (21), (62), (2) p^{-1} ;

(52) $a^{-1}a\left(abcb^{-1}\right)^2a^{-1}a = \left(abcb^{-1}\right)^2$;

(53) $b^{-1}a\left(abcb^{-1}\right)^2a^{-1}b = a^2\left(cb^{-1}ab\right)^2a^2$ by p Lemma 5 (11), (68), (2) p^{-1} ;

(54) $c^{-1}a\left(abcb^{-1}\right)^2a^{-1}c = a^{-1}\left(b^{-1}abc\right)^2a$ by (34);

(55) $aa^2\left(cb^{-1}ab\right)^2a^2a^{-1} = a^{-1}\left(cb^{-1}ab\right)^2a$;

(56) $ba^2\left(cb^{-1}ab\right)^2a^2b^{-1} = a\left(abcb^{-1}\right)^2a^{-1}$ by (53);

(57) $ca^2\left(cb^{-1}ab\right)^2a^2c^{-1} = a\left(cb^{-1}ab\right)^2a^{-1}$ by p Lemma 5 (17), (50), (70), (44),

$\qquad\qquad\qquad\qquad\qquad$ (72), (62), (2), p^{-1} ;

(58) $a^2a^2\left(cb^{-1}ab\right)^2a^2a^2 = \left(cb^{-1}ab\right)^2$;

(59) $b^2a^2\left(cb^{-1}ab\right)^2a^2b^2 = a^{-1}\left(b^{-1}abc\right)^2a$ by (41);

(60) $c^2a^2\left(cb^{-1}ab\right)^2a^2c^2 = a^{-1}\left(b^{-1}abc\right)^2a$ by (42);

(61) $a^{-1}a^2\left(cb^{-1}ab\right)^2a^2a = a\left(cb^{-1}ab\right)^2a^{-1}$;

(62) $b^{-1}a^2\left(cb^{-1}ab\right)^2a^2b = a\left(cb^{-1}ab\right)^2a^{-1}$ by p Lemma 5 (17), (50), (72), (62),

$\qquad\qquad\qquad\qquad\qquad$ (2), p^{-1} ;

(63) $c^{-1}a^2\left(cb^{-1}ab\right)^2a^2c = a\left(abcb^{-1}\right)a^{-1}$ by (48);

(64) $aa\left(cb^{-1}ab\right)^2a^{-1}a^{-1} = a^2\left(cb^{-1}ab\right)^2a^2$;

(65) $ba\left(cb^{-1}ab\right)^2a^{-1}b^{-1} = a^2\left(cb^{-1}ab\right)^2a^2$ by (62);

(66) $ca\left(cb^{-1}ab\right)^2a^{-1}c^{-1} = a^{-1}\left(b^{-1}abc\right)^2a$ by (45);

(67) $a^2a\left(cb^{-1}ab\right)^2a^{-1}a^2 = a^{-1}\left(cb^{-1}ab\right)^2a$;

(68) $b^2a(cb^{-1}ab)^2a^{-1}b^2 = a(abcb^{-1})^2a^{-1}$ by (50);

(69) $c^2a(cb^{-1}ab)^2a^{-1}c^2 = a(abcb^{-1})^2a^{-1}$ by (51);

(70) $a^{-1}a(cb^{-1}ab)^2a^{-1}a = (cb^{-1}ab)^2$;

(71) $b^{-1}a(cb^{-1}ab)^2a^{-1}b = a^{-1}(b^{-1}abc)^2a$ by (38);

(72) $c^{-1}a(cb^{-1}ab)^2a^{-1}c = a^2(cb^{-1}ab)^2a^2$ by (57).

Proof. The term p Lemma $x(y)p^{-1}$ on the right-hand side means that the appropriate formula stated on the left-hand side is implied by applying the automorphism induced by p on $\langle a, b, c \rangle$ on the formula Lemma $x(y)$.

By Lemma 5 we have

$$ab(ca)^2b^{-1}a^{-1} = (ca)^2 .$$

Applying p , we obtain

(*) $$(bcb^{-1}a)^2 = (b^{-1}abc)^2$$

which yields (4), (5), (7).

LEMMA 8. *The following rules of transformation apply:*

(1) $a(abcb)^2a^{-1} = (c^{-1}b^{-1}a^{-1}b^{-1})^2a^{-1}(b^{-1}abc)^2a \ (abcb^{-1})^2$;

(2) $b\ (abcb)^2b^{-1} = (babc)^2$;

(3) $c(abcb)^2c^{-1} = (c^{-1}b^{-1}a^{-1}b^{-1})^2 \ (b^{-1}abc)^2 \ a^{-1}(cb^{-1}ab)^2a$;

(4) $a^2(abcb)^2a^{-2} = (b^{-1}c^{-1}b^{-1}a^{-1})^2 \ (b^{-1}abc)^2 \ a(abcb^{-1})^2a^{-1}$;

(5) $b^2(abcb)^2b^{-2} = (b^{-1}c^{-1}b^{-1}a^{-1})^2 \ (abcb^{-1})^2 \ (cb^{-1}ab)^2$;

(6) $c^2(abcb)^2c^{-2} = (b^{-1}c^{-1}b^{-1}a^{-1})^2 \ (abcb^{-1})^2 \ (cb^{-1}ab)^2$;

(7) $a^{-1}(abcb)^2a = (babc)^2$;

(8) $b^{-1}(abcb)^2b = (c^{-1}b^{-1}a^{-1}b^{-1})^2(b^{-1}abc)^2a^{-1}(cb^{-1}ab)^2a$;

(9) $c^{-1}(abcb)^2c = (babc)^2$;

(10) $a\ (babc)^2a^{-1} = (abcb)^2$;

(11) $b(babc)^2b^{-1} = (b^{-1}c^{-1}b^{-1}a^{-1})^2(cb^{-1}ab)^2 \ (abcb^{-1})^2$;

(12) $c(babc)^2c^{-1} = (abcb)^2$;

(13) $a^2(babc)^2a^{-2} = (c^{-1}b^{-1}a7b^{-1})^2a^{-1}(b^{-1}abc)^2a \ (abcb^{-1})^2$;

(14) $b^2(babc)^2 b^{-2} = (c^{-1}b^{-1}a^{-1}b^{-1})^2 (b^{-1}abc)^2 a^{-1}(cb^{-1}ab)^2 a$;

(15) $c^2(babc)^2 c^{-2} = (c^{-1}b^{-1}a^{-1}b^{-1})^2 (b^{-1}abc)^2 a^{-1}(cb^{-1}ab)^2 a$;

(16) $a^{-1}(babc)^2 a = (b^{-1}c^{-1}b^{-1}a^{-1})^2 (b^{-1}abc)^2 a(abcb^{-1})^2 a^{-1}$;

(17) $b^{-1}(babc)^2 b = (abcb)^2$;

(18) $c^{-1}(babc)^2 c = (b^{-1}c^{-1}b^{-1}a^{-1})^2 (cb^{-1}ab)^2 (abcb^{-1})^2$.

Proof. Transform the formulae of Lemma 6 by p and use Lemma 7 and Lemma 1 to conclude Lemma 8. Transform the normal subgroup generated by $(ac)^2$ in H by p to obtain the normal subgroup generated by $(abcb^{-1})^2$. Hence this subgroup is also elementary abelian. The formulae of Lemma 5 imply that $(abcb^{-1})^2$ commutes with all eight conjugates of $(ac)^2$. Hence

$N := \langle (ac)^2, (ca)^2, a(ac)^2 a^{-1}, a^{-1}(ca)^2 a, c(ca)^2 c^{-1}, c^{-1}(ac)^2 c, a^{-1}c^{-1}(ac)^2 ca,$
$\qquad c^{-1}a^{-1}(ca)^2 ac, (abcb^{-1})^2, (b^{-1}abc)^2, a^{-1}(cb^{-1}ab)^2 a, (cb^{-1}ab)^2, a^{-1}(b^{-1}abc)^2 a,$
$\qquad\qquad\qquad\qquad a(abcb^{-1})^2 a^{-1}, a^2(cb^{-1}ab)^2 a^2, a(cb^{-1}ab)^2 a^{-1}\rangle$

is an elementary abelian normal subgroup in H . The formulae of Lemma 5, Lemma 7 show that $(c^{-1}a)^2$ and $(ac^{-1})^2$ centralize the group N . Transforming this statement by p implies that $(abcb)^2$ and $(babc)^2$ centralize N .

Moreover Lemma 6 yields the commutator relation

$(a^{-1}c)^2 \cdot (ac^{-1})^2 \cdot (c^{-1}a)^2 \cdot (ca^{-1})^2 = (ac)^2 \cdot (ca)^2 \cdot a(ac)^2 a^{-1} \cdot a^{-1}(ac)^2 a \cdot c^{-1}(ac)^2 c$
$\qquad\qquad\qquad\qquad \cdot c(ca)^2 c^{-1} \cdot a^{-1}c^{-1}(ac)^2 ca \cdot c^{-1}a^{-1}(ca)^2 ac$.

Transform by p obtaining

$(abcb)^{-2}(babc)^2(abcb)^2 (babc)^{-2}$
$= (abcb^{-1})^2 \cdot (b^{-1}abc)^2 \cdot a^{-1}(cb^{-1}ab)^2 a \cdot (cb^{-1}ab)^2 \cdot a^{-1}(b^{-1}abc)^2 a \cdot a(abcb^{-1})^2 a^{-1}$
$\qquad\qquad\qquad\qquad \cdot a^2(cb^{-1}ab)^2 a^{-2} a(cb^{-1}ab)^2 a^{-1}$.

LEMMA 9. *In* $\langle a, b, c, F\rangle$ *we have the following relations:*

(1) $ac^2 b^{-1}a^{-1}b^{-1}ca^2 bc^{-1}b = c^{-1}a^{-1}(ca)^2 ac \cdot (ac)^2 \cdot c(ca)^2 c^{-1}$
$\qquad\qquad\qquad\qquad \cdot a^{-1}c^{-1}(ac)^2 ca(ca)^2 a^{-1}(ca)^2 a$;

(2) $b^{-1}a^{-1}b^{-1}cba^2 cba = (ac)^2 c^{-1}a^{-1}(ca)^2 ac$;

(3) $(c^{-1}b^{-1})^2 a^{-1}c^{-1}(bc)^2 ca = (ca)^2 c^{-1}(ac)^2 c$;

(4) $\left(bc^2a\right)^2 = \left(a^{-1}c^{-1}bc\right)^2$;

(5) $\left(ab^2a^{-1}b^2\right)^2 = \left(bc^2b^{-1}c^2\right)^2 = 1$;

(6) $\left(abc^{-1}b\right)^4 = 1$;

(7) $\left(bab^2abc^{-1}a\right)^2 = (ac)^2a^{-1}(ca)^2aa^{-1}c^{-1}(ac)^2ca$;

(8) $\left(bcabc^{-1}b^2a^{-1}\right)^2 = 1$;

(9) $b\cdot a^2b^2a^2\cdot b^{-1}\cdot a^2b^2a^2 = a^{-1}c^{-1}(ac)^2ca\cdot c(ca)^2c^{-1}\cdot a(ac)^2a^{-1}\cdot(ca)^2$;

(10) $\left(babcbaba^{-1}\right)^2 = (ac)^2a^{-1}(ca)^2aa^{-1}c^{-1}(ac)^2ca$;

(11) $\left(abca^{-1}b^{-1}c^{-1}\right)^2 = 1$;

(12) $\left(abcb^{-1}\right)^2a^{-1}\left(b^{-1}abc\right)^2a = a^{-1}(ca)^2aa^{-1}c^{-1}(ac)^2ca$;

(13) $\left(bcbc^{-1}\right)^2 = \left(b^{-1}abc\right)^2\cdot a^{-1}\left(cb^{-1}ab\right)^2a\left(c^{-1}b^{-1}a^{-1}b^{-1}\right)^2\cdot\left(c^{-1}a\right)^2$;

(14) $\left(abab^{-1}\right)^2 = (ca)^2a(ac)^2a^{-1}\left(ca^{-1}\right)^2\left(b^{-1}c^{-1}b^{-1}a^{-1}\right)^2$;

(15) $Fbcb^{-1}aF = a^{-1}(ca)^2a\cdot c(ca)^2c^{-1}\cdot b^{-1}ca^2ba^{-1}$;

(16) $Fab^{-1}cb^{-1}F = a^{-1}c^{-1}abca^2bca^{-1}$;

(17) $Fbc^{-1}bcF = a(ac)^2a^{-1}c(ca)^2c^{-1}c^{-1}a^{-1}(ca)^2aca^{-1}c^{-1}(ac)^2ca$

$\qquad a\left(cb^{-1}ab\right)^2a^{-1}a\left(abcb^{-1}\right)^2a^{-1}\left(a^{-1}c\right)^4\left(c^{-1}b^{-1}a^{-1}b^{-1}\right)^2\cdot ac^{-1}b^{-1}a^{-1}b^{-1}c$;

(18) $FacF = (ca)^2(ac)^2a^{-1}c^{-1}(ac)^2cac^{-1}(ac)^2c\left(a^{-1}c\right)^4\left(c^{-1}b^{-1}a^{-1}b^{-1}\right)^2$

$\qquad\qquad\qquad\qquad\qquad\qquad\qquad a^{-1}c^{-1}abcab^{-1}c$;

(19) $Fabcb^{-1}F = (ac)^2a^{-1}c^{-1}(ac)^2ca\cdot a^{-1}b^{-1}a^2c^{-1}b^2c^2b^{-1}$;

(20) $Fab^2cF = a(ac)^2a^{-1}c^{-1}a^{-1}(ca)^2ac\left(b^{-1}abc\right)^2a^{-1}\left(cb^{-1}ab\right)^2a\left(a^{-1}c\right)^4(babc)^2$

$\qquad\qquad\qquad\cdot a^{-1}b^{-1}a^2c^{-1}b^2c^2b^{-1}ac^{-1}b^{-1}a^{-1}b^{-1}c$;

(21) $c^{-1}a^2ba^{-1}cb^{-1}a^2c^{-1}b^2c^2b^{-1}ac^{-1}b^{-1} = a(ac)^2a^{-1}c^{-1}(ac)^2c$;

(22) $FcaF = a(ac)^2a^{-1}a^{-1}(ca)^2ac(ca)^2c^{-1}c^{-1}a^{-1}(ca)^2ac(ca)^2c^{-1}(ac)^2ca^{-1}\left(cb^{-1}ab\right)^2a$

$\qquad\left(cb^{-1}ab\right)^2a^2\left(cb^{-1}ab\right)^2a^2a\left(cb^{-1}ab\right)^2a^{-1}\left(c^{-1}a\right)^4\left(c^{-1}b^{-1}a^{-1}b^{-1}\right)^2abcab^{-1}a^{-1}$.

Proof. (1) By Lemma 2 (12) we have

$$abc^{-1}b^{-1}a^{-1}D^2D^{-1}bc^{-1}bca^{-1}b^{-1}DD^2b^{-1}ca = 1$$.

Lemma 2 (5), Lemma 3 (2), (8), (1) and Lemma 4 (1) imply

$$c^{-1}b^{-1}a^2c^{-1}a^2ca^{-1}b^{-1}ac^2ac^{-1}a^2caca^{-1}bc^{-1}bca^{-1} = 1 \ .$$

Hence

$$bc^{-1}bac^2 \cdot c\left(ca^{-1}\right)^2 c^{-1} \cdot b^{-1}a^{-1} \cdot \left(a^{-1}c^{-1}\right)^2 \left(ca^{-1}\right)^2 \cdot b^{-1} \cdot ac^2ac^{-1}a^2caca^{-1} = 1 \ .$$

Lemmas 5 and 6 yield (1).

(2) By Lemma 2 (11) we have

$$b^{-1}D^{-1}b^{-1}cb^{-1}D^{-1}abca^{-1}Dbc^{-1}bDbac^{-1}b^{-1}a^{-1} = 1 \ .$$

Lemma 3 (4), (7), (1), Lemma 4 (1) and Lemma 5 imply

$$b^{-1}D^{-1}b^{-1}ac^{-1}ab^2Dbac^{-1}b^{-1}a^{-1} = 1 \ .$$

Lemma 3 (4), (1), (3), (7), (2) and Lemma 4 (1) imply

$$b^{-1}a^{-1}c^{-1}b^{-1}a^2c^{-1}ac^{-1}a^2b^{-1}c^2a^{-1}ca^{-1}b \cdot bab \cdot cac^{-1} = 1 \ .$$

Use Lemma 9 (1) to replace bab and apply Lemmas 5 and 6 obtaining Lemma 9 (2).

(3) By Lemma 2 (25), (26) we have

$$c^{-1}bcbca^{-1}c^2b^{-1}a^2 \cdot b^{-1}D^{-1}Fbca^{-1}Fa^{-1}c^{-1}Fac^{-1} \cdot b^{-1}FDba \cdot c^2b^{-1}a^{-1} = 1 \ .$$

By Lemma 2 (27),

$$c^{-1}bcbca^{-1}c^2b^{-1}a^{-1}c^{-1}a^{-1}c^{-1}c^{-1}b^{-1}a^{-1} = 1 \ .$$

Hence Lemma 5 yields (3).

(4) By Lemma 2 (25), (28) we have

$$bc^2a^{-1}b^{-1}D^{-1}Fbca^{-1}F \cdot bc^2a^{-1}b^{-1}D^{-1}Fbca^{-1}F = a^{-1}c^{-1}bca^{-1}c^{-1}bc \ .$$

Hence Lemma 2 (26) implies (4).

(5) By Lemma 2 (13) we have

$$\left(D^{-1}abca^{-1}Db^{-1}c^{-1}\right)^2 = 1 \ .$$

By Lemma 2 (5), Lemma 3 (2), (4) we have

$$\left(ab^2cab^2ac^2b^2a^{-1}b^2c^{-1}\right)^2 = 1 \ .$$

Using Lemma 4 (1) we have

$$\left(a^{-1}c^{-1}b^2a^{-1}b^2c^{-1}\right)^2 = 1 \ .$$

Then Lemma 5 implies the first part of (5). When we apply to $\left(ab^2a^{-1}b^2\right)^2 = 1$ the automorphism induced by p on $\langle a, b, c \rangle$, we obtain the second part of (5).

(6) Lemma 2 (14) yields

$$\left(D^2D^{-1}ba^2bDba^2b\right)^2 = 1 \ .$$

Lemma 3 (10), (5), (4), Lemma 4 (1) imply

$$\left(abca^2b \cdot caca \cdot abca^2b\right)^2 = 1 \ .$$

Use Lemma 5 to conclude

$$\left(abca^2b\right)^4 = 1 \ .$$

Apply to this equation the automorphism induced by p on $\langle a, b, c \rangle$ obtaining (6).

(7) By Lemma 2 (15) we have

$$\left(Dba^2bc^{-1}bc^{-1}b\right)^2 = 1 \ .$$

Lemma 3 (10), (5), (4), (8), (2), Lemma 4 (1) imply

$$babca^2bcac^2a^{-1}bac^{-1}a^2c^2a^{-1}bac^{-1}a^2ca^2bca^2bc^{-1}bc^{-1} = 1 \ .$$

From Lemma 5 follows

$$\left(bab \cdot ca^2bc^{-1}bc^{-1}\right)^2 = (ac)^2(ca)^2a^{-1}(ca)^2aa^{-1}c^{-1}(ac)^2cac^{-1}a^{-1}(ca)^2ac \ .$$

Replacing $ca^2bc^{-1}b$ by Lemma 9 (1) and using Lemma 5 yields (7).

(8) and (11) By the definition of a, b, c we have

$$bcabc^{-1}b^2a^{-1} = \left(pqp^2q^{-1}\right)^4 \ ,$$

$$abca^{-1}b^{-1}c^{-1} = \left(qp^{-1}q^{-1}p^2\right)^4 \ .$$

Then (xvi) and (xvii), respectively, imply (8) and (11), respectively.

(9) By Lemma 2 (17) we have

$$ac^{-1}D^{-1}abcac^{-1}D^{-1}abcb^{-1} \cdot b^2 \cdot bc^{-1}b^{-1}a^{-1}Dca^{-1}c^{-1}b^{-1}a^{-1}Dca^{-1}b^2 = 1 \ .$$

Lemma 2 (5), Lemma 3 (2) imply

$$ac^{-1}D^{-1}abcb^{-1}bac^{-1}a^{-1}c^{-1}b^2caca^{-1}c^{-1}b^{-1}a^{-1}Dca^{-1}b^2 = 1 \ .$$

By Lemma 2 (5), Lemma 3 (5), (6), (2), (8), Lemma 4 (1), Lemma 5 follows

$$(ca)^2 \cdot a(ac)^2a^{-1}b \cdot (ca)^2a^2(ac)^2a(ac)^2a^{-1}c(ca)^2c^{-1}$$
$$c^{-1}a^{-1}(ca)^2aca^{-1}c^{-1}(ac)^2cab^2ac^{-1}a^2\left(a^{-1}c\right)^2ac^{-1}b^{-1}ca^{-1}b^2ac^{-1} = 1 \ .$$

Applying again Lemmas 5 and 6, implies Lemma 9 (9).

(10) By Lemma 2 (18) we have

$$\left(D^2D^{-1}baDbc^{-1}bDbac^{-1}\right)^2 = 1 \ .$$

By Lemma 3 (10), (5), Lemma 4 (1), Lemma 5 we have

$$\left(bac^{-1}abca^2b^2D\right)^2 = 1 \ .$$

By Lemma 3 (10), (5), (3), (7), (8), (1), Lemma 4 (1) we have

$$ac^{-1}abca^2b^2abc^{-1}acab^{-1}a^{-1}c^{-1}b^{-1}ca^2c^{-1}a^2b^{-1}c^{-1}a^2ca^{-1}b^2abc = 1 \ .$$

Hence

$$ac^{-1}ab\cdot\left(ca^{-1}\right)^2\cdot a(ac)^2a^{-1}\cdot a^2cb^2abc^2\cdot(ca)^2\cdot ba^2c\cdot(ac)^2a^{-1}b\left(ca^{-1}\right)^2a(ac)^2a^{-1}$$
$$\cdot b^{-1}\left(c^{-1}a\right)^2\cdot a^{-1}(ca)^2a\cdot ab^2abc = 1$$

By Lemmas 5 and 6 follows

$$\left(c^2\cdot ba^2cb^2ab\right)^2 = (ca)^2a(ac)^2a^{-1}a^{-1}(ca)^2ac^{-1}(ac)^2c\cdot c(ca)^2c^{-1} \ .$$

Replace cba^2cb by Lemma 9 (2) and use Lemma 5 to obtain (10).

(12) By Lemma 9 (10) we have

$$(babc)^2c^{-1}a^{-1}(babc)^2c^{-1}a^{-1} = (ac)^2\cdot a^{-1}(ca)^2aa^{-1}c^{-1}(ac)^2ca \ .$$

Applying Lemma 8 we obtain Lemma 9 (12).

(13) By Lemma 9 (3) follows

$$bcbc^{-1}\cdot\left(c^{-1}a\right)^2\cdot bcb\cdot(abcb)^{-2}\cdot c^{-1} = (ca)^2\cdot c^{-1}(ac)^2c \ .$$

Use Lemma 5, Lemma 6, Lemma 7, Lemma 8 to conclude Lemma 9 (13).

(14) Apply to Lemma 9 (13) the automorphism induced by p^{-1} on $\langle a, b, c \rangle$ obtaining Lemma 9 (14).

(15) By Lemma 9 (29) we have

$$Dbc^{-1}bD^2\cdot D^{-1}bac^{-1}b^{-1}a^{-1}b^{-1}D\cdot D^2ca^{-1}b^{-1}D^{-1}Fbcb^{-1}aF = 1 \ .$$

Using Lemma 3 (10), (5), (8), Lemma 4 (1) we obtain

$$D^2D^{-1}bc^{-1}babca^2a^{-1}b^{-1}a^{-1}c^{-1}b^{-1}a^{-1}ca^{-1}b^{-1}DD^2Fbcb^{-1}aF = 1 \ .$$

By Lemma 3 (10), (8), (3), (9), (4), (5), (7), (1), Lemma 4 (1) follows

$$abcb^{-1}\cdot(ca)^2\cdot bc^{-1}babac^{-1}acb\cdot a(ac)^2a^{-1}\left(ac^{-1}\right)^2$$
$$\cdot b^{-1}a^2c^{-1}a^{-1}c^{-1}bcaba^{-1}c^{-1}a^{-1}cb^{-1}a^{-1}Fbcb^{-1}aF = 1$$

By Lemmas 6 and 7 follows

$$ab\cdot(ca)^2\cdot bab\cdot ac^{-1}a^2c^{-1}a(ac)^2a^{-1}(ca)^2bcaba^{-1}c^{-1}a^{-1}cb^{-1}a^{-1}Fbcb^{-1}aF = 1 \ .$$

Replace bab using Lemma 9 (1) and apply Lemma 4 (1), Lemmas 6 and 7 to obtain

$$a \cdot ba^2cb \cdot c(ca)^2c^{-1} \cdot (ac)^2c^{-1}(ac)^2cc^{-1}a^{-1}(ca)^2aca^{-1}(ca)^2a(ca)^2 \cdot ab^{-1}c^2b^{-1}a^{-1}Fbcb^{-1}aF = 1 \ .$$

Replace ba^2cb using Lemma 9 (2), and apply Lemmas 6 and 7 and replace again using Lemma 9 (1) to obtain (15).

(16) follows immediately by Lemma 2 (23), (33).

(17) By Lemma 2 (31) we have

$$Fc^{-1}b^{-1}cb^{-1}Fac^{-1}b^{-1}a^{-1}c^{-1}bD^2 \cdot D^{-1}bac^{-1}b^{-1}a^{-1}b^{-1}DD^2c^{-1}b^{-1} = 1 \ .$$

By Lemma 3 (10), (5), (8), Lemma 4 (1) we have

$$Fbc^{-1}bcF = ac^{-1}b^{-1}a^{-1}c^{-1}ba \cdot bc^2a^{-1}b^{-1} \cdot a^{-1}c^{-1}b^{-1}a^{-1}c^{-1}b^{-1} \ .$$

Replace $bc^2a^{-1}b^{-1}$ using Lemma 9 (1) and apply Lemma 6 to obtain

$$Fbc^{-1}bcF = a(ac)^2a^{-1}a^{-1}(ca)^2ac(ca)^2c^{-1}c^{-1}(ac)^2cc^{-1}a^{-1}(ca)^2ac \cdot ac^{-1}a^{-1}b^{-1}a^{-1}$$
$$\cdot (abab^{-1})^2 \cdot a^{-1}c^{-1}b^{-1} \ .$$

Replace $(abab^{-1})^2$ using Lemma 9 (14) and apply Lemmas 6, 7, 8, 5 to obtain

$$Fbc^{-1}bcF = c^{-1}(ac)^2ca^{-1}c^{-1}(ac)^2ca(ac)^2(a^{-1}c)^2a^{-1}(b^{-1}abc)^2a(abcb^{-1})^2 \cdot a(cb^{-1}ab)^2a^{-1}$$
$$\cdot a(abcb^{-1})^2a^{-1}(c^{-1}b^{-1}a^{-1}b^{-1})^2ac^{-1}a^{-1}b^{-1}c^{-1}a^2 \cdot a(ac)^2a^{-1}(ac^{-1})^2 \ b^{-1} \ .$$

Apply Lemmas 5 and 6, replace $b^{-1}c^{-1}a^2b^{-1}$ using Lemma 9 (2), and apply Lemmas 5 and 6 and Lemma 9 (12) to obtain (17).

(18) follows by Lemma 9 (16), (17) using Lemmas 5 to 8.

(19) follows by Lemma 2 (20), Lemma 9 (15) using Lemma 5.

(20) follows by Lemma 9 (17), (19) using Lemmas 5, 7, 8.

(21) Lemma 2 (24) implies

$$abca^{-1} \cdot Fbc^{-1}bc \cdot c^{-1}a^{-1} \cdot abcb^{-1}F \cdot ac^{-1}b^{-1}a^{-1}b^2 = 1 \ .$$

By Lemma 9 (17), (18), (19), Lemma 4 (1) we have

$$a^{-1}c^{-1}b^{-1}a^{-1}cb^{-1}a^2c^{-1}b^2c^2b^{-1}ac^{-1}b^{-1}a^{-1}b^2 = (ac)^2 \cdot a^{-1}c^{-1}(ac)^2ca \ .$$

Lemma 5, Lemma 4 (1) imply (21).

(22) By Lemma 9 (20), Lemma 2 (21) and Lemmas 5 to 8 we have

$$FcaF = (c^{-1}a)^4 \cdot (c^{-1}b^{-1}a^{-1}b^{-1})^2a^{-1}(cb^{-1}ab)^2a \cdot (b^{-1}abc)^2(cb^{-1}ab)^2 \cdot a^2(cb^{-1}ab)^2a^2$$
$$\cdot a(cb^{-1}ab)^2a^{-1}a(abcb^{-1})^2a^{-1}c^{-1}a^{-1}(ca)^2aca(ac)^2a^{-1} \cdot c^{-1}ba \cdot bca^{-1}bc^2b^2ca^2bc^{-1} \cdot c^{-1}b^2a^{-1} \ .$$

Applying Lemma 9 (21), Lemma 5, Lemma 9 (12) we obtain

$$FcaF = a(ac)^2a^{-1}a^{-1}(ca)^2ac(ca)^2c^{-1}c^{-1}a^{-1}(ca)^2ac(ca)^2c^{-1}(ac)^2ca^{-1}(cb^{-1}ab)^2a(cb^{-1}ab)^2$$
$$a^2(cb^{-1}ab)^2a^2a(cb^{-1}ab)^2a^{-1} \cdot c^{-1} \cdot bac^{-1}a^2b \cdot a^{-1}c^{-1}b^2a^{-1}$$

Then follows (22) by Lemma 4 (1), Lemma 2 (2).

LEMMA 10. *We have*

(1) $q(ac)^2q^{-1} = (ca)^2$;

(2) $q(ca)^2q^{-1} = (ac)^2$;

(3) $(c^{-1}a)^4 = (ca)^2a(ac)^2a^{-1}$, $(ac^{-1})^4 = (ac)^2a^{-1}(ca)^2a$;

(4) $(abcb)^4 = (b^{-1}abc)^2 \cdot a^{-1}(cb^{-1}ab)^2a$, $(babc)^4 = (abcb^{-1})^2 \cdot (cb^{-1}ab)^2$;

(5) $\langle (ac)^2, (ca)^2 \rangle$, $\langle (abcb^{-1})^2, (b^{-1}abc)^2 \rangle$, $\langle (abcb)^2, (babc)^2 \rangle$,

   $\langle (c^{-1}a)^2, (ac^{-1})^2 \rangle$ *are four elementary abelian normal subgroups of*
   exponent 2 *in* $\langle a, b, c \rangle$. *They centralize each other elementwise.*
   Each of them is centralized by H_g . *Any element of* $H \backslash H_g$ *permutes,*
   in each case, the two generators.

Proof. (1) By Lemma 1 we have

$$qacacq^{-1} = Dbc^{-1}b^{-1}a^{-1}Dbc^{-1}b^{-1}a^{-1} .$$

Lemma 2 (5) and Lemma 3 (10) imply (1).

(2) By Lemma 1 we have

$$qcacaq^{-1} = adbc^{-1}b^{-1}a^{-1}Dbc^{-1}b^{-1}a^{-1}da^{-1} .$$

By Lemma 2 (5) we have

$$qcacaq^{-1} = a(abcb^{-1})^2a^{-1}aD^2a^{-1} .$$

Lemma 5, Lemma 3 (10) imply (2).

(3) By Lemma 2 (30), Lemma 3 (2), (8), (6), (3), (5), (10) and Lemma 4 (1) we
have

$$Fbcb^{-1}aFa^{-1}c^{-1}b^{-1}a^2c^{-1}a^2ca^{-1}b^{-1}ac^{-1}acba^{-1}cac^{-1}a^2cacb^2c^{-1}a^{-1}cb^{-1}a^{-1} = 1 .$$

By Lemma 9 (15) we have

$$a^{-1}(ca)^2a \cdot c(ca)^2c^{-1} \cdot b^{-1}ca^2b \cdot a^2c^{-1}b^{-1}a^2c^{-1}a^2ca^{-1}b^{-1}ac^{-1}acba^{-1}cac^{-1}a^2cacb^2$$
$$\cdot c^{-1}a^{-1}cb^{-1}a^{-1} = 1$$

Use Lemma 9 (2) to replace $b^{-1}ca^2b$ obtaining

$ba^2 \cdot b^{-1}a^{-1}b^{-1} \cdot ac^{-1}acba^{-1}cac^{-1}a \cdot (ac)^2 \cdot (ca)^2 a(ac)^2 a^{-1}a^{-1}(ca)^2 a$

$$\cdot c(ca)^2 c^{-1}c^{-1}(ac)^2 ca^{-1}c^{-1}(ac)^2 cac^{-1}a^{-1}(ca)^2 ac = 1 \ .$$

Replace $b^{-1}a^{-1}b^{-1}$ using Lemma 9 (2) to obtain

$ba\, b^{-1}c^{-1}a^2 b^{-1} \cdot c^{-1}ac^{-1}acba^{-1}cac^{-1}a$

$$\cdot (ac)^2 (ca)^2 a^{-1}(ca)^2 ac^{-1}(ac)^2 cc(ca)^2 c^{-1}c^{-1}a^{-1}(ca)^2 ac = 1 \ .$$

By Lemma 9 (2), Lemma 5 we have

$c^{-1}a^2 b^{-1}c^{-1}b^{-1}a^{-1}c^{-1}bc^{-1}ac^{-1}acba^{-1}cac^{-1}a \cdot (ca)^2 a^{-1}(ca)^2 ac^{-1}(ac)^2 cc(ca)^2 c^{-1} = 1 \ .$

By Lemma 9 (3) and Lemma 5 we have

$c^{-1}a^2 ca^{-1}c^2 b^{-1}c^{-1} \cdot (c^{-1}a)^2 \cdot cba^{-1}cac^{-1}a \cdot (ca)^2$

$$\cdot a(ac)^2 a^{-1}a^{-1}(ca)^2 ac^{-1}(ac)^2 cc(ca)^2 c^{-1}c^{-1}a^{-1}(ca)^2 ac \ .$$

By Lemmas 5 and 6 we have

$$(c^{-1}a)^4 = (ac)^2 (ca)^2 a(ac)^2 a^{-1}a^{-1}(ca)^2 ac(ca)^2 c^{-1}a^{-1}c^{-1}(ac)^2 ca \ .$$

Conjugate the relation Lemma 9 (12) with a^2 and compare the result with Lemma 9 (12), then we obtain

$$(ac)^2 c(ca)^2 c^{-1}a^{-1}(ca)^2 aa^{-1}c^{-1}(ac)^2 ca = 1 \ .$$

Using this relation in the above yields the first part of (3). Applying Lemma 5 yields the second part.

(4) Apply the automorphism induced by p on $\langle a, b, c \rangle$ to the formulae of (3) to obtain (4).

(5) Apply to (3) the automorphism induced by p^2 on $\langle a, b, c \rangle$ and compare the result with the original formula

(I)
$$\begin{cases} c(ca)^2 c^{-1} = (ac)^2 (ca)^2 a(ac)^2 a^{-1} \ , \\ c^{-1}(ac)^2 c = (ac)^2 (ca)^2 a^{-1}(ca)^2 a \ , \\ a^{-1}c^{-1}(ac)^2 ca = (ca)^2 a(ac)^2 a^{-1}a^{-1}(ca)^2 a \ , \\ c^{-1}a^{-1}(ca)^2 ac = (ac)^2 a(ac)^2 a^{-1}a^{-1}(ca)^2 a \ . \end{cases}$$

Conjugate with b, a^2, b^{-1} and use Lemma 5 to obtain the second, third and fourth formula by the first formula.

Apply to (I) the automorphism induced by p to obtain

$$(\text{II})\quad\begin{cases} a^{-1}(b^{-1}abc)^2 a = (abcb^{-1})^2\cdot(b^{-1}abc)^2\cdot a^{-1}(cb^{-1}ab)^2 a \ , \\ a(abcb^{-1})^2 a^{-1} = (abcb^{-1})^2\cdot(b^{-1}abc)^2\cdot(cb^{-1}ab)^2 \ , \\ a^2(cb^{-1}ab)^2 a^2 = (b^{-1}abc)^2(cb^{-1}ab)^2 a^{-1}(cb^{-1}ab)^2 a \ , \\ a(cb^{-1}ab)^2 a^{-1} = (abcb^{-1})^2(cb^{-1}ab)^2 a^{-1}(cb^{-1}ab)^2 a \ . \end{cases}$$

Using Lemma 10 (2) and Lemma 1 we have

$$qa^{-1}(ca)^2 aq^{-1} = aD^{-1}d(ac)^2 d^{-1}Da^{-1} \ .$$

By Lemma 2 (36), (42) follows

$$qa^{-1}(ca)^2 aq^{-1} = aD^{-1}(Dbabca^{-1}Fbc^{-1}a^{-1}b^{-1}Fac^{-1}b^{-1}a^{-1}b^{-1}D^{-1})^2 Da^{-1}$$
$$= ababca^{-1}F(ca)^2 Fac^{-1}b^{-1}a^{-1}b^{-1}a^{-1} \ .$$

Use Lemma 10 (3), (4) and (I) and (II) in Lemma 9 (22), then follows

$$FcaF = (ac)^2(ca)^2 a(ac)^2 a^{-1}a^{-1}(ca)^2 a(babc)^2 abcab^{-1}a^{-1} \ .$$

Insert this in the above formula to obtain, using Lemma 5,

$$qa^{-1}(ca)^2 aq^{-1} = ababca^{-1}\cdot((babc)^2 abcab^{-1}a^{-1})^2\cdot ac^{-1}b^{-1}a^{-1}b^{-1}a^{-1} \ .$$

Apply Lemma 8, then follows

$$qa^{-1}(ca)^2 aq^{-1} = ababca^{-1}\cdot a^{-1}(b^{-1}abc)^2 a\cdot(abcb^{-1})^2\cdot a(cb^{-1}ab)^2 a^{-1}a(abcb^{-1})^2 a^{-1}(cb^{-1}ab)^2$$
$$\cdot a^2(cb^{-1}ab)^2 a^2 ab(ca)^2 b^{-1}a^{-1}\cdot ac^{-1}b^{-1}a^{-1}b^{-1}a^{-1}$$

Lemma 5 and (II) imply

$$qa^{-1}(ca)^2 aq^{-1} = ababca^{-1}a^{-1}(b^{-1}abc)^2 a(abcb^{-1})^2(ca)^2 ac^{-1}b^{-1}a^{-1}b^{-1}a^{-1} \ .$$

Lemma 9 (12), Lemma 5 and (I) imply

$$qa^{-1}(ca)^2 aq^{-1} = (ca)^2 \ .$$

A comparison with Lemma 10 (1) shows

$$a^{-1}(ca)^2 a = (ca)^2 \ .$$

Applying the automorphism induced by p^2 yields

$$c^{-1}(ac)^2 c = (ac)^2 \ .$$

Then the statement on $\langle(ac)^2(ca)^2\rangle$ follows by Lemma 5. To obtain the statement on $\langle(abcb^{-1})^2(b^{-1}abc)^2\rangle$ conjugate with p . One obtains the relations

$$a^{-1}(cb^{-1}ab)^2 a = (b^{-1}abc)^2 \ ,$$

$$(cb^{-1}ab)^2 = (abcb^{-1})^2 \ .$$

Use Lemmas 6 and 8 and Lemma 10 (3), (4) to conclude the remaining statements.

LEMMA 11. *In* $\langle a, b, c \rangle$ *we have the following relations*

(1) $\left(b^{-1}cbc\right)^2 = \left(c^{-1}a\right)^2 \cdot (babc)^2$;

(2) $\left(a^{-1}bab\right)^2 = \left(ac^{-1}\right)^2 \cdot (abcb)^2$;

(3) $\left(bcab^{-1}ca\right)^2 = \left(ac^{-1}\right)^2 \cdot \left(c^{-1}a\right)^2$;

(4) $cabacb^{-1}a^{-1}c^{-1} = bacb^{-1}$

(5) $\left(b^{-1}a^2ba^2\right)^2 = 1$;

(6) $\left(b^2ca\right)^2 = \left(ac^{-1}\right)^2$;

(7) $\left(cb^{-1}a^{-1}b\right)^2 = (babc)^2$;

(8) $bcbc^{-1}a^{-1}b^{-1}c^{-1}b^{-1}a^{-1}c^{-1} = \left(c^{-1}a\right)^2$;

(9) $\left(bab^{-1}c^{-1}\right)^2 = \left(ac^{-1}\right)^2 \cdot \left(c^{-1}a\right)^2 (babc)^2$;

(10) $b^{-1}a^{-1}b^{-1} \cdot ac \cdot bab \cdot a^{-1}c^{-1} = (ca)^2 \left(c^{-1}a\right)^2 (abcb)^2 \cdot (babc)^2$;

(11) $b^{-1}a^{-1}b^{-1} \cdot ca \cdot bab \cdot c^{-1}a^{-1} = \left(c^{-1}a\right)^2 \left(ac^{-1}\right)^2 (abcb)^2 \cdot (babc)^2$;

(12) $b^2acb^2c^{-1}a^{-1} = (ac)^2 \left(ac^{-1}\right)^2 (abcb)^2 \cdot (babc)^2$;

(13) $(abcb)^2 = (babc)^2$, $\left(c^{-1}a\right)^2 = \left(ac^{-1}\right)^2$;

(14) $(abcab)^4 = 1$;

(15) $b^{-1}cabca^{-1}b^{-1}cabca^{-1} = 1$;

(16) $b^{-1}caba^2b^{-1}caba^2 = \left(c^{-1}a\right)^2$;

(17) $\left(a^{-1}b^{-1}ab\right)^2 = (ca)^2 \cdot \left(abcb^{-1}\right)^2$, $\left(b^{-1}c^{-1}bc\right)^2 = \left(b^{-1}abc\right)^2 \cdot (ca)^2$;

(18) $(ab)^4 = (ba)^4$;

(19) $(ab)^4 \in Z(\langle a, b, c \rangle)$;

(20) $ba^2c^2b^{-1}c^2a^2 = (ac)^2(babc)^2 \cdot (ab)^4$;

(21) $ba^2b^{-1}acba^2b^{-1}ac = \left(c^{-1}a\right)^2$;

(22) $(ba)^4 = (cb)^4$, $(ab)^4 = (bc)^4$;

(23) $b^2cb^{-1}ac^{-1}b^{-1}a^{-1} = \left(ac^{-1}\right)^2(ac)^2(ca)^2 \cdot (ab)^4$;

(24) $\quad ba^{-1}c^{-1}b^{-1}c^{-1}a^{-1} = \left(c^{-1}a\right)^2 (babc)^2 \cdot \left(abcb^{-1}\right)^2 \cdot \left(b^{-1}abc\right)^2 \cdot (ab)^4 \; ;$

(25) $\quad (ab)^2 = (ca)^2 \cdot \left(abcb^{-1}\right)^2 \cdot \left(b^{-1}abc\right)^2 \cdot (abcb)^2 \cdot (ab)^4 \cdot (ba)^2 \; ,$

$\qquad (bc)^2 = (ac)^2 (ca)^2 \left(b^{-1}abc\right)^2 \left(ac^{-1}\right)^2 \cdot (ab)^4 \cdot (cb)^2 \; ;$

(26) $\quad (ab)^4 = \left(abcb^{-1}\right)^2 \cdot \left(b^{-1}abc\right)^2 \cdot (babc)^2 \cdot \left(c^{-1}a\right)^2 \; ;$

(27) $\quad (ac)^2 (ca)^2 = \left(abcb^{-1}\right)^2 \left(b^{-1}abc\right)^2 \; ;$

(28) $\quad (ac)^2 = (ca)^2 \; , \quad \left(abcb^{-1}\right)^2 = \left(b^{-1}abc\right)^2 \; .$

Proof. (1) By Lemma 9 (2), Lemma 10 (5) we have

$$b^{-1}cbc = (ca)^2 \left(a^{-1}c\right)^2 aba^{-1}b^{-1}a^2 \; .$$

Then (1) follows by Lemma 9 (14).

(2) We use the automorphism induced by p^{-1} in Lemma 11 (1). Then Lemma 11 (2) follows.

(3) By Lemma 2 (2) and Lemma 4 (1),

$$bcab^{-1}ca \cdot (ca)^2 \left(ca^{-1}\right)^2 \cdot bca \cdot (ca)^2 \left(ca^{-1}\right)^2 b^{-1}ca = 1 \; .$$

Hence Lemma 10 (5) implies Lemma 11 (3).

(4) By Lemma 2 (25) we have

$$bcFac^{-1}b^{-1}FDbac^2b^{-1}a^{-1}c^{-1} = c^{-1}b^{-1}cabc^2a^{-1}b^{-1}D^{-1}Fbca^{-1}F \; .$$

Insert this in Lemma 2 (28); then

$$Fac^{-1}b^{-1}FDbac^2b^{-1}Fac^{-1}b^{-1}FDba = c^{-1}b^{-1}cac^{-1}b^{-1}cabc^2 \; .$$

A comparison with Lemma 2 (26) yields

$$c^2a^{-1}b^{-1}cabc^2a = bacb^{-1} \; .$$

Then Lemma 11 (4) follows by Lemma 9 (3).

(5) By Lemma 9 (1), (2), Lemma 10 (5)

$$bab = ca^2bc^{-1}bac^2 = cba^2cba \cdot (ac)^2 (ca)^2 \; .$$

Using Lemma 10 (5) we obtain

$$b^{-1}a^2ba^2 = (ca)^2 \left(ca^{-1}\right)^2 c^{-1} \cdot bc^2b^{-1}c^2 \cdot c \; .$$

Then Lemma 9 (5) implies Lemma 11 (5).

(6) follows immediately by Lemma 4 (1) and Lemma 10 (5).

(7) Applying the automorphism induced by p yields Lemma 11 (7).

(8) follows immediately by Lemma 9 (3) and Lemma 10 (5).

(9) By Lemma 10 (5),

$$bab^{-1}c^{-1} \cdot a^{-1} \cdot ba^{-1}b^{-1}a^{-1} \cdot aca = (ca)^2 bab^{-1} \cdot ac \cdot ba^{-1}b^{-1}ca \ .$$

Replace $ba^{-1}b^{-1}a^{-1}$ using Lemma 9 (14) and bab^{-1} on the right-hand side using Lemma 9 (2), then

$$\left(a^{-1}c\right)^2 (babc)^2 \left(bab^{-1}c^{-1}\right)^2 = c^{-1}a^2c \cdot \left(c^{-1}b^{-1}\right)^2 \cdot ca \cdot (bc)^2 \cdot c^{-1}a^2c^2a \ .$$

Then Lemma 9 (3) and Lemma 10 (5) imply Lemma 11 (9).

(10) By Lemma 4 (1) we have

$$ababac^{-1}a^2b^{-1}a^{-1}b^{-1}a^2c^{-1} = abab^{-1} \cdot a^{-1}c^{-1} \cdot ba^{-1}b^{-1}a^{-1} \cdot a^{-1}c^{-1} \ .$$

Replacing $abab^{-1}$ using Lemma 9 (14) yields, by Lemma 10 (5),

$$ababac^{-1}a^2b^{-1}a^{-1}b^{-1}a^2c^{-1} = \left(ac^{-1}\right)^2 ba^{-1}b^{-1} \cdot ac \cdot bab^{-1} \cdot a^{-1}c^{-1} \ .$$

Replacing bab^{-1} using Lemma 9 (2) yields, by Lemma 10 (5),

$$ababac^{-1}a^2b^{-1}a^{-1}b^{-1}a^2c^{-1} = a^{-1}b^{-1}cba^{-1} \cdot c^{-1}b^{-1}c^{-1}b \cdot c^{-1} \ .$$

Replacing $c^{-1}b^{-1}c^{-1}b$ using Lemma 11 (1) yields, by Lemma 10 (5),

$$ababac^{-1}a^2b^{-1}a^{-1}b^{-1}a^2c^{-1} = \left(ac^{-1}\right)^2 (abcb)^2 \left(a^{-1}b^{-1}cb\right)^2 \ .$$

Then Lemma 11 (9) and Lemma 10 (5) imply Lemma 11 (10).

(11) Applying to Lemma 11 (10) the automorphism induced by p^2 yields Lemma 11 (11) by Lemma 10 (5).

(12) By Lemma 11 (8) and Lemma 4 (1)

$$b^2acb^2c^{-1}a^{-1} = \left(ca^{-1}\right)^2 \cdot \left(c^{-1}a\right)^2 \cdot bc^{-1}b \cdot cab^{-1}cb^{-1}c^{-1}a^{-1} \ .$$

Replacing $bc^{-1}b$ using Lemma 9 (1) yields, by Lemma 10 (5),

$$b^2acb^2c^{-1}a^{-1} = a^2c^{-1} \cdot babacb^{-1}a^{-1}b^{-1} \cdot ca^2c^{-1}a^{-1} \ .$$

Use Lemma 11 (11) to conclude Lemma 11 (12).

(13) We have

$$abcb^{-1}c \cdot abc^{-1}b^{-1} \cdot a^2c^{-1} = ab^2 \cdot b^{-1}cb^{-1} \cdot ca \cdot bc^{-1}bb^2a^2c^{-1} \ .$$

Replacing $abc^{-1}b^{-1}$ using Lemma 11 (7) and $bc^{-1}b$ using Lemma 9 (1) yields, by Lemma 10 (5),

$$(babc)^2abc \cdot b^{-1}cbc \cdot b^{-1}ac^{-1} = \left(c^{-1}a\right)^2 \left(ac^{-1}\right)^2 ab^2 \cdot ac^2 \cdot b^{-1}a^{-1}b^{-1} \cdot ac \cdot bab \cdot c^2a^{-1}ba^2c^{-1} \ .$$

Replacing $b^{-1}cbc$ using Lemma 11 (1) and applying Lemma 11 (10) to the right hand
side yields

$$\left(c^{-1}a\right)^2\left(ac^{-1}\right)^2 = ab^2acb^2a^2c^{-1} \ .$$

A comparison with Lemma 11 (12) furnishes the first part of Lemma 11 (13). The second
part is obtained by applying the automorphism induced by p .

(14) By Lemma 9 (1), (13)

$$ababc = (ac)^2\left(c^{-1}a\right)^2(babc)^2a^{-1}b^{-1}cb^{-1}a^{-1} \ .$$

Use Lemma 10 (5) to conclude

$$cab\cdot\left(c^{-1}b^{-1}a^{-1}b^{-1}a^{-1}\right)^4b^{-1}a^{-1}c^{-1} = \left(ac^{-1}\right)^2cabc^{-1}b^{-1}a^{-1}c^{-1}bc^2\cdot c^{-1}a^{-1}b^{-1}a^{-1}c^{-1}bab^{-1}a^{-1}c^{-1}$$

$$= \left(ac^{-1}\right)^2cabc^{-1}b^{-1}a^{-1}c^{-1}bc^2b^{-1}a^{-1}\cdot c^{-1}bc^{-1}b^{-1}\cdot a^{-1}c^{-1}$$

by Lemma 11 (4)

$$= (babc)^2c\cdot abc^{-1}b^{-1}\cdot a^{-1}c^{-1}bc^2\cdot b^{-1}a^{-1}bc\cdot b^{-1}ca^{-1}c^{-1}$$

by Lemma 11 (1)

$$= \left(a^{-1}c\right)^2(babc)^2cbcb^{-1}a^2c^{-1}\cdot bcb^{-1}a^{-1}\cdot a^{-1}c^2$$

by Lemma 11 (7)

$$= cbcb\cdot b^2acb^2b^{-1}c^{-1}b^{-1}a^{-1}c^2 \text{ by Lemma 11 (7).}$$

Then Lemma 11 (12), Lemma 9 (3), Lemma 10 (5) imply Lemma 11 (14).

(15) As in the proof of Lemma 11 (14)

$$\left(c^{-1}b^{-1}a^{-1}b^{-1}a^{-1}\right)^4 = \left(ac^{-1}\right)^2c^{-1}b^{-1}a^{-1}c^{-1}bca^{-1}b^{-1}a^{-1}c^{-1}ba \ .$$

A comparison with Lemma 11 (14) yields Lemma 11 (15).

(16) By $(abc)^4 = 1$ and Lemma 4 (1)

$$b^{-1}cabcab^{-1}a^{-1}c^{-1}ba^{-1}c^{-1} = 1 \ .$$

A comparison with Lemma 11 (15) yields Lemma 11 (16).

(17) Use Lemma 11 (10) to conclude

$$b^2\cdot babca\cdot a^{-1}b^{-1}abc = \left(b^{-1}abc\right)^2 \ ,$$

$$b^2c^{-1}a^{-1}b^2\cdot b^{-1}aba^{-1}b^{-1}abc = \left(c^{-1}a\right)^2\left(b^{-1}abc\right)^2 \ .$$

Then we obtain (17) by Lemma 11 (12) and Lemma 10 (5). Transforming by p yields the
second part.

(18) From (viii) follows $(ba)^8 = 1$. Then use Lemma 9 (14) to conclude

$$(ab)^4(ba)^4 = abab \cdot abab^{-1} \cdot b^{-1}aba \cdot baba$$

$$= abab \cdot ba^{-1}b^{-1}a \cdot ab^{-1}a^{-1}b^2aba$$

$$= (ac)^2(b^{-1}abc)^2 a \cdot baba^{-1} \cdot bab^{-1}ab^{-1}a^{-1}b^2aba \quad \text{by Lemma 11 (17)}$$

$$= (ac)^2(b^{-1}abc)^2(ac^{-1})^2(abcb)^2a^2b^{-1} \cdot b^{-1}ab^{-1}a^{-1} \cdot b^2aba \quad \text{by Lemma 11 (2)}$$

$$= (ac)^2(b^{-1}abc)^2a^{-1}(a^{-1}b^{-1}ab)^2a \quad \text{by Lemma 11 (2);}$$

hence Lemma 11 (18) by Lemma 11 (17).

(19) The relations $a(ab)^4a^{-1} = (ab)^4$ and $b(ab)^4b^{-1} = (ab)^4$ follow by Lemma 11 (18), $c(ab)^4c^{-1} = (ab)^4$ follows by Lemma 11 (10).

(20) By Lemma 9 (1), (2)

$$(ab)^2 = aca^2bc^{-1}bac^2 ,$$

$$(ab)^2 = (ac)^2(ca)^2acba^2cba .$$

Hence

$$(ab)^4 = (ca)^2(c^{-1}a)^2acba^2c \cdot bcbc^{-1} \cdot bac^2 .$$

Then Lemma 11 (20) follows by Lemma 9 (13) and Lemma 10 (5).

(21) By Lemma 11 (11), (12)

$$b^{-1}cab = ab^{-1}acba^{-1} \cdot (ac)^2(ac^{-1})^2 .$$

By Lemma 11 (10) and Lemma 4 (1)

$$b^{-1}cab = a^{-1}b^{-1}acba .$$

A comparison yields Lemma 11 (21).

(22) Use Lemma 9 (2), Lemma 10 (5) to conclude

$$(ba)^4 = bab \cdot a \cdot bab \cdot a ,$$

$$(ba)^4 = cbcb \cdot b^{-1}a^2bca \cdot aba^2cba^2 .$$

By Lemma 11 (16) and Lemma 4 (1)

$$(ba)^4 = (ca^{-1})^2(ca)^2 \cdot (cb)^2 \cdot cab^{-1}a^{-1} \cdot a^{-1}bab \cdot a^2cba^2$$

$$= (ca)^2(abcb)^2(cb)^2 \cdot c \cdot ab^{-1}a^{-1}b^{-1} \cdot a^{-1}b^{-1}a^{-1}cba^2 \quad \text{by Lemma 11 (2)}$$

$$= (ca)^2(cb)^3aba^2b^{-1}c^{-1}a^{-1} \cdot a^2ba^2 \quad \text{by Lemma 11 (2).}$$

Then the first part of Lemma 11 (22) follows by Lemma 11 (21), (5). The second part follows by Lemma 11 (18), (19).

(23) Use Lemma 11 (1) to conclude

$$b^2cb^{-1}ac^{-1}b^{-1}a^{-1} = (ac^{-1})^2 b \cdot bcb^{-1}c \cdot a^{-1}b^{-1}a^{-1}$$

$$= (ac)^2(abcb)^2ba \cdot a^{-1}c^{-1}bacb^{-1} \cdot c^{-1}a^2b^{-1}c^{-1}a^{-1} \quad \text{by Lemma 9 (2)}$$

$$= (ac)^2(ca)^2(abcb)^2babc^{-1}a^{-1}b^{-1}a^{-1}b^{-1} \cdot bc^2a^2b^{-1}c^{-1}a^{-1}$$

$$\text{by Lemma 11 (4).}$$

By Lemma 11 (13), (11)

$$b^2cb^{-1}ac^{-1}b^{-1}a^{-1} = (ac)^2(ca)^2(abcb)^2a^{-1}c^{-1}bc^2a^2b^{-1}c^{-1}a^{-1} \ ,$$

hence

$$b^2cb^{-1}ac^{-1}b^{-1}a^{-1} = (abcb)^2a^{-1}c^{-1}ba^2c^2b^{-1}c^{-1}a^{-1} \ .$$

Then Lemma 11 (23) follows by Lemma 11 (20).

(24) Apply the automorphism induced by p^{-1} to Lemma 11 (23) and use Lemma 10 (5), Lemma 11 (22) to conclude Lemma 11 (24).

(25) By Lemma 9 (2)

$$(ab)^2 = (ac)^2 \cdot c^{-1}a^{-1}ba^{-1}c^{-1}b^{-1} \cdot (ba)^2 \cdot (ca^{-1})^2(ac)^2(ca)^2 \ .$$

Then Lemma 11 (25) follows by Lemma 11 (24). The second part is obtained by transformation with p .

(26) By Lemma 3 (4), Lemma 4 (1) and Lemma 10 (5)

$$D^{-1}abD = (ca)^2 \cdot bc^{-1} \cdot abc \cdot b^{-1} \ .$$

Replacing abc using Lemma 11 (23) yields

$$D^{-1}abD = (ac^{-1})^2(ac)^2(ab)^4 \cdot bc^{-1}b^2ca \cdot a^{-1}b^{-1}ab^{-1} \ .$$

Use Lemma 4 (1), Lemma 11 (2), Lemma 10 (5) to conclude

$$D^{-1}abD = (ac^{-1})^2(abcb)^2(ac)^2(ca)^2(ab)^4 \cdot ba \cdot b^{-1}a^{-1}ba \ .$$

By Lemma 11 (17)

$$D^{-1}abD = (ac^{-1})^2(abcb)^2(abcb^{-1})^2(ac)^2(ab)^4 \cdot ab \ .$$

This computation can also be carried through in the following way:

$$D^{-1}abD = (ca)^2bc^{-1}a \cdot bcbc \cdot c^{-1}ba^{-1} \cdot ab \ .$$

By Lemma 11 (25)

$$D^{-1}abD = (ab)^4 \cdot (ac)^2(abcb^{-1})^2(ac^{-1})^2bc^{-1}ac \cdot bcbc^{-1} \cdot ba^{-1} \cdot ab \ .$$

By Lemma 9 (13)

$$D^{-1}abD = (ab)^4 \cdot (ac)^2 \cdot (babc)^2(abcb^{-1})^2 \cdot (c^{-1}a)^2 \cdot ba^{-1}c^{-1}b^{-1}c^{-1}a^{-1} \cdot ab \ .$$

By Lemma 11 (24)

$$D^{-1}abD = (ac)^2 \left(b^{-1}abc\right)^2 \cdot ab .$$

Then Lemma 11 (26) follows by comparing the formulae for $D^{-1}abD$.

(27) Apply to Lemma 11 (26) the automorphism induced by p and use Lemma 11 (22) to deduce Lemma 11 (27).

(28) By $(abc)^4 = 1$, Lemma 4 (1), Lemma 11 (24), (26), (27) we have

$$abcab^{-1} \cdot b^2 cabcabc = 1 ,$$

$$\left(c^{-1}a\right)^2 (ac)^2 \cdot ac^{-1}aca^{-1} \cdot bcab^{-1}c = 1 .$$

Use Lemma 11 (24), (27), (26) to conclude the first part of Lemma 11 (28). The second part is obtained by transformation with p .

Using Lemma 10 (5), the relations Lemma 11 (13), (25), (28) show that

$$\langle (ac)^2, \left(abcb^{-1}\right)^2, \left(ac^{-1}\right)^2, (abcb)^2, (ab)^4\rangle$$

is an elementary abelian normal subgroup of exponent 2 in the centre of $\langle a, b, c\rangle$. It is of order 2^4 at most, and is generated by $(ac)^2$, $\left(abcb^{-1}\right)^2$, $\left(ac^{-1}\right)^2$, $(abcb)^2$. In the following we shall use the simplified relations obtained by Lemma 11 (26), (28) without any special indication.

LEMMA 12. *We have the transformation formulae:*

(1) $D^{-1}a^2D = (ca)^2 \cdot \left(ac^{-1}\right)^2 \cdot \left(abcb^{-1}\right)^2 \cdot (abcb)^2 \cdot a^2$;

(2) $D^{-1}b^2D = (ca)^2 \cdot \left(ac^{-1}\right)^2 \left(abcb^{-1}\right)^2 \cdot (abcb)^2 \cdot b^2$;

(3) $D^{-1}c^2D = (ca)^2 \left(ac^{-1}\right)^2 \left(abcb^{-1}\right)^2 \cdot (abcb)^2 \cdot c^2$;

(4) $D^{-1}abD = (ac)^2 \cdot \left(b^{-1}abc\right)^2 \cdot ab$;

(5) $D^{-1}acD = ac$;

(6) $D^{-1}bcD = \left(c^{-1}a\right)^2 (abcb)^2 \cdot bc$;

(7) $D^2 = (ca)^2 \cdot \left(abcb^{-1}\right)^2$;

(8) $Fa^2F = a^2$;

(9) $Fca^{-1}F = (babc)^2 \cdot \left(c^{-1}a\right)^2 \cdot ca^{-1}$;

(10) $FacF = (abcb)^2 \cdot \left(c^{-1}a\right)^2 \cdot ac$;

(11) $Fbcb^{-1}a^{-1}F = (abcb)^2 \cdot \left(c^{-1}a\right)^2 \cdot bcb^{-1}a^{-1}$;

(12) $Fabcb^{-1}F = (abcb)^2 (c^{-1}a)^2 abcb^{-1}$;

(13) $Fb^2F = b^2$;

(14) $Fab^2a^{-1}F = ab^2a^{-1}$;

(15) $Fabab^{-1}F = abab^{-1}$;

(16) $Fbab^{-1}a^{-1}F = bab^{-1}a^{-1}$;

(17) $D^{-1}aD = ab^{-1}a^{-1}FabF$;

(18) $D^{-1}bD = (abcb)^2 \cdot (c^{-1}a)^2 \cdot a^{-1}FabF$;

(19) $D^{-1}cD = (abcb)^2 \cdot (c^{-1}a)^2 \cdot (ac)^2 (abcb^{-1})^2 \cdot cb^{-1}a^{-1}FabF$;

(20) $D^{-1}FD = F$.

Proof. (1) By Lemma 3 (1), Lemma 4 (1), Lemma 10 (5) follows immediately
$$D^{-1}a^2D = (ca)^2 (ac^{-1})^2 bc^{-1}b^{-1}a^{-1} \cdot a^{-1}bcb^{-1} .$$
Use Lemma 11 (7) to conclude Lemma 12 (1).

(2) By Lemma 3 (2), Lemma 4 (1), Lemma 10 (5)
$$D^{-1}b^2D = (ca)^2 (ac^{-1})^2 \cdot bc^{-1}b^{-1}a^{-1}b^2 \cdot bc^{-1}b^{-1}a^{-1} \cdot (abcb^{-1})^2 .$$
Hence Lemma 12 (2).

(3) By Lemma 3 (3), Lemma 4 (1), Lemma 10 (5)
$$D^{-1}c^2D = (ca)^2 (c^{-1}a)^2 \cdot bc^{-1}b^{-1}a^{-1}c^2 ac^{-1} \cdot cbcb^{-1} .$$
Lemma 11 (1), Lemma 10 (5) imply
$$D^{-1}c^2D = (c^{-1}a)^2 (babc)^2 (ca)^2 \cdot (ac)^2 \cdot (bc^{-1}b^{-1}c)^2 \cdot c^2 .$$
Use Lemma 11 (17) to conclude Lemma 12 (3).

(4) See the proof of Lemma 11 (26).

(5) By Lemma 3 (6), Lemma 4 (1), Lemma 10 (5)
$$D^{-1}acD = bc^{-1}b^{-1}cabcb^{-1} .$$
Use Lemma 4 (1) and Lemma 11 (24) to conclude Lemma 12 (5).

(6) By Lemma 3 (8), Lemma 4 (1)
$$D^{-1}bcD = bc^{-1}b^{-1}a^{-1}bcab^{-1} \cdot b^2 \cdot cb^{-1}c^{-1}b^{-1} \cdot bc \cdot (ac^{-1})^2 .$$
By Lemma 9 (13), Lemma 11 (24)

$$D^{-1}bcD = b \cdot c^{-1}b^{-1}cb \cdot b^2cbc^{-1} \cdot bc \cdot (abcb)^2 \cdot (ac)^2 \ .$$

By Lemma 11 (17) follows

$$D^{-1}bcD = (abcb)^2 \cdot \left(abcb^{-1}\right)^2 \cdot c^{-1} \cdot bcbc \cdot c^{-1} \cdot bcbc^{-1} \cdot bc \ .$$

Use Lemma 11 (25), Lemma 9 (13) to conclude Lemma 12 (6).

(7) follows by Lemma 3 (10) and Lemma 4 (1).

(8) is Lemma 2 (19).

(9) follows by Lemma 9 (22), Lemma 11 (24), Lemma 12 (8).

(10) follows by Lemma 9 (18), Lemma 11 (28), (24).

(11) By Lemma 9 (15), Lemma 12 (8) follows

$$Fbcb^{-1}a^{-1}F = b \cdot b^2cab^2 \cdot b \cdot baba \ .$$

By Lemma 4 (1), Lemma 11 (25), (27), (28) follows

$$Fbcb^{-1}a^{-1}F = (ab)^4bcb^{-1}a^{-1} \ .$$

Hence Lemma 12 (11) follows by Lemma 11 (28).

(12) By Lemma 9 (19), Lemma 10 (5) follows

$$Fabcb^{-1}F = (ac)^2a^{-1}b^{-1}a^{-1} \cdot cab^2 \cdot c^2b^{-1}$$

implying

$$Fabcb^{-1}F = \left(ac^{-1}\right)^2(ac)^2a^{-1}b^{-1}a^{-1}b^{-1} \cdot b^{-1}a^{-1}cb^{-1}$$

by Lemma 4 (1). Then Lemma 12 (12) follows by Lemma 11 (25), (28).

(13) By Lemma 2 (22), (23) and $(abc)^4 = 1$ follows

$$Fb^2ca^{-1}F = (ac)^2 \cdot bcb^{-1}cabb^2cb^{-1}$$

implying

$$Fb^2ca^{-1}F = \left(ac^{-1}\right)^2b^2 \cdot b^{-1}a^{-1}b^{-1}a^{-1} \cdot ab^{-1}cb^{-1} \quad \text{by Lemma 11 (24)}$$
$$= (ca)^2 \cdot b^2a^{-1}b^{-1}a^{-1}b^{-1}ab^{-1} \cdot cb^{-1} \quad \text{by Lemma 11 (25).}$$

Then Lemma 12 (13) follows by Lemma 11 (2), (24) and Lemma 12 (9).

(14) By Lemma 9 (20) and Lemma 12 (10) follows

$$Fab^2a^{-1}F = \left(c^{-1}a\right)^2 \cdot a^{-1}b^{-1}a^{-1} \cdot a^{-1}c^{-1}b^2c^2b^{-1}ac^{-1}b^{-1}a^{-1}b^{-1}a^{-1}$$
$$= (ca)^2a^{-1}b^{-1}a^{-1}b^{-1} \cdot b^{-1}a^{-1}cb^{-1}ac^{-1}b^{-1}a^{-1}b^{-1}a^{-1} \quad \text{by Lemma 4 (1)}$$
$$= (abcb)^2\left(abcb^{-1}\right)^2 \cdot b \cdot c^{-1}b^{-1}c^{-1}b^{-1} \cdot a^{-1}b^{-1}a^{-1} \quad \text{by Lemma 11 (25), (26).}$$

Hence Lemma 12 (14) follows by Lemma 11 (25), (24).

(15) By Lemma 12 (14) and Lemma 2 (22)

$$Fabab^{-1}F = ab \cdot ba^2c^2b^{-1}a^{-1}c^2b^{-1} \; .$$

Then Lemma 12 (15) follows by Lemma 11 (20) and Lemma 10 (5).

(16) By Lemma 2 (22), (20)

$$FbabccaF = bc^2a^{-1}ba^{-1} \; .$$

By Lemma 12 (8), (9), (14)

$$Fbab^{-1}a^{-1}F = bc^2a^{-1}ba^2c^2b^{-1}b^{-1}a^{-1} \; .$$

Lemma 11 (20) and Lemma 10 (5) imply Lemma 12 (16).

(17) By Lemma 2 (32), Lemma 12 (8), (7), (9)

$$
\begin{aligned}
D^{-1}aD &= a^{-1}b^{-1}aFac^{-1}abacc^2F \\
&= a^{-1}b^{-1}aFc^2 \cdot abcb^{-1} \cdot bcF \quad \text{by Lemma 11 (24), Lemma 12} \\
&= \left(abcb^{-1}\right)^2 \cdot a^{-1}b^{-1}aFc^2bc^{-1} \cdot b^{-1}a^{-1}bc \cdot F \quad \text{by Lemma 12} \\
&= \left(abcb^{-1}\right)^2\left(abcb\right)^2a^{-1}b^{-1}ac^2bc^2b^{-1} \cdot FabF \quad \text{by Lemma 2 (20), Lemma 12.}
\end{aligned}
$$

Then Lemma 12 (17) follows by Lemma 11 (24), (1), (17).

(18) follows by Lemma 12 (18), (2), (4).

(19) follows by Lemma 12 (18), (6), (3).

(20) By Lemma 2 (34), Lemma 12

$$D^{-1}FD = \left(ac^{-1}\right)^2 \cdot \left(abcb\right)^2 \cdot \left(cbcb^{-1}\right)^2F \; .$$

Then Lemma 12 (20) follows by Lemma 11 (1).

We shall now describe the normal subgroup which is generated by the fourth powers in G .

DEFINITION.

$$v_1 = b^{-1}a^{-1}Dba = \left(p^2q\right)^4 \; ,$$

$$v_2 = ab^{-1}a^{-1}Db \; ,$$

$$v_3 = c^{-1}b^{-1}cb^{-1}D^{-1}cb^{-1}Fac^{-1}b^{-1}a^{-1}b^{-1}Fb \; ,$$

$$v_4 = (ac)^2 = \left(qpq^{-1}p\right)^4 \; ,$$

$$v_5 = \left(abcb^{-1}\right)^2 \; ,$$

$$v_6 = \left(a^{-1}c\right)^2 ,$$

$$v_7 = (abcb)^2 ,$$

$$v_8 = bc^{-1}b^{-1}cb^{-1}D^{-1}cb^{-1}Fac^{-1}b^{-1}a^{-1}b^{-1}F ;$$

$$V := \langle v_1, \ldots, v_8 \rangle .$$

We shall show that V is a normal subgroup in G generated by fourth powers.

LEMMA 13. *We have the transformation formulae*

(1) $pv_1p^{-1} = v_3$;

(2) $pv_2p^{-1} = v_8$;

(3) $pv_3p^{-1} = v_4v_2$;

(4) $pv_4p^{-1} = v_5$;

(5) $pv_5p^{-1} = v_4$;

(6) $pv_6p^{-1} = v_7$;

(7) $pv_7p^{-1} = v_6$;

(8) $pv_8p^{-1} = v_4v_1$;

(9) $qv_1q^{-1} = v_4v_2$;

(10) $qv_2q^{-1} = v_6v_1$;

(11) $qv_3q^{-1} = v_7v_3$;

(12) $qv_4q^{-1} = v_4$;

(13) $qv_5q^{-1} = v_5$;

(14) $qv_6q^{-1} = v_7$;

(15) $qv_7q^{-1} = v_6$;

(16) $qv_8q^{-1} = v_5v_8$;

(17) V *is an elementary abelian normal subgroup of exponent* 2 *in* G ;

(18) $v_8 = v_1v_2v_3$;

(19) $v_7 = v_4v_5v_6$.

Proof. (1) follows immediately by Lemma 1 and Lemma 2 (47).

(2) follows immediately by Lemma 1 and Lemma 2 (47).

(3) By Lemmas 12, 1, and the definition of v_i

$$p^{-1}v_4 \cdot v_2 p = p^{-1} \cdot \left(abcb^{-1}\right)^2 aDb^{-1}a^{-1}b \cdot p$$

$$= (ca)^2 c^{-1}b^{-1}a^{-1}c^{-1}e^2 cbca .$$

If one computes p^4ep^{-4} using Lemma 1, then

$$ca \neq e .$$

Hence Lemma 13 (3) follows by Lemma 11 (24), Lemma 2 (47), Lemma 12.

(4) to (7) follow by Lemma 1, Lemma 11 (28), Lemma 9 (13).

(8) Use the definition of v_i to conclude

$$pv_8p^{-1} = pb \cdot v_3b^{-1} \cdot p^{-1} \; .$$

Then Lemma 13 (8) follows by Lemma 1, Lemma 13 (3), Lemma 11 (24), Lemma 12.

(12) follows by Lemma 10 (1) and Lemma 11 (28).

(13) By Lemma 1

$$q\left(abcb^{-1}\right)^2q^{-1} = \left(deb^{-1}dbc^{-1}e^{-1}a^{-1}\right)^2 \; .$$

But

$$\begin{aligned}
deb^{-1}dbc^{-1}ec &= q^2p^2 \cdot q^{-1}p^{-1}q^{-1}p^{-1} \cdot q^2 \cdot pq^{-1}pq^{-1} \cdot pq \\
&= p^{\pm 1}q^{-1}p^{-1}q^2 \cdot pq^{-1}pq^{-1} \cdot p \\
&= q^{-1}\left(q^{-1}p^{-1}\right)^4q = 1 \quad \text{by (iv), (iii)}
\end{aligned}$$

implying

$$q\left(abcb^{-1}\right)^2q^{-1} = c^{-1}e^{-2}a^{-1}c^{-1}e^{-2}a^{-1} \; .$$

Using $ca \not\smile e$ as in the proof of Lemma 13 (3) yields

$$q\left(abcb^{-1}\right)^2q^{-1} = (ac)^2ae^4a^{-1} \; .$$

We apply the automorphism induced by p on Lemma 12 (7) obtaining

$$e^4 = (ca)^2 \cdot \left(b^{-1}abc\right)^2 \; .$$

Hence the proof of Lemma 13 (13) is complete using Lemma 10 (5).

(14) By Lemma 1

$$\begin{aligned}
q\left(ac^{-1}\right)^2q^{-1} &= dbcb^{-1}d^{-1}a^{-1}db \cdot cb^{-1}d^{-1} \cdot a^{-1} \\
&= dbcb^{-1}d^{-1}a^{-1} \cdot dbdb \cdot c^{-1}a^{-1} \quad \text{by Lemma 2 (49)} \\
&= dbcb^{-1} \cdot d^{-1}a^{-1}b^{-1}d^{-1} \cdot b^{-1}d^{-1}c^{-1}a^{-1} \quad \text{by Lemma 2 (49)} \\
&= dbcab^{-1} \cdot d^{-1}c^{-1}a^{-1} \quad \text{by Lemma 2 (49).}
\end{aligned}$$

Then by Lemma 2 (39), (42), (36),

$$q\left(ac^{-1}\right)^2q^{-1} = cb^{-1} \cdot Fac^{-1}b^{-1}a^{-1}c^{-1}bca^{-1}Fbc^2a^{-1} \; .$$

Hence Lemma 13 (14) follows by Lemma 11 (24), Lemma 12, Lemma 11 (28).

(15) By Lemma 13 (6)

$$q(abcb)^2 q^{-1} = qpq^{-1}p^{-1} \cdot pq(ac^{-1})^2 q^{-1}p^{-1} \cdot pqp^{-1}q^{-1}$$
$$= ap(abcb)^2 p^{-1} \cdot a^{-1} .$$

Hence Lemma 13 (15) follows by Lemma 13 (7) and Lemma 10 (5).

(9) We have $v_1 = (p^2q)^2$; hence $p^2 v_1 p^{-2} = qv_1 q^{-1}$. Then Lemma 13 (9) follows by Lemma 13 (3).

(10) By Lemma 1

$$qv_2 q^{-1} = dbe^{-1}d^{-1}c^{-1}e^{-1}a^{-1}c^{-1}b^{-1}a^{-1} .$$

The definition of d, e and (v) implies immediately

$$(de)^2 = 1 .$$

This implies

$$qv_2 q^{-1} = dbd^{-1} \cdot \mathcal{D} \cdot ec^{-1}e^{-1} \cdot a^{-1}c^{-1}b^{-1}a^{-1} .$$

By Lemma 2 (39), (43),

$$qv_2 q^{-1} = Dbc^{-1}a^{-1}c^{-1}b^{-1}a^{-1} .$$

Hence

$$qv_2 q^{-1} = Dbc^{-1}a^{-1}c^{-1} \cdot b^{-1}a^{-1}b^{-1}a^{-1} \cdot Dba \cdot v_1^{-1}$$

implying

$$qv_2 q^{-1} = D^2 D^{-1} \cdot (a^{-1}a)^2 \cdot a^{-1}b^{-1}Dba \cdot v_1^{-1}$$

by Lemma 11 (25). Then Lemma 13 (10) follows by Lemma 12, Lemma 10 (5).

(11) By Lemma 13 (1), (9), (8)

$$qv_3 q^{-1} = v_5 \cdot apv_2 p^{-1}a^{-1}$$
$$= v_5 \cdot pc^{-1}b^2 a^{-1}Dbabcp^{-1}$$
$$= v_5 p(c^{-1}a)^2 \cdot (abcb)^2(ac)^2(b^{-1}abc)^2 \cdot c^{-1}b^2 \cdot a^{-1}bab \ cb^{-1}a^{-1}bap^{-1} \cdot v_3^{-1}$$

by Lemma 12, Lemma 13 (1)

$$= v_5 \cdot p(ac^{-1})^2 \cdot (ca)^2 c^{-1}b \cdot a^{-1}b^{-1}a^{-1}b^{-1} \cdot c^{-1}bp^{-1} \cdot v_3^{-1}$$

by Lemma 11 (2), Lemma 11 (24), Lemma 4 (1).

Hence Lemma 13 (11) follows by Lemma 11 (25), Lemma 11 (24) and Lemma 13 (5), (6).

(16) By Lemma 13 (2), (10), Lemma 1 and $v_1^2 = 1$

$$qv_8q^{-1} = v_7 \cdot pc^{-1}b^{-1}a^{-1}a^{-1}b^{-1}D^{-1}ababcp^{-1}$$

$$= v_7 \cdot p(ac)^2 \cdot (c^{-1}a)^2 \cdot c^{-1}b^{-1}a^{-1}baD^{-1}cab^{-1}a^{-1}Dbp^{-1}v_8^{-1}$$

by Lemma 12 (4), Lemma 11 (25).

Then Lemma 13 (12) follows by Lemma 12 (1), (3), (5), (4) and Lemma 13 (4), (6).

(17) Use the definition of v_i and the transformation formulae to conclude

$$v_1^2 = v_2^2 = \ldots = v_8^2 = 1 .$$

Lemma 10 (5), Lemma 11 (28), Lemma 12 show

$$\langle v_4, v_5, v_6, v_7 \rangle \leq Z(V) .$$

$(v_1v_2)^2 = (v_1v_3)^2 = (v_1v_8)^2 = 1$ follows by $v_1 = (p^2q)^4$ and Lemma 13 (1)-(16).

The other commutator formulae are obtained by transforming the above formulae by p and q.

(18) By the definition of v_1, v_3 and Lemma 12

$$v_1 \cdot v_3 = (abcb)^2 \cdot b^{-1}a^{-1}bac^{-1}b^{-1}cb^{-1}cb^{-1} \cdot abab \cdot caFa^{-1}Fb$$

$\Rightarrow$
$$v_1v_3 = (abcb)^2(ac)^2 \cdot b^{-1}a^{-1}ba \cdot c^{-1}b^{-1}a^{-1}c^{-1}Fa^{-1}Fb \quad \text{by Lemma 11 (25), (24)}$$

$\Rightarrow$
$$v_1v_3 = (abcb^{-1})^2 \cdot (c^{-1}a)^2 \cdot (ca)^2 \cdot a^{-1}cb^{-1}c^{-1}a^{-1} \cdot a^{-1}Fa^{-1}Fb \quad \text{by Lemma 11 (17), (7)}$$

$\Rightarrow$
$$v_1v_3 = (ac^{-1})^2 \cdot (abcb^{-1})^2 \cdot (abcb)^2 \cdot c^2ababFa \cdot (a^{-1}FabF)^2 \cdot Fb^{-1}a^{-1}b$$

by Lemma 11 (24), (28).

Raise the formula Lemma 12 (18) to the second power and compare with Lemma 12 (2) to obtain

$$(a^{-1}FabF)^2 = (ac)^2(ac^{-1})^2 \cdot (abcb^{-1})^2 \cdot (abcb)^2 \cdot b^2 .$$

We insert this in the above formulae obtaining, by Lemma 12,

$$v_1 \cdot v_3 = (ac)^2 \cdot c^2ababab a^{-1}bFa^{-1}Fa$$

$\Rightarrow$
$$v_1v_3 = (ca)^2 \cdot (abcb)^2 \cdot (c^{-1}a)^2 \cdot c^2b^{-1}a^2b \cdot Fa^{-1}Fa \quad \text{by Lemma 11 (28).}$$

By the definition of v_2, v_8 and Lemma 12

$$v_2 v_8 v_1 \cdot v_3 = (ca)^2 \cdot ab^{-1}a^{-1}b^2 c^{-1}b^{-1}cb^{-1}cb^{-1}a \cdot c^{-1}b^{-1}a^{-1}b^{-1} \cdot c^2 b^{-1}a^2 b$$

$$= (abcb)^2 \cdot (c^{-1}a)^2 \cdot ab^{-1}a^{-1}b^2 c^{-1}b^{-1}c \cdot b^{-1}cab \cdot ac^{-1}b^{-1}a^2 b \quad \text{by Lemma 11 (25)}$$

$$= (abcb)^2 \cdot (ca)^2 \cdot ab^{-1}a^{-1}b^{-1} \cdot b^{-1}c^{-1}b^{-1}c^{-1}b^{-1}a^2 b$$

$$\text{by Lemma 11 (24), Lemma 4 (1)}$$

$$= (abcb)^2 \cdot (ca)^2 ab^{-1}a^{-1}c^{-1}b^{-1}c^{-1}ba^2 b \quad \text{by Lemma 11 (22), (26), (2)}$$

$$= (babc)^2 \cdot (ca)^2 \cdot ab \cdot babaab \quad \text{by Lemma 11 (24).}$$

Then Lemma 13 (18) follows by Lemma 11 (25), (28).

(19) Lemma 13 (19) is obtained by applying to Lemma 13 (18) the automorphism induced by q .

V is also an elementary abelian normal subgroup of G . It is generated by $v_1, v_2, \ldots, v_6$. By definition the elements $v_1, \ldots, v_5$ are contained in G^4 .

By Lemma 13 (10),

$$V \leq G^4 .$$

The definition of v_1 shows that V is the normal subgroup in G generated by $(p^2 q)^4$.

We shall now describe the normal subgroup generated by $(pq^2)^4$. Let

$$\alpha : G \to G$$

be the involutory automorphism of G which just interchanges p and q . Obviously, (V) is then the normal subgroup in G generated by $(pq^2)^4$. Here are the formulae for the operation of G on V :

$$pv_1 p^{-1} = v_3 \qquad\qquad qv_1 q^{-1} = v_4 v_2$$

$$pv_2 p^{-1} = v_1 v_2 v_3 \qquad\qquad qv_2 q^{-1} = v_6 v_1$$

$$pv_3 p^{-1} = v_4 v_2 \qquad\qquad qv_3 q^{-1} = v_4 v_5 v_6 \, v_3$$

$$pv_4 p^{-1} = v_5 \qquad\qquad qv_4 q^{-1} = v_4$$

$$pv_5 p^{-1} = v_4 \qquad\qquad qv_5 q^{-1} = v_5$$

$$pv_6 p^{-1} = v_4 v_5 v_6 \qquad\qquad qv_6 q^{-1} = v_4 v_5 v_6$$

$$av_1 a^{-1} = v_2 \qquad bv_1 b^{-1} = v_2 v_4 v_6 \qquad cv_1 c^{-1} = v_2$$

$$av_2 a^{-1} = v_1 \qquad bv_2 b^{-1} = v_1 v_4 v_6 \qquad cv_2 c^{-1} = v_1$$

$$av_3a^{-1} = v_1v_2v_3v_4v_6 \qquad\qquad bv_3b^{-1} = v_1v_2v_3 \qquad\qquad cv_3c^{-1} = v_1v_2v_3v_4v_6$$

$$av_4a^{-1} = v_4 \qquad\qquad bv_4b^{-1} = v_4 \qquad\qquad cv_4c^{-1} = v_4$$

$$av_5a^{-1} = v_5 \qquad\qquad bv_5b^{-1} = v_5 \qquad\qquad cv_5c^{-1} = v_5$$

$$av_6a^{-1} = v_6 \qquad\qquad bv_6b^{-1} = v_6 \qquad\qquad cv_6c^{-1} = v_6$$

$$dv_1d^{-1} = v_1v_4v_6 \qquad\qquad\qquad\qquad ev_1e^{-1} = v_1v_5v_6$$

$$dv_2d^{-1} = v_2v_4v_6 \qquad\qquad\qquad\qquad ev_2e^{-1} = v_2v_5v_6$$

$$dv_3d^{-1} = v_3v_5v_6 \qquad\qquad\qquad\qquad ev_3e^{-1} = v_3v_4v_6$$

$$dv_4d^{-1} = v_4 \qquad\qquad\qquad\qquad ev_4e^{-1} = v_4$$

$$dv_5d^{-1} = v_5 \qquad\qquad\qquad\qquad ev_5e^{-1} = v_5$$

$$dv_6d^{-1} = v_6 \qquad\qquad\qquad\qquad ev_6e^{-1} = v_6 \ .$$

To enable a more detailed study of the normal subgroup generated in G by the fourth powers, it will be convenient to somewhat refine some of the formulae, Lemma 2 (36)-(48).

LEMMA 14.

(1) $ece^{-1} = v_3v_4v_6 \cdot bcb^{-1}F$;

(2) $de\,b\,e^{-1}d^{-1} = v_4(cb)^2 \cdot bD^{-1}$;

(3) $ede^{-1}d^{-1} = v_3v_5v_6 \cdot bcb^{-1}c^{-1}D^{-1}$;

(4) $e^2 = v_3v_4v_6bcb^{-1}c^{-1}$.

Proof. (1) By Lemma 2 (43) and the definition of v_3 follows

$$ece^{-1} = bc \cdot v_3 \cdot b^{-1}F \ .$$

Then Lemma 14 (1) follows by Lemma 13.

(2) By Lemma 2 (41) and Lemma 12,

$$de \cdot b \cdot e^{-1}d^{-1} = c^{-1}b^{-1} \cdot c^{-1}b^{-1}cb^{-1} \cdot ac^{-1}b^{-1}a^{-1}b^{-1} \cdot D^{-1}$$

$$= v_6 \cdot c^2 bc \cdot abab \cdot cD^{-1} \quad \text{by Lemma 9 (13)}$$

$$= v_6 \cdot ccbcb \cdot c^{-1}b \cdot D^{-1} \quad \text{by Lemma 11 (25), (24).}$$

Then Lemma 14 (2) follows by Lemma 11 (25) and Lemma 13 (19).

(3) By Lemma 2 (45), Lemma 12, Lemma 13,

$$ede^{-1}d^{-1} = v_3 v_5 v_6 babca^{-1}bc^{-1}b^{-1}c \cdot b^2 ca \cdot bca^{-1} \cdot bc^{-1}bc \cdot bc \cdot D^{-1}$$

$$= v_3 v_6 \cdot babca^{-1}b \cdot c^{-1}b^{-1}cb^{-1} \cdot cb^{-1}c^2 D^{-1}$$

$$\qquad\qquad\qquad\qquad \text{by Lemma 4 (1), Lemma 11 (24), Lemma 9 (13)}$$

$$= v_3 v_5 v_6 \cdot c^{-1}a^{-1}b^{-1}a^{-1}c^{-1}b^{-1}cb^{-1}cb^{-1}c^2 \cdot D^{-1} \quad \text{by Lemma 9 (13), Lemma 11 (25)}$$

$$= v_3 v_4 v_5 \cdot c^{-1}cb \cdot bcb^{-1}c^{-1} \cdot c^2 b^{-1}c^2 D^{-1} \quad \text{by Lemma 11 (24)}$$

$$= v_3 v_4 v_6 \cdot c^{-1} \cdot bc^{-1}b^{-1}c^{-1} \cdot c^{-1}D^{-1} \quad \text{by Lemma 11 (17), (25).}$$

Then Lemma 14 (3) follows by Lemma 11 (1).

(4) By Lemma 2 (47) and the definition of v_3 follows

$$e^2 = bcv_3 b^{-1}c^{-1} .$$

Hence Lemma 14 (4) follows by Lemma 13.

DEFINITION.

$$u_i = \alpha(v_i) , \quad i = 1, 2, \ldots, 6 ,$$

$$U := \langle u_1, \ldots, u_6 \rangle .$$

From Lemma 13 it follows that U is an elementary abelian normal subgroup in G which is generated by fourth powers. U is the normal subgroup generated by $(q^2 p)^4$, obviously it centralizes V.

LEMMA 15. *The elements u_i can be computed in terms of the v_i* :

(1) $u_1 = v_6 \cdot bFb^{-1}$;

(2) $u_2 = v_6 \cdot a^{-1}b^{-1}Fba$;

(3) $u_3 = v_2 \cdot ab^2 cF$;

(4) $u_4 = v_6 \cdot b^{-1}a^{-1}FabF$;

(5) $u_5 = v_4 v_5 v_6 \cdot b^{-1}a^{-1}FabF$;

(6) $u_6 = v_6$;

(7) $u_4 u_5 = v_4 v_5$.

The following transformation formulae apply:

$$(8) \begin{cases} pu_1 p^{-1} = u_4 u_2 \\ pu_2 p^{-1} = u_6 u_1 \\ pu_3 p^{-1} = u_4 u_5 u_6 u_3 \\ pu_4 p^{-1} = u_4 \\ pu_5 p^{-1} = u_5 \\ pu_6 p^{-1} = u_4 u_5 u_6 \end{cases} \qquad (9) \begin{cases} qu_1 q^{-1} = u_3 \\ qu_2 q^{-1} = u_1 u_2 u_3 \\ qu_3 q^{-1} = u_4 u_2 \\ qu_4 q^{-1} = u_5 \\ qu_5 q^{-1} = u_4 \\ qu_6 q^{-1} = u_4 u_5 u_6 \end{cases}$$

Proof. The following formulae follow immediately by the definition of $a, b, \ldots, e$ and Lemma 1:

$$\alpha(a) = pqp^{-1}q^{-1} = a^{-1} \ ;$$

$$\alpha(b) = qa^{-1}q^{-1} = ad^{-1} \ ;$$

$$\alpha(c) = q^2 a^{-1} q^{-2} = dec \ ;$$

$$\alpha(d) = pq^{-1}p^{-1} \cdot a^{-1} = b^{-1}a^{-1} \ ;$$

$$\alpha(e) = qb^{-1}a^{-1}q^{-1} = abe^{-1}d^{-1} \ ;$$

$$(1) \qquad u_1 = (q^2 p)^4 = p^{-1}Fp = p^{-1} \cdot ebe^{-1}a^{-1}ba^{-1}p$$
$$= dad^{-1}abca^2bc$$

which follows by the definition of F and Lemma 1. Moreover Lemma 2 (36) and Lemma 12 imply

$$u_1 = v_4 v_5 v_6 \cdot D^{-1}babca^2c^2D \cdot D^{-1}cD \cdot Fabca^2bc$$

$$\Rightarrow \qquad u_1 = v_4 v_5 v_6 \cdot bab \cdot ca^2 c^{-1} a^2 \cdot a^2 b^{-1} a^{-1} Fababca^2bc \quad \text{by Lemma 12, Lemma 13 (19)}$$

$$\Rightarrow \qquad u_1 = v_5 \cdot baba^{-1}bca \cdot abcb \cdot Fb^{-1} \quad \text{by Lemma 12.}$$

Then Lemma 15 (1) follows by Lemma 11 (24) and Lemma 4 (1).

(2) $\alpha(v_2) = \alpha\left(av_1 a^{-1}\right) = a^{-1}v_6 bFb^{-1}a$. Lemma 15 (2) then follows by Lemma 13 and Lemma 12.

(4), (5), (6). Obviously,

$$\alpha\left((ab)^4\right) = D^2 = v_4 v_5$$

and

$$\alpha\left((ac)^2\right) = a^{-1}de \cdot ca^{-1} \cdot e^{-1}d^{-1}c \ .$$

By Lemma 2 (44), (38),

$$\alpha\big((ac)^2\big) = a^{-1}Dbc^{-1}bDbabc$$

$\Rightarrow$
$$\alpha\big((ac)^2\big) = v_4 v_5 \cdot a^{-1} \cdot bc^{-1}b^{-1}c \cdot c^{-1}b^{-1}abcb^{-1}a^{-1}FabF \quad \text{by Lemma 12}$$

$\Rightarrow$
$$\alpha\big((ac)^2\big) = v_4 v_6 \cdot a^{-1}c^2 \cdot b^{-1}cab \cdot cb^{-1}a^{-1}FabF \quad \text{by Lemma 11 (17), (25).}$$

Then Lemma 15 (4) follows by Lemma 11 (24), Lemma 4 (1) and Lemma 15 (5) follows by Lemma 11 (26), Lemma 13 (19) and the formula for $\alpha\big((ab)^4\big)$.

$\alpha\big((ac^{-1})^2\big) = a^{-1}c^{-1}da^{-1}c^{-1}d^{-1}$ follows by the definition and $(ed)^2 = 1$ and $ca \neq e$. By Lemma 2 (36), (42)

$$\alpha\big((ac^{-1})^2\big) = a^{-1}c^{-1}Dbacb^{-1}D^{-1} .$$

Then Lemma 15 (6) follows by Lemma 11 (24), Lemma 12 and Lemma 13 (19).

(8), (9). These formulae follow by Lemma 13 applying α .

(3) Use Lemma 14 (4) to conclude

$$v_3 = e^2 cbc^{-1}b^{-1} v_4 v_6 .$$

Apply α , then we have by the statements already proven

$$u_3 = ab \cdot de \cdot abca \cdot e^{-1}d^{-1} \cdot ded^{-1}e^{-1} \cdot ec^{-1}e^{-1} \cdot a^{-1} \cdot u_4 \cdot u_6 .$$

By Lemma 14 (1)-(3), Lemma 2 (38), (41), Lemma 13, Lemma 12 follows

$$u_3 = v_1 v_2 v_6 \cdot b^{-1} \cdot D^{-1} cbcb^{-1}c^{-1}a^{-1} \cdot b^{-1} cbc^2 b^{-1}a^{-1}Fb^{-1}a^{-1}FabF$$

$\Rightarrow$
$$u_3 = v_1 v_2 \cdot b^{-1}D^{-1}cbcb^{-1}a \cdot b^{-1}c^{-1}bc \cdot b^{-1}a^{-1}Fb^{-1}a^{-1}FabFv_5 v_6 \quad \text{by Lemma 11 (1), (24)}$$

$\Rightarrow$
$$u_3 = v_1 v_2 v_4 v_5 \cdot b^{-1}cbcb^{-1}a^2 b^{-1}c^{-1}D^{-1}c^{-1}D \cdot D^{-1} \cdot Fb^{-1}a^{-1}FabF$$

$$\text{by Lemma 11 (17), (24), Lemma 12}$$

$\Rightarrow$
$$u_3 = v_1 v_2 \cdot b^{-1}cbcb^{-1}a^2 b^{-1}c^2 ab \cdot v_1 \cdot a^{-1}b^{-1} \cdot F$$

$$\text{by Lemma 12 and the definition of } v_1$$

$\Rightarrow$
$$u_3 = v_2 \cdot b^{-1}cbcb^{-1}ac^{-1}b^{-1}cba^{-1}b^{-1}F \quad \text{by Lemma 11 (24)}$$

$\Rightarrow$
$$u_3 = v_2 v_4 v_6 b^{-1}cbcb^{-1}cb^{-1}c^{-1}F \quad \text{by Lemma 11 (9)}$$

$\Rightarrow$
$$u_3 = v_2 c^{-1}b^2 c^{-1}F \quad \text{by Lemma 11 (1).}$$

Conjugating this relation with c^2 and using Lemma 13, Lemma 15 (7), (8) yields Lemma 15 (3).

By Lemma 15 follows immediately the collection of formulae given hereafter.

The formulae for the operation of G on U

$$au_1a^{-1} = u_2 \qquad\qquad\qquad bu_1b^{-1} = u_2u_4u_6$$

$$au_2a^{-1} = u_1 \qquad\qquad\qquad bu_2b^{-1} = u_1u_4u_6$$

$$au_3a^{-1} = u_1u_2u_3u_4u_6 \qquad\quad bu_3b^{-1} = u_1u_2u_3u_4u_5$$

$$au_4a^{-1} = u_4 \qquad\qquad\qquad bu_4b^{-1} = u_4$$

$$au_5a^{-1} = u_5 \qquad\qquad\qquad bu_5b^{-1} = u_5$$

$$au_6a^{-1} = u_6 \qquad\qquad\qquad bu_6b^{-1} = u_6$$

$$cu_1c^{-1} = u_2u_4u_5 \qquad\qquad du_1d^{-1} = u_1u_4u_6$$

$$cu_2c^{-1} = u_1u_4u_5 \qquad\qquad du_2d^{-1} = u_2u_4u_6$$

$$cu_3c^{-1} = u_1u_2u_3u_5u_6 \qquad\; du_3d^{-1} = u_3u_4u_6$$

$$cu_4c^{-1} = u_4 \qquad\qquad\qquad du_4d^{-1} = u_4$$

$$cu_5c^{-1} = u_5 \qquad\qquad\qquad du_5d^{-1} = u_5$$

$$cu_6c^{-1} = u_6 \qquad\qquad\qquad du_6d^{-1} = u_6$$

$$eu_1e^{-1} = u_1u_5u_6$$

$$eu_2e^{-1} = u_2u_5u_6$$

$$eu_3e^{-1} = u_3u_5u_6$$

$$eu_4e^{-1} = u_4$$

$$eu_5e^{-1} = u_5$$

$$eu_6e^{-1} = u_6 \; .$$

The join of V and U is again an elementary abelian normal subgroup in G , because U and V centralize each other. Obviously $V \cap U$ is just the group fixed by α . Hence we have

$$\langle v_6 , \; v_4 \cdot v_5 \rangle \leq V \cap U \; .$$

It is easy to check that the representation of G on the $\mathbb{F}_2$ vector space $V \cdot U$ is compatible with the relations (i)-(xvi). The factor group $G/V \cdot U$ is of exponent 4 as mentioned in part A. We shall now collect some relations in G/G^4 .

LEMMA 16.

(1) $F = u_2 u_4$;

(2) $b^2 c^2 = u_2 u_3 v_1 v_6$;

(3) $a^2 b^2 = u_1 u_3 u_5 v_1 v_2 v_3 v_5$;

(4) $a^2 c^2 = u_1 u_2 u_5 u_6 v_2 v_3 v_5$;

(5) $aca^{-1} c^{-1} = u_1 u_2 u_4 v_2 v_3 v_4$;

(6) $(ab)^2 d^{-2} = u_1 u_3 v_1 v_3 v_4 v_5$;

(7) $(ab)^2 (bc)^{-2} = u_5 v_1 v_2 v_4$;

(8) $d^2 e^{-2} = u_1 u_2 u_6$;

(9) $a^2 b a^2 b^{-1} = u_5 u_6 v_1 v_2$;

(10) $b^2 a b^2 a^{-1} = u_5 v_1 v_2 v_5$;

(11) $b^2 c b^2 c^{-1} = u_4 u_6 v_1 v_2$;

(12) $c^2 b c^2 b^{-1} = u_4 v_1 v_2 v_5$;

(13) $d^{-2} a d^2 a^{-1} = u_4 u_6$;

(14) $d^{-2} b d^2 b^{-1} = u_5 u_6$;

(15) $d^{-2} c d^2 c^{-1} = u_4 u_6$;

(16) $dad^{-1} a^{-1} d^{-2} = u_3 v_1 v_3$;

(17) $dbd^{-1} b^{-1} d^{-2} = u_2 u_4 u_6$;

(18) $dcd^{-1} c^{-1} d^{-2} = u_1 u_2 u_3 \, v_1 v_3$;

(19) $eae^{-1} a^{-1} d^{-2} = u_1 v_4 v_5$;

(20) $ebe^{-1} b^{-1} d^{-2} = u_1 u_2 u_3 v_2 v_3 v_5$;

(21) $ece^{-1}c^{-1}d^{-2} = u_2 u_5 u_6$;

(22) $da^2 d^{-1} a^2 = u_1 u_2 v_5 v_6$;

(23) $db^2 d^{-1} b^2 = u_1 u_2$;

(24) $dc^2 d^{-1} c^2 = u_1 u_2 u_6 v_4$;

(25) $ea^2 e^{-1} a^2 = u_1 u_2 u_5 u_6$;

(26) $eb^2 e^{-1} b^2 = u_1 u_2 u_4 v_5$;

(27) $ec^2 e^{-1} c^2 = u_1 u_2 u_4 u_6$;

(28) $D^2 = v_4 v_5$.

Proof. (1) follows by Lemma 15 (1).

(2) follows by Lemma 15 (3), if one replaces F using Lemma 16 (1).

(3) is Lemma 16 (2) transformed by p^{-1} .

(4), (5) follow by Lemma 16 (2), (3).

(6) Use the definition of v_1 to conclude

$$v_1 v_4 v_6 = aba^{-1} b^2 a^{-1} b^{-1} a^{-1} \cdot (ab)^2 \cdot D^{-1} .$$

Lemma 16 (6) follows by Lemma 16 (3).

(7) By Lemma 11 (25), (18), (26)

$$(ab)^2 \cdot (bc)^{-2} = babacbcb \cdot v_4 v_5 .$$

Lemma 16 (7) follows by Lemma 11 (24) and Lemma 16 (2).

(8) Transform Lemma 16 (6) by p and use Lemma 16 (7) to obtain Lemma 16 (8).

(9)-(12) follow by Lemma 16 (2)-(4) under consideration of Lemmas 13 and 15.

(13)-(15) follow by Lemma 16 (6), (7) under consideration of Lemma 11 (25), (26), (22).

(16) By Lemma 2 (36)

$$dad^{-1} a^{-1} D^{-1} = ab \cdot b^{-1} a^{-1} Dba \cdot bca^{-1} \cdot F \cdot bc^{-1} a^{-1} b^{-1} \cdot D^{-1} a^{-1} D \cdot D^{-2} .$$

Lemma 16 (16) follows by Lemma 11 (24), Lemma 16 (1), (13), (3), (4).

(17) By Lemma 2 (39) follows

$$dbd^{-1}b^{-1}D^{-1} = cb \cdot b^2 \cdot F \cdot a^2 \cdot a^{-1}c^{-1}b^{-1} \cdot a \cdot a^2 b^2 \cdot b \cdot D^{-1}b^{-1}D \cdot D^{-2} \ .$$

Lemma 16 (17) follows by Lemma 11 (24), Lemma 16 (3), (14).

(18)   By Lemma 2 (42)

$$dcd^{-1}c^{-1}D^{-1} = ab \cdot b^{-1}a^{-1}Dba \cdot a^{-1}b^{-1} \cdot F \cdot abab \cdot \left(b^{-1}a^{-1}b^{-1}c^{-1}\right)^2 \cdot cD^{-1}c^{-1}D \cdot D^{-2} \ .$$

Lemma 16 (18) follows by Lemma 11 (25), Lemma 16 (3), (15).

(19)-(21) follow by Lemma 16 (16)-(18) on transformation by p under consideration of Lemma 16 (18).

(22)-(24) follow by raising Lemma 16 (19)-(21) to the second power under consideration of Lemma 12 (10).

REMARK. The formulae in Lemma 16 imply that a^2, b^2 and c^2 commute pairwise.

CHAPTER 2

A FINITENESS PROOF FOR Γ

We shall now study the following group and we show that it is of finite order:

$\Gamma = \langle Y, Z \mid$ (i) $Y^4 = Z^8 = 1$;

(ii) $\left(YZ^{-1}\right)^2 = 1$;

(iii) $(YZ)^4 = 1$;

(iv) $\left(Y^2Z^2\right)^8 = 1$;

(v) $\left(YZ^2\right)^8 = 1$;

(vi) $\left(Z^4Y^2\right)^8 = 1$;

(vii) $\left(Z^4YZ^2Y\right)^8 = 1$;

(viii) $\left(Y^2ZYZ^{-3}\right)^8 = 1$;

(ix) $\left(Z^4Y^2Z^2Y^2Z^4Y^2Z^{-2}Y^2\right)^8 = 1$;

(x) $\left(Z^4Y^2Z^2Y^2\right)^8 = 1$;

(xi) $\left(YZ^2YZ^3Y^2Z^3\right)^8 = 1$;

(xii) $\left(Z^3Y\right)^8 = 1$;

(xiii) $\left(Z^3Y^2\right)^8 = 1$;

(xiv) $\left(Z^4Y^2Z^2Y^2ZY\right)^8 = 1$;

(xv) $\left(Z^4Y^2ZY\right)^8 = 1$;

(xvi) $\left(Z^4YZY^2\right)^8 = 1$;

(xvii) $\left(Z^3Y^2Z^2Y^2\right)^8 = 1$;

(xviii) $\left(Z^3Y^2Z^{-2}Y^2\right)^8 = 1$;

(xix) $\left(Z^2Y^2ZY\right)^8 = 1$;

(xx) $\left(Z^4YZ^2Y^2ZY\right)^8 = 1$;

(xxi) $\left(Z^4Y^2Z^2Y^2Z^2Y^2Z^{-2}Y^2Z^4Y^2Z^2Y^2Z^{-2}Y^2Z^{-2}Y^2\right)^8 = 1$;

$$(\text{xxii}) \quad \left(Z^4 Y^2 Z^2 Y^2 Z^2 Y^2 Z^{-2} Y^2\right)^8 = 1 \; ;$$

$$(\text{xxiii}) \quad \left(Z^4 Y^2 Z^4 Y^2 Z^4 Y^2 Z^{-2} Y^2\right)^8 = 1 \; ;$$

$$(\text{xxiv}) \quad \left(Z^4 Y^2 Z^4 Y^2 Z^2 Y^2 Z^2 Y^2\right)^8 = 1 \; ;$$

$$(\text{xxv}) \quad \left(Z^4 Y^2 Z^{-2} Y^2 Z^2 Y^2 Z^2 Y^2\right)^8 = 1) \; .$$

A. The main theorems

We shall carry through our computations in a subgroup of Γ . For this we need the

DEFINITION.

$$p := (ZY)^2 Z^{-2} ,$$

$$q := (YZ)^2 Z^{-2} ,$$

$$x := (ZY)^2 ,$$

$$\Gamma_1 := \langle p, q, x \rangle \leq \Gamma .$$

THEOREM 1. Γ_1 *is a normal subgroup of* Γ , *and we have*

$$|\Gamma : \Gamma_1| = 2^3$$

and

$$\Gamma^4 = \Gamma_1^2 .$$

Proof. The formulae in Lemma 1 (Part B) show that Γ_1 is normal in Γ . The quotient Γ/Γ_1 is obviously isomorphic to the dihedral group of order 8 . The presentation of $B(2, 4)$ in Coxeter and Moser (1957, p. 81), proves that Γ^4 is the normal subgroup of Γ , which is generated by Z^4 and $\left(Y^2 Z^2\right)^4$. Then Lemmas 1, 3-8 prove $\Gamma^4 = \Gamma_1^2$.

For our further computations, an automorphism of Γ_1 induced by an inner automorphism of Γ will be of importance.

DEFINITION. Let α be the automorphism induced by $ZY^2 Z^{-1} = ZYZY^{-1} = xY^2$ on Γ_1 .

THEOREM 2. α *operates on the generators of* Γ_1 *in the following way:*

$$\alpha : p \rightarrow q$$
$$q \rightarrow p$$
$$x \rightarrow x .$$

Proof. See the transformation formulae in Lemma 1, Part B.

Next we shall show some relations in Γ_1 between p, q and x . Then we shall find (see Theorem 3 (1)-(15)) that the subgroup of Γ generated by p and q is a factor group of the group studied in Chapter 1. Hence we have in particular the relations shown in Chapter 1, Lemma 16, between p and q . When we obviously use these relations, we shall not indicate it separately.

THEOREM 3. *In Γ_1 we have the following relations:*

(1) $p^4 = q^4 = 1$;

(2) $(qp)^4 = 1$;

(3) $(pq^{-1})^4 = 1$;

(4) $(p^2q^2)^4 = 1$;

(5) $(pqp^{-1}q^{-1})^4 = 1$;

(6) $(qp^{-1}qpq^{-1}p)^4 = 1$;

(7) $(p^2qpq^2pq)^2 = 1$;

(8) $(q^2pq^2p^{-1})^8 = 1$, $(p^2qp^2q^{-1})^8 = 1$;

(9) $(p^2q)^8 = (q^2p)^8 = 1$;

(10) $(pqpq^{-1})^8 = (qpqp^{-1})^8 = 1$;

(11) $(p^2qpq^{-1}p^2qp^{-1}q^{-1})^8 = (q^2pqp^{-1}q^2pq^{-1}p^{-1})^8 = 1$;

(12) $(qpq^{-1}p^2)^8 = (pqp^{-1}q^2)^8 = 1$;

(13) $(p^2q^2p^2q^{-1})^8 = (q^2p^2q^2p^{-1})^8 = 1$;

(14) $(pqp^2q^2)^8 = (qpq^2p^2)^8 = 1$;

(15) $(pqp^2q^{-1})^8 = (qpq^2p^{-1})^8 = 1$;

(16) $x^2 = 1$;

(17) $(px)^4 = (qx)^4 = 1$;

(18) $(pq^{-1}x)^2 = 1$;

(19) $\left(xpxpq^2\right)^4 = \left(xqxqp^2\right)^4 = 1$;

(20) $(pxqx)^4 = 1$;

(21) $\left(p^{-1}xqx\right)^4 = 1$;

(22) $(qxqxp)^4 = (pxpxq)^4 = 1$;

(23) $\left(p^2qx\right)^4 = \left(q^2px\right)^4 = 1$;

(24) $\left(xpxp^{-1}\right)^4 = \left(xqxq^{-1}\right)^4 = 1$;

(25) $\left(q^{-1}p^2qpxpx\right)^4 = \left(p^{-1}q^2pqxqx\right)^4 = 1$;

(26) $\left(p^2qxqx\right)^4 = \left(q^2pxpx\right)^4 = 1$;

(27) $\left(qp^2q^{-1}xpxp\right)^4 = \left(pq^2p^{-1}xqxq\right)^4 = 1$;

(28) $\left(qpq^{-1}xpxp\right)^4 = \left(pqp^{-1}xqxq\right)^4 = 1$;

(29) $\left(qp^2q^{-1}p^2xp^{-1}x\right)^4 = \left(pq^2p^{-1}q^2xq^{-1}x\right)^4 = 1$;

(30) $\left(qpq^{-1}pxp^{-1}x\right)^4 = \left(pqp^{-1}qxq^{-1}x\right)^4 = 1$;

(31) $\left(q^{-1}px\right)^4 = 1$;

(32) $\left(xp^2q^2\right)^4 = 1$, $\left(q^2p^2x\right)^4 = 1$;

(33) $\left(pqpqxqpqpx\right)^2 = 1$;

(34) $\left(pqp^{-1}q^{-1}xpxqxp^{-1}x\right)^4 = \left(qpq^{-1}p^{-1}xqxpxq^{-1}x\right)^4 = 1$.

Proof. See Part B.

The list in Theorem 3 is not a complete presentation of Γ_1 , because in our computations we use also the conjugates of relations (1)-(34) under Y or Z .

It is now necessary to pass to a subgroup of Γ_1 .

DEFINITION. Let Γ_2 be the normal subgroup of Γ_1 which is generated by the elements

$$pqp^{-1}q^{-1}, \quad xpxp, \quad xqxq .$$

We want to state a system of generators of the group Γ_2 . So assume that

$$a := qpq^{-1}p^{-1} , \qquad\qquad s := pxpx ,$$

$$b := pqpq^{-1}p^{-1}p^{-1} , \qquad t := ppxpxp^{-1} ,$$

$$c := p^2qpq^{-1}p^{-1}p^{-2} , \qquad u := p^{-1}qpxpxq^{-1}p ,$$

$$d := q^2pq^{-2}p^{-1} , \qquad\qquad v := qpxpxq^{-1} ,$$

$$e := pq^2pq^{-2}p^{-1}p^{-1} , \qquad w := pqpxpxq^{-1}p^{-1} ,$$

$$r := xpxp , \qquad\qquad\qquad z := p^{-1}q^2pxpxq^{-2}p .$$

Obviously we have $\langle a, b, c, d, e, r, \ldots, z \rangle \le \Gamma_2$.

THEOREM 4.

$$\Gamma_2 = \langle a, b, c, d, e, r, s, t, u, v, w, z \rangle ;$$

$$|\Gamma_1 : \Gamma_2| \le 2^5 .$$

Proof. Lemma 3 shows

$$\langle a, b, c, d, e, r, s, t, u, v, w, z \rangle \trianglelefteq \Gamma_1 \Rightarrow \Gamma_2 \le \langle a, b, c, d, e, r, s, t, u, v, w, z \rangle .$$

The other inclusion is obvious. Γ_1/Γ_2 is isomorphic to the semidirect product of an abelian group of order 16 with a group of order 2 .

DEFINITION.

$$\Gamma_3 := \langle a, b, c, d, e, r, s, t, u, v, w \rangle ,$$

$$\Gamma_4 := \langle a, b, c, d, e, r, t, u, v, w \rangle ,$$

$$\Gamma_5 := \langle a, b, c, d, e, r, t, v, w \rangle ,$$

$$\overline{\Gamma}_4 := \langle a, b, c, d, e, s, t, u, v, w \rangle ,$$

$$\overline{\Gamma}_5 := \langle a, b, c, d, e, s, t, u, v \rangle .$$

Now we deduce sufficient relations between $a, b, c, d, e, r, s, t, u, v, w, z$ to prove the following theorem.

THEOREM 5. *For the above-defined groups* Γ_i *we have:*

(1) $|\Gamma_2 : \Gamma_3| \le 2$;

(2) $|\Gamma_3 : \Gamma_4| \le 2$;

(3) $|\Gamma_4 : \Gamma_5| \le 2$;

(4) $|\Gamma_3 : \overline{\Gamma}_4| \le 2$;

(5) $|\overline{\Gamma}_4 : \overline{\Gamma}_5| \le 2$.

In particular

$$\Gamma_5 = \langle a, b, c, d, e, r, t, v, w \rangle$$

and

$$\overline{\Gamma}_5 = \langle a, b, c, d, e, s, t, u, v \rangle$$

have finite indices in Γ .

Γ_5 *and* $\overline{\Gamma}_5$ *are subnormal*

subgroups of Γ .

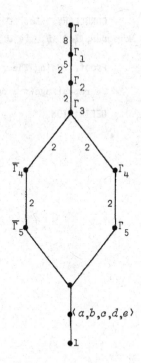

Proof. Theorem 5 is proved by Lemmas 8, 9, 10, 9',
10'.

DEFINITION. If G is a group and $S \subset G$ is a subset, then we write $\langle\langle S \rangle\rangle_G$
for the normal subgroup generated by S in G and $C_G(S)$ for the centralizer of S
in G .

For our following argumentation the simple theorem stated hereafter is of
importance.

THEOREM 6. *Let* G *be a group and* $H_1 \trianglelefteq H \trianglelefteq G$ *where*

(1) H_1 *is finite,*

(2) G/H *is finite;*

then $\langle\langle H_1 \rangle\rangle_G$ *is finite.*

Proof. If $g_1, \ldots, g_n$ is a system of coset representatives of G/H , then

$$\langle\langle H_1 \rangle\rangle_G = \left\langle g_1 H_1 g_1^{-1}, \ldots, g_n H_1 g_n^{-1} \right\rangle .$$

Since we have

$$g_i H_1 g_i^{-1} \trianglelefteq H \quad \text{for all} \quad i = 1, \ldots, n ;$$

hence $\langle\langle H_1 \rangle\rangle_G$ is generated by finitely many normal finite subgroups of H .

COROLLARY. *Let* G *be a group with a subnormal subgroup* H *of finite index. Moreover, let* H_1 *be a finite normal subgroup of* H. *Then* $\langle\langle H_1 \rangle\rangle_G$ *is also finite.*

Proof. Using Theorem 6 several times yields the corollary.

We repeat again a definition of Chapter 1.

DEFINITION.
$$D := d^2 = \left(q^2 pq^{-2}p^{-1}\right)^2 ;$$
$$u_1 := \left(q^2 p\right)^4 ;$$
$$u_2 := pqp^{-1}q^{-1}\left(q^2 p\right)^4 qpq^{-1}p^{-1} ;$$
$$u_3 := q\left(q^2 p\right)^4 q^{-1} ;$$
$$v_1 := \left(p^2 q\right)^4 ;$$
$$v_2 := qpq^{-1}p^{-1}\left(p^2 q\right)^4 pqp^{-1}q^{-1} ;$$
$$v_3 := p\left(p^2 q\right)^4 p^{-1} ;$$
$$v_4 := \left(qpq^{-1}p\right)^4 ;$$
$$u_4 := \left(pqp^{-1}q\right)^4 ;$$
$$v_6 := \left(pqp^{-1}q^{-1}p^2 qpq^{-1}p\right)^2 ;$$
$$N := \left\langle\left\langle u_1,\, a^2,\, de,\, D\right\rangle\right\rangle_\Gamma .$$

Next we shall show that N is a finite group. In order to prove it, we take elements $\theta \in \langle a, b, c, d, e \rangle$ and show that θ generates a finite normal subgroup either in Γ_5 or in $\overline{\Gamma}_5$. In both cases Theorem 5 and the corollary yield that θ generates also in Γ a finite normal subgroup $\langle\langle \theta \rangle\rangle_\Gamma$. But in the factor group $\Gamma/\langle\langle \theta \rangle\rangle_\Gamma$ the Theorems 1-5 and the Lemmas 1-12 hold. Then, in the factor group, one finds new elements in $\langle a, b, c, d, e \rangle$, whose normal closure in Γ_5 or $\overline{\Gamma}_5$ can be computed. After some steps one exhausts N by this method.

THEOREM 7. N *is a finite group.*

Proof. See Lemmas 13-17.

If we use the same symbols for the images of the elements $a, b, c, d, e, r, s, t, u, v, w$ in Γ/N, then Chapter 1, Lemma 16 yields that $\langle a, b, c, d, e \rangle$ in Γ/N is an elementary abelian group of exponent 2. Moreover, we show in Γ/N that

$$\langle a, b, c, d, e, r, s, t \rangle \trianglelefteq \langle a, b, c, d, e, r, s, t, u, v, w \rangle$$

and that the index is 8 at most.

We use this in Lemma 19. Then Lemma 20 shows that

(1) $$\left| \langle a, b, c, d, e, r, s, t \rangle : \langle a, b, c, d, e, s, t \rangle \right| \leq 2 ,$$

(2) $$\left| \langle a, b, c, d, e, s, t \rangle : \langle a, b, c, d, e, s \rangle \right| \leq 2 ,$$

in Γ/N . Hence it suffices to show the finiteness of $\langle a, b, c, d, e, s \rangle$ in Γ/N . We have now deduced so many relations in Γ/N that it is easy to show that

(3) $$\langle a, b, c, d, e, s \rangle \leq \Gamma/N$$

is metabelian at most,

(4) $$\langle a, b, c, d, e, s \rangle \leq \Gamma/N$$

is finite. Thus we have proved

THEOREM 8. Γ *is finite.*

Hence, if we use the shortened form

$$U := YZ^{-1}$$

we see that the group

$$\Gamma^* := \langle U, Y \mid U^2 = Y^4 = \left(Y^2 U \right)^4 = 1, W^8 = 1 \rangle$$

is a factor group of Γ . Hence we get

THEOREM 9. Γ^* *is finite.*

B. The proofs

LEMMA 1. *We have the rules of conjugation*

(1) $YpY^{-1} = qp^{-1}xq^{-1}$,   (2) $ZpZ^{-1} = p^{-1}xqx^{-1}p$,

   $YqY^{-1} = xq^{-1}$, $ZqZ^{-1} = p$,

   $YxY^{-1} = qp^{-1}x$, $ZxZ^{-1} = p^{-1}xq$;

and the formulae

(3) $Z^4 = (xp)^2$;   (4) $\left(Y^2 Z^2 \right)^4 = x \left(q^{-1} p^{-1} \right)^2 x^{-1}$;

(5) $p^2 = ZYZ^4 Y^{-1} Z^{-1}$;   (6) $q^2 = YZ^4 Y^{-1}$;

(7) $(pq)^2 = ZY \left(Z^2 Y^2 \right)^4 Y^{-1} Z^{-1}$;   (8) $(qx)^2 = Y^2 Z^4 Y^{-2}$.

Proof. All these formulae are obvious consequences of the relations (i)-(iii).

Proof of Theorem 3. By Lemma 1 we have

$$xY^2 : p \to q$$
$$q \to p$$
$$x \to x \ .$$

Hence, if we have proved one formula of Theorem 3, we have also shown the formula which is implied by the first one when putting q for p and p for q .

In our computations we shall not indicate separately obvious consequences of the relations (i)-(iii).

(1)

$$q^4 = \left(YZYZ^{-1}\right)^4 = YZ^8Y^{-1} = 1$$

by (ii), (i).

(2)

$$(pq)^4 = \left(ZYZYZ^{-2}YZYZ^{-1}\right)^4$$
$$= ZY^{-1}\left(Y^2Z^2\right)^8YZ^{-1}$$
$$= 1$$

by (iv).

(3)

$$\left(pq^{-1}\right)^4 = \left(ZYZYZ^{-1}Y^{-1}Z^{-1}Y^{-1}\right)^4$$
$$= \left(Y^{-1}Z^{-1}Y^{-1}Z^{-1}YZYZ^2YZYZ^{-1}Y^{-1}Z^{-1}Y^{-1}\right)^2$$
$$= \left(Y^2Z\left(YZ^2\right)^4Z^{-1}Y^2\right)^2$$
$$= 1$$

by (v).

(4)

$$\left(p^2q^2\right)^4 = \left(ZYZYZ^{-1}YZYZ^{-2}YZYZ^{-1}YZYZ^{-1}\right)^4$$
$$= Z \cdot \left(YZ^3YZ^{-2}YZ^3Y\right)^4Z^{-1}$$
$$= ZY^{-1}\left(Y^2Z^4\right)^8YZ^{-1}$$
$$= 1$$

by (vi).

(5)

$$\left(pqp^{-1}q^{-1}\right)^4 = \left(ZYZYZ^{-2}YZYZ^2YZYZY^{-1}Z^{-1}Y^{-1}\right)^4$$
$$= \left(ZYZYZ^{-3}Y^{-1}Z^{-1}Y^2Z^{-1}Y^2Z^{-1}Y^{-1}\right)^4$$
$$= \left(ZYZYZ^{-3}Y^2Z^3Y^2\right)^4 = Y^{-1}Z^{-1}Y^{-1}\left(Z^4Y^2\right)^8YZY = 1$$

by (vi).

(6)

$$(qp^{-1}qpq^{-1}p)^4 = (yzyz^2yzy^2zy^{-1}z^{-1}yzyz^2yzy^2zy^{-1}z^{-1})^4$$
$$= (yzyz^2yzy^2zy^{-1}z^{-1})^8$$
$$= (yzyz^2yzy^{-1}z^{-2})^8$$
$$= (z^{-1}y^{-1}z^{-2}y^{-1}z^{-3})^8$$
$$= z^{-1}(y^{-1}z^{-2}y^{-1}z^4)^8 z = 1$$

by (vii).

(7) By Lemma 1 we have

$$y^{-1}py = qp^2x .$$

Hence

$$p \cdot y^{-1}py = pqp^2x .$$

Moreover, we have

$$y^2py^{-1}pyy^{-2} = xqpq^2 .$$

We multiply these formulae obtaining

$$(py^{-1})^8 = (pqp^2qpq^2)^2 ,$$
$$(py^{-1})^8 = (zyzyz^{-2}y^{-1})^8$$
$$= zy^{-1}(y^2zyz^{-3})^8 yz^{-1} = 1$$

by (viii).

(8)

$$q^2pq^2p^{-1} = yzyz^{-1}yzy^2zyz^{-2}yzyz^{-1}yzyz^2yzy$$
$$= yz^3y^2zyz^{-2}yz^3yz^2yzy$$
$$= yz^3y^2z^2y^2z^4yzy^{-1}z^{-1}y^{-1}z^{-1}$$
$$= yz^{-1}z^4y^2z^2y^2z^4y^2z^{-2}y^2zy^{-1} .$$

Hence we have $(q^2pq^2p^{-1})^8 = 1$ by (ix).

(9)

$$q^2p = yzyz^{-1}yzy^2zyz^{-2}$$
$$= yz^3y^2zyz^{-2}$$
$$= zy^{-1}z^4y^2z^2y^2zyz^{-1} .$$

Hence we have $(q^2p)^8 = 1$ by (x).

(10)

$$pqpq^{-1} = ZYZYZ^{-2}YZY^2ZYZ^{-1}Y^{-1}Z^{-1}Y^{-1}$$
$$= Y^{-1}Z^{-1}Y^{-1}Z^{-3}Y^2Z^{-2}Y^2Z^{-1}Y^{-1}Z^2Z^2Y^2YZY$$
$$= Y^{-1}Z^{-1}Y^{-1}Z^{-3}Y^2Z^{-2}Y^{-1}Z^{-2}Y^{-1}YZY \ .$$

Hence we have $\left(pqpq^{-1}\right)^8 = 1$ by (xi).

(11)

$$\left(q^2pqp^{-1}q^2pq^{-1}p^{-1}\right)^8 = \left(YZ^3Y^2ZYZ^{-2}YZYZ^2YZY^2Z^3Y^2ZYZ^{-1}Y^{-1}Z^{-1}Y^{-1}Z^3YZY\right)^8$$
$$= \left(YZ^3Y^2Z^2Y^2Z^2Y^2Z^{-2}Y^2Z^4Y^2Z^2Y^2Z^{-1}Y^{-1}Z^3YZY\right)^8$$
$$= YZ^{-1}\left(Z^4Y^2Z^2Y^2Z^2Y^2Z^{-2}Y^2Z^4Y^2Z^2Y^2Z^{-2}Y^2Z^{-2}Y^2\right)^8ZY^{-1} =$$

by (xxi).

(12)

$$\left(pqp^{-1}q^2\right)^8 = Z\left(YZYZ^{-2}YZYZ^2YZY^2Z^3Y\right)^8Z^{-1}$$
$$= ZY\left(Z^2Y^2Z^2Y^2Z^{-2}Y^2Z^4Y^2\right)^8Y^{-1}Z^{-1} = 1$$

by (xxii).

(13)

$$\left(p^2q^2p^2q^{-1}\right)^8 = \left(ZYZ^3YZ^{-2}YZ^3Y^2Z^3YZ^{-1}Y^{-1}Z^{-1}Y^{-1}\right)^8$$
$$= ZY\left(Z^4Y^2Z^4Y^2Z^4Y^2Z^{-2}Y^2\right)^8Y^{-1}Z^{-1} = 1$$

by (xxiii).

(14)

$$\left(pqp^2q^2\right)^8 = \left(ZYZYZ^{-2}YZY^2Z^3YZ^{-2}YZ^3YZ^{-1}\right)^8$$
$$= ZY\left(Z^2Y^2Z^2Y^2Z^4Y^2Z^4Y^2\right)^8Y^{-1}Z^{-1} = 1$$

by (xxiv).

(15)

$$\left(qpq^2p^{-1}\right)^8 = \left(YZY^2ZYZ^{-2}YZ^3YZ^2YZY\right)^8$$
$$= \left(YZY^2Z^2Y^2Z^4YZ^2YZY\right)^8$$
$$= YZ\left(Y^2Z^2Y^2Z^4Y^2Z^{-2}Y^2Z^2\right)^8Z^{-1}Y^{-1} = 1$$

by (xxv).

(16) $x^2 = (ZYZY)^2 = (ZY)^4 = 1$ by (iii).

(17) $Y (1) Y^{-1} \Rightarrow (17)$.

(18) Y (16) $Y^{-1} \rightarrow$ (18).

(19)

$$\left(xpxpq^2\right)^4 = \left(ZYZYZYZYZ^{-2}ZYZYZYZYZ^{-2}YZYZ^{-1}YZYZ^{-1}\right)^4$$
$$= \left(Z^4YZ^3YZ^{-1}\right)^4$$
$$= Z\left(Z^3Y\right)^8 Z^{-1} = 1$$

by (xii).

(20) Y (2) $Y^{-1} \rightarrow$ (20).

(21) Z (3) $Z^{-1} \rightarrow$ (21).

(22)

$$(xpxqp)^4 = \left(Z^{-1}YZY^2ZY^2ZYZ^{-2}\right)^4$$
$$= \left(Z^{-1}Y^2Z^{-3}Y^2Z^{-2}\right)^4$$
$$= Z^{-1}\left(Y^2Z^{-3}\right)^8 Z$$
$$= 1$$

by (xiii).

(23)

$$\left(q^2px\right)^4 = \left(YZ^3Y^2ZYZ^{-1}YZY\right)^4$$
$$= Y\left(Z^3Y^2\right)^8 Y^{-1}$$
$$= 1$$

by (xiii).

(24)

$$\left(xpxp^{-1}\right)^4 = \left(Z^{-1}YZYZ^3YZY\right)^4$$
$$= Y^{-1}Z^{-1}\left(Z^3Y\right)^8 ZY$$
$$= 1$$

by (xii).

(25)   We have $ZYp^2qY^{-1}Z^{-1} = xp^{-1}xp^{-1}q^{-1}x$. Hence

$$\left(p^2qZY\right)^8 = \left(p^2qZYp^2qY^{-1}Z^{-1}x\right)^4$$
$$= \left(p^2qpxpxq^{-1}\right)^4$$
$$= 1$$

by (17) and because

$$\left(p^2qzy\right)^8 = \left(zyzyz^{-1}yzyz^{-2}yzy^2\right)^8$$
$$= \left(zyz^4y^2z^2y^2\right)^8$$
$$= 1$$

by (xiv).

(26) We have $zy^{-1}p^2yz^{-1} = xq^{-1}xq^{-1}$. Hence

$$\left(p^2zy^{-1}\right)^8 = \left(p^2zy^{-1}p^2yz^{-1}\right)^4$$
$$= \left(p^2qxqx\right)^4$$
$$= 1$$

because

$$\left(p^2zy^{-1}\right)^8 = \left(zyzyz^{-1}yzyz^{-1}y^{-1}\right)^8$$
$$= \left(zyz^3yz^{-1}y^{-1}\right)^8$$
$$= \left(zyz^4y^2\right)^8$$
$$= 1$$

by (xv).

(27) We have by Lemma 1,

$$zyqp^2y^{-1}z^{-1} = q^{-1}p^{-1}xp^{-1} \Rightarrow \left(qp^2zy\right)^8 = \left(qp^2zyqp^2y^{-1}z^{-1}x\right)^4$$
$$= \left(qp^2q^{-1}p^{-1}xp^{-1}x\right)^4$$
$$= 1$$

because

$$\left(qp^2zy\right)^8 = \left(yzy^2zyz^{-1}yzyz^{-1}y\right)^8$$
$$= \left(yzy^2z^4\right)^8$$
$$= 1$$

by (xvi).

(28) Essentially is (22).

(29) We have $zy^2p^2xy^2z^{-1} = q^{-1}xp^{-1}$. Hence

$$\left(p^2xzy^2\right)^8 = \left(p^2xzy^2p^2xy^2z^{-1}zy^2zy^2\right)^4$$
$$= \left(p^2xq^{-1}xp^{-1}zy^{-1}z^{-1}y^{-1}\right)^4$$
$$= \left(p^2xq^{-1}xp^{-1}q^{-1}\right)^4$$
$$= 1$$

because

$$\left(p^2xZY^2\right)^8 = \left(ZYZYZ^{-1}YZYZ^{-1}YZYZY^2\right)^8$$
$$= \left(ZYZ^5YZY^2\right)^8$$
$$= ZY\left(Z^{-3}Y^2Z^{-2}Y^2\right)^8Y^{-1}Z^{-1}$$
$$= 1$$

by (xvii).

(30) We have $ZY^2pxY^2Z^{-1} = p^{-1}xqp$. Hence

$$\left(pxZY^2\right)^8 = \left(pxZY^2pxY^2Z^{-1}ZY^2ZY^2\right)^4$$
$$= \left(pxp^{-1}xqpZY^{-1}Z^{-1}Y^{-1}\right)^4$$
$$= \left(pxp^{-1}xqpq^{-1}\right)^4$$
$$= 1$$

because

$$\left(pxZY^2\right)^8 = \left(ZYZYZ^{-1}YZYZY^2\right)^8$$
$$= \left(ZYZ^3YZY^2\right)^8$$
$$= ZY\left(Z^3Y^2Z^{-2}Y^2\right)^8Y^{-1}Z^{-1}$$
$$= 1$$

by (xviii).

(31) We have $ZY^{-1}pYZ^{-1} = xq^{-1}$. Hence

$$\left(pZY^{-1}\right)^8 = \left(pZY^{-1}pYZ^{-1}\right)^4$$
$$= \left(pxq^{-1}\right)^4$$
$$= 1$$

because

$$\left(pZY^{-1}\right)^8 = \left(ZYZYZ^{-1}Y^{-1}\right)^8$$
$$= ZY\left(Z^2Y^2ZY\right)^8Y^{-1}Z^{-1}$$
$$= 1$$

by (xix).

(32) We have $ZYpq^{-1}Y^{-1}Z^{-1} = xp^{-1}xq$. Hence

$$\left(pq^{-1}ZY\right)^8 = \left(pq^{-1}ZYpq^{-1}Y^{-1}Z^{-1}x\right)^4$$
$$= \left(pq^{-1}xp^{-1}xqx\right)^4$$
$$= x\left(qp^2xq\right)^4x$$
$$= 1$$

because

$$(pq^{-1}zy)^8 = (zyzyz^{-1}y^{-1}z^{-1}y^{-1}zy)^8$$
$$= (zy^2z^{-2}y^2z^2y)^8$$
$$= zy^2(z^{-2}y^2z^2yzy^2)^8\,y^2z^{-1}$$
$$= zy^2(z^{-1}y^{-1}zyz^2yzy^2)^8\,y^2z^{-1}$$
$$= zy^2(z^{-1}y^2z^{-1}y^{-1}z^{-1}y^{-1}zyzy^2)^8\,y^2z^{-1}$$
$$= zy^2(z^{-1}y^2z^{-1}y^2yz^{-1}z^{-1}y^{-1}y^2zy^2)^8\,y^2z^{-1}$$
$$= zy^2(z^{-1}y^2z^{-1}y^2zyz^{-1}z^{-1}y^{-1}y^{-1}y^{-1}zy^2)^8\,y^2z^{-1}$$
$$= zy^2(z^{-1}y^2z^{-1}y^2zyz^{-1}z^{-1}y^{-1}z^{-1}y^{-1})^8\,y^2z^{-1}$$
$$= zy^2(z^{-1}y^2z^{-1}y^2z^2yz)^8\,y^2z^{-1}$$
$$= zy^2z^{-1}y^{-1}z^{-2}(yz^{-1}y^2z^{-2}y^2)^8\,z^2yzy^2z^{-1}$$
$$= zy^2z^{-1}y^{-1}z^{-2}(y^{-1}z^{-1}y^2z^{-2})^8\,z^2yzy^2z^{-1}$$
$$= 1$$

by (xix).

(33)   We have $y^{-1}q^{-1}py = xpqp^2x$.   Hence

$$(q^{-1}py^{-1})^8 = (q^{-1}py^{-1}q^{-1}pyy^2q^{-1}py^{-1}q^{-1}pyy^2)^2$$
$$= (q^{-1}pxpqp^2xy^2q^{-1}pxpqp^2xy^2)^2$$
$$= (q^{-1}pxpqp^2xxp^{-1}qxqpq^2)^2$$
$$= q^{-1}p(xpqpqxqpqp)^2p^{-1}q$$
$$= 1$$

because

$$(q^{-1}py^{-1})^8 = (zy^{-1}z^{-1}y^{-1}zyzyz^{-2}y^{-1})^8$$
$$= yz^2(z^4y^{-1}zyzy)^8z^{-2}y^{-1}$$
$$= yz^2(z^3y^2zy)^8z^{-2}y^{-1}$$
$$= yz^2(z^3y^{-1}z^{-1}y^2)^8z^{-2}y^{-1} = 1$$

by (viii).

(34)   We have $zy^2pqp^{-1}y^2z^{-1} = q^{-1}xpxqxp^{-1}xq$.   Hence

$$(pqp^{-1}zy^2)^8 = (pqp^{-1}zy^2pqp^{-1}y^2z^{-1}zy^2zy^2)^4$$
$$= (pqp^{-1}q^{-1}xpxqxp^{-1}xqq^{-1})^4$$
$$= 1$$

because

$$\left(pqp^{-1}zy^2\right)^8 = \left(zyzyz^{-2}yzyz^2yzyzy^2\right)^8$$
$$= y^{-1}\left(z^{-1}y^{-1}z^{-3}yzyzy^{-1}z^{-1}\right)^8y$$
$$= y^{-1}z^{-1}\left(y^{-1}z^4y^{-1}z^{-1}y^2z^{-2}\right)^8zy$$
$$= 1 \quad \text{by (xx).}$$

Next we shall express some elements of Γ_2 in terms of $a, b, \ldots, t, z$.

LEMMA 2.

(1) $p^{-1}xpxpp = rst$;

(2) $p^{-1}qxpxq^{-1}p^2 = uc^{-1}b^{-1}a^{-1}vawb$;

(3) $pq^2xpxq^2 = wbtsuzb^{-1}d^{-1}$;

(4) $p^2q^2xpxq^2p^{-1} = zc^{-1}ecvatrva$;

(5) $p^{-1}q^2xpxq^2p^2 = dbruzswb$;

(6) $xqxq = tb^{-1}wa^{-1}vabcuzc^{-1}ec$;

(7) $pxqxqp^{-1} = sabcudbruzswb$;

(8) $p^2xqxqp^2 = zc^{-1}ecvat$;

(9) $p^{-1}xqxqp = dbruz$;

(10) $qxqx = as$;

(11) $qqxqxq^{-1} = da^{-1}v$;

(12) $pqqxqxq^{-1}p^{-1} = eb^{-1}w$.

Proof. (1) follows by using Theorem 3 (17) several times.

(2) follows by using Theorem 3 (17) several times.

(3) By Theorem 3 (3), (17) follows

$$wbtsuzb^{-1}d^{-1} = pq^2\left(q^{-1}xp^{-1}x\right)^2\left(xp^{-1}\right)^2\left(qpxpx\right)^2qpq .$$

By Theorem 3 (17), (20), (22) follows (3).

(4) follows obviously by Theorem 3 (20), (21), (17).

(5)

$$dbruzswb = q^{-1}p^{-1}(qpxpx)^3q^2p^2xpxpqpxpxpq^{-1}p^2$$
$$= q^{-1}xpxqp\left(xp^{-1}xq\right)^2q^2p^2 \quad \text{by Theorem 3 (17), (22)}$$
$$= p^{-1}(xq)^4q^2xpxq^2p^2 \quad \text{by Theorem 3 (21), (18).}$$

Hence we have shown (5).

(6)

$$tb^{-1}wa^{-1}vabcuzc^{-1}ec = p^2 xpxpqxp^{-1}xqpxpxpq^2$$
$$= p(xp^{-1}xq)^3 q \quad \text{by Theorem 3 (17)}$$
$$= pq^{-1}xpxq \quad \text{by Theorem 3 (21)}$$
$$= xqxq \quad \text{by Theorem 3 (18).}$$

(7) Use (5) to conclude

$$sabcudbruzswb = p(xpxq)^3 qp^2$$
$$= pq^{-1}xp^{-1}xqp^{-1}p^{-1} \quad \text{by Theorem 3 (20).}$$

Hence we have proved (7).

(8) Use (4) to conclude

$$zc^{-1}ecvat = p^2 q^{-1}(q^{-1}xpx)^3 p$$
$$= p^2 q^{-1}xp^{-1}xqp^{-1}p^2 \quad \text{by Theorem 3 (21).}$$

Hence the proof of (8) is complete.

(9) follows obviously by Theorem 3 (22), (18).

(10)

$$as = qpq^{-1}xpx$$
$$= qxqx \quad \text{by Theorem 3 (18).}$$

(11)

$$da^{-1}v = q^2 pq^{-1}xpxq^{-1}$$
$$= q^2 xqxq^{-1} \quad \text{by Theorem 3 (18).}$$

(12)

$$eb^{-1}w = pq^2 pq^{-1}xpxq^{-1}p^{-1}$$
$$= pq^2 xqxq^{-1}p^{-1} \quad \text{by Theorem 3 (18).}$$

Next we shall sum up the transformation formulae which are necessary to prove Theorem 4.

LEMMA 3.

(1) $pap^{-1} = b$;

(2) $pbp^{-1} = c$;

(3) $pcp^{-1} = c^{-1}b^{-1}a^{-1}$;

(4) $pdp^{-1} = e$;

(5) $pep^{-1} = b^{-1}db$;

(6) $prp^{-1} = s$;

(7) $\ psp^{-1} = t$;

(8) $ptp^{-1} = rst$;

(9) $pup^{-1} = v$;

(10) $pvp^{-1} = w$;

(11) $pwp^{-1} = cuc^{-1}b^{-1}a^{-1}vawb$;

(12) $pzp^{-1} = dwbtsuzb^{-1}d^{-1}$;

(13) $qaq^{-1} = da^{-1}$;

(14) $qbq^{-1} = aeb^{-1}a^{-1}$;

(15) $qcq^{-1} = adbc^{-1}b^{-1}a^{-1}$;

(16) $qdq^{-1} = c^{-1}e^{-1}a^{-1}$;

(17) $qeq^{-1} = ad^{-1}abcb^{-1}a^{-1}$;

(18) $qrq^{-1} = abcuc^{-1}b^{-1}a^{-1}$;

(19) $qsq^{-1} = v$;

(20) $qtq^{-1} = awa^{-1}$;

(23) $quq^{-1} = abczc^{-1}b^{-1}a^{-1}$;

(24) $qvq^{-1} = dwbtsuzb^{-1}d^{-1}$;

(25) $qwq^{-1} = aezc^{-1}ecvatrv$;

(26) $qzq^{-1} = abcruzc^{-1}b^{-1}a^{-1}$;

(27) $xax^{-1} = tb^{-1}wa^{-1}vabcuzc^{-1}ecr$;

(28) $xbx^{-1} = rdbruzts$;

(29) $xcx^{-1} = sa^{-1}vc^{-1}e^{-1}czs$;

(30) $xdx^{-1} = ecsvdczuc^{-1}b^{-1}a^{-1}c^{-1}e^{-1}b^{-1}a^{-1}$;

(31) $xex^{-1} = rdbruc^{-1}ecdbdbruzsa^{-1}vabcub^{-1}d^{-1}bcr$;

(32) $xrx^{-1} = s$;

(33) $xsx^{-1} = r$;

(34) $xtx^{-1} = str$;

(35) $xux^{-1} = abcudbruzswbtb^{-1}wa^{-1}vabcuzc^{-1}b^{-1}wszurb^{-1}d^{-1}uc^{-1}b^{-1}a^{-1}$;

(36) $xvx^{-1} = ecsvdwbtsuzrstb^{-1}wd^{-1}vsc^{-1}e^{-1}$;

(37) $xwx^{-1} = rdbruc^{-1}ecvate^{-1}a^{-1}dbruzr$;

(38) $xzx^{-1} = abcudbruzswbd^{-1}abczc^{-1}ecvatrvawbtsuzb^{-1}eb^{-1}wbts$;

(39) $\alpha(a) = a^{-1}$;

(40) $\alpha(b) = ad^{-1}$;

(41) $\alpha(c) = dec$;

(42) $\alpha(d) = b^{-1}a^{-1}$;

(43) $\alpha(e) = abde$;

(44) $\alpha(r) = tb^{-1}wa^{-1}vabcuzc^{-1}ec$;

(45) $\alpha(s) = as$;

(46) $\alpha(t) = da^{-1}v$;

(47) $\alpha(u) = c^{-1}e^{-1}sabcudbruzswbec$;

(48) $\alpha(v) = bt$;

(49) $\alpha(w) = aeb^{-1}wa^{-1}$;

(50) $\alpha(z) = dbaezc^{-1}ecvate^{-1}a^{-1}b^{-1}d^{-1}$.

Proof. (1)-(5) follow by Chapter 1, Lemma 1.

(6)-(12) follow by Lemma 2 (1), (2), (3).

(13)-(17) follow by Chapter 1, Lemma 1.

(18)-(25) follow by Lemma 2 (1), (2), (4).

(26) Because

$$qzq^{-1} = qp^{-1}q^2pqp^{-1}(pq^{-1}x)^2\,xqxq\,p\,p^{-1}qpq^{-1}$$

this formula is implied by Lemma 2 (9) and Theorem 3.

(27) Theorem 3 (18) yields

$$xax = xqxqp^{-1}xp^{-1}x$$
$$= xqxq\,xpxp \quad \text{by Theorem 3 (17).}$$

Then Lemma 2 (6) implies (27).

(28) By Theorem 3 (18) we have

$$xbx = (xp)^2\,p^{-1}(xq)^2p\,p^2xp^2x \ .$$

Then Lemma 2 (2), (9) imply (28).

(29) By Theorem 3 (18) we have

$$xcx = (xp)^2\,p^{-1}(xp)^2p\,p^2(xq)^2p^{-2}(px)^2 \ .$$

Then (29) follows by using Lemma 2 (1), (8).

(30) We write out the right-hand side of the formula obtaining

$$pq^{-1}p^{-1}\,(qpxpx)^2\,qpqp^{-1}q^{-1}xpxq^{-1}(px)^2\,pqp^{-1}q^{-1}pq^{-1} = xdx$$

using Theorem 3 (17), (18), (22).

(31) By Lemma 2 (5) we have

$$p^{-1}zp = b^{-1}dbdbruaswb \ .$$

Since we have

$$xex = r \ p^{-1}xdxpr \ ,$$

then (31) follows by Lemma 3 (30), (1)-(12) and the formula for $p^{-1}zp$ stated above.

(32)-(34) are obvious.

(35) follows by Lemma 2 (7), (6) and Theorem 3.

(36)

$$ecsvd \ wbtsuzb^{-1}d^{-1} \ db \ rst \ b^{-1}wd^{-1}vsc^{-1}e^{-1}$$
$$= pq^{-1}p^{-1}(qpxpx)^4 \ xp^{-1}xp^2 \ (qxpx)^2 \ qp^{-1} \left(q^{-1}pxpx\right)^2 q^{-1}pqp^{-1}$$

by Lemma 2 (3), (1). Use Theorem 3 (17), (18), (20), (22), to conclude (36).

(37) Use Lemma 2 (9), (8) to obtain (37).

(38) follows immediately by Lemma 2 (3), (4), (7), (12).

(39)-(43) are the formulae contained in Chapter 1, Lemma 15.

(44) follows by Lemma 2 (6).

(45) follows by Lemma 2 (10).

(46) follows by Lemma 2 (11).

(47) follows by Lemma 2 (7).

(48) follows by Theorem 3 (18).

(49) follows by Lemma 2 (12).

(50) follows by Lemma 2 (8).

LEMMA 4. *In addition to the relations proved in Chapter 1, we have in the group* $G = \langle p, q \rangle \le \Gamma_1$,

(1) $u_4 = u_5$, (2) $v_4 = v_5$, (3) $D^2 = 1$.

Proof. Theorem 3 (6) yields

$$\left(qp^{-1}qpq^{-1}p\right)^4 = 1 \ .$$

Hence

$$(qpc)^4 = 1 \ .$$

Now we use the relations of Chapter 1 to obtain (1).

(2) and (3) follow by (1) using Chapter 1, Lemma 16.

Next we collect some relations between the elements $a, b, c, d, e, r, s, t, u,$

v, w, z which are immediate consequences of Theorem 3.

LEMMA 5.

(1) $u^2 = v^2 = w^2 = r^2 = s^2 = t^2 = z^2 = 1$;

(2) $(uaebc)^2 = 1$;

(3) $\left(vda^{-1}\right)^2 = 1$;

(4) $\left(web^{-1}\right)^2 = 1$;

(5) $(rabc)^2 = 1$;

(6) $(sa)^2 = 1$;

(7) $(tb)^2 = 1$;

(8) $\left(zc^{-1}ecdb\right)^2 = 1$;

(9) $(rst)^2 = 1$;

(10) $(rt)^2 = 1$;

(11) $(rsts)^2 = 1$;

(12) $(uw)^2 = 1$;

(13) $\left(cuc^{-1}b^{-1}a^{-1}vawb\right)^2 = 1$;

(14) $\left(vcuc^{-1}b^{-1}a^{-1}vawb\right)^2 = 1$;

(15) $\left(wbcuc^{-1}b^{-1}\right)^2 = 1$;

(16) $\left(abcuc^{-1}b^{-1}a^{-1}vawa^{-1}v\right)^2 = 1$;

(17) $(crstas)^2 = 1$;

(18) $(trabcb)^2 = 1$;

(19) $\left(ta^{-1}rab^2\right)^2 = 1$;

(20) $rsa^{-1}bsr = abc^2$;

(21) $stb^{-1}cts = a^{-1}c^{-1}b^{-1}a^{-1}$;

(22) $absbctb^{-1}a^{-1}rstc^{-1}b^{-1}r = 1$;

(23) $arstc^{-1}b^{-1}a^{-1}tcsbr = 1$;

(24) $(vabt)^2 = 1$;

(25) $(bcus)^2 = 1$;

(26) $\left(ta^{-1}vae\right)^2 = 1$;

(27) $\left(sbcuc^{-1}b^{-1}a^{-1}da^{-1}\right)^2 = 1$;

(28) $\left(tb^{-1}a^{-1}vabcu\right)^2 = 1$;

(29) $\left(c^{-1}wbrst\right)^2 = 1$;

(30) $(crst)^2 = 1$;

(31) $(zr)^2 = 1$;

(32) $(zur)^2 = 1$;

(33) $\left(dbuc^{-1}b^{-1}a^{-1}vabts\right)^2 = 1$;

(34) $(ecvawbcrs)^2 = 1$;

(35) $\left(zc^{-1}ectb^{-1}wa^{-1}vabcu\right)^2 = 1$;

(36) $\left(dwbtsuzb^{-1}d^{-1}\right)^2 = 1$;

(37) $\left(zc^{-1}ecvat\right)^2 = 1$;

(38) $\left(zc^{-1}ecvate^{-1}a^{-1}\right)^2 = 1$;

(39) $(dbruz)^2 = 1$;

(40) $\left(zc^{-1}ecvatrvae\right)^2 = 1$;

(41) $\left(zc^{-1}ecvatrvaet\right)^2 = 1$;

(42) $\left(zswbe^{-1}a^{-1}dbru\right)^2 = 1$;

(43) $(zswdbdbru)^2 = 1$;

(44) $\left(zc^{-1}ecvatrvad^{-1}abc\right)^2 = 1$;

(45) $\left(tb^{-1}wa^{-1}vabcuzc^{-1}\right)^2 = 1$;

(46) $b^{-1}d^{-1}seb^{-1}td^{-1}abcrstaebcr = 1$;

(47) $rste^{-1}a^{-1}rda^{-1}sc^{-1}e^{-1}tdec^{-1}e^{-1} = 1$;

(48) $sa^{-1}dstb^{-1}esrc^{-1}b^{-1}dbtsabecr = 1$;

(49) $dbruc^{-1}ecdbc^{-1}ecabcurabc = 1$;

(50) $aeb^{-1}a^{-1}vabcub^{-1}d^{-1}becabcuc^{-1}b^{-1}a^{-1}da^{-1}v = 1$;

(51) $\left(deb^{-1}wbts\right)^2 = 1$.

Proof. (1) Since $u, \ldots, z$ are conjugates of

$$s = (px)^2 ,$$

(1) follows by Theorem 3 (17).

(6)

$$(sa)^2 = \left(pxpxqpq^{-1}p^{-1}\right)^2$$
$$= pqp^{-1}(xq)^4pq^{-1}p^{-1} = 1$$

by Theorem 3 (17).

(7) p (6) $p^{-1} \Rightarrow$ (7) .

(5) p^2 (7) $p^{-2} \Rightarrow$ (5) .

(3) q (6) $q^{-1} \Rightarrow$ (3) .

(2) p^{-1} (3) $p \Rightarrow$ (2) .

(4) p (3) $p^{-1} \Rightarrow$ (4) .

(8) q (2) $q^{-1} \Rightarrow$ (8) .

(9)

$$(rst)^2 = \left(xpxp \cdot pxpx \cdot p^2xpxp^{-1}\right)^2$$
$$= p^{-1}(xp)^4p = 1$$

using Theorem 3 (17) several times.

(10) $(rt)^2 = \left(xpxp^{-1}\right)^4 = 1$ by Theorem 3 (24).

(11)

$$(rsts)^2 = \left(xpxp \cdot pxpx \cdot p^2xp^2x\right)^2$$
$$= \left(p^{-1}xpx\right)^4 = 1$$

by Theorem 3 (17), (24).

(12)

$$(uw)^2 = \left(p^{-1}qpxpxq^{-1}p^2qpxpxq^{-1}p^{-1}\right)^2$$
$$= pq\left(q^{-1}p^2qpxpx\right)^4q^{-1}p^{-1} = 1$$

by Theorem 3 (25).

(14) p (12) $p^{-1} \Rightarrow$ (14).

(15) q (10) $q^{-1} \Rightarrow$ (15).

(13) p (15) $p^{-1} \Rightarrow$ (13).

(16) q (11) $q^{-1} \Rightarrow$ (16).

(17)

$$(crstas)^2 = \left(p^2 qxqxp^2 qxqx\right)^2$$
$$= \left(p^2 qxqx\right)^4 = 1$$

by Theorem 3 (26).

(18) p (17) $p^{-1} \Rightarrow$ (18).

(19) By Theorem 3 (27) we have

$$1 = \left(qp^2 q^{-1} xpxp\right)^4$$
$$= \left(qp^2 q^{-1} p^2 p^2 xpxpp^{-2} p^2 qp^2 q^{-1} xpxp\right)^2$$
$$= \left(abtb^{-1}a^{-1}r\right)^2 \ .$$

Then (19) follows using (7).

(20) By Theorem 3 (28) we have

$$1 = \left(qpq^{-1}p^{-1}pxpxp\right)^4$$
$$= (asp)^4 = aspasp^{-1} p^2 asp^{-2} p^{-1}asp$$
$$= asbtcrstc^{-1}b^{-1}a^{-1}r \ .$$

Then (20) follows using (5), (6), (7).

(21) p (20) $p^{-1} \Rightarrow$ (21).

(22) By Theorem 3 (29) we have

$$1 = \left(qp^2 q^{-1}p^{-2}pxpxp\right)^4$$
$$= (absp)^4 = abs\ pabsp^{-1} p^2 absp^{-2}p^{-1}absp$$
$$= absbctb^{-1}a^{-1}rstc^{-1}b^{-1}r \ .$$

Thus we have proved (22).

(23) By Theorem 3 (30) we have

$$1 = \left(qpq^{-1}pxp^{-1}x\right)^4 = \left(pqpq^{-1}p^2xpxpp^{-1}\right)^4$$
$$= \left(brp^{-1}\right)^4 = br \cdot p^{-1}brp \cdot p^2brp^{-2}pbrp^{-1}$$
$$= brarstc^{-1}b^{-1}a^{-1}tcs \ .$$

Thus we have proved (23).

(24)

$$(vabt)^2 = \left(q \cdot pxpx \cdot q^{-1}qp^2q^{-1}p^2p^2xpxp^{-1}\right)^2$$
$$= \left(qxp^{-1} \cdot xpq^{-1}x \cdot pxp^{-1}\right)^2$$
$$= \left(qxp^{-1}\right)^4 = 1$$

by Theorem 3 (17), (18), (31).

(25) p^{-1} (24) $p \Rightarrow$ (25).

(26)

$$\left(ta^{-1}vae\right)^2 = p^2 \cdot \left(xpxqxpx \cdot q^{-1}p^{-1}q^{-1}p^{-1} \cdot q\right)^2 \cdot p^{-2}$$
$$= p^2\left(xp \cdot xqp^{-1}x \cdot p^{-1}xqpq^2\right)^2 p^{-2}$$
$$= p^2\left(xp^2q^{-1}p^{-1}xqpq^2\right)^2 p^{-2}$$
$$= p^2\left(xp^2q^2\right)^4 p^{-2} = 1$$

by Theorem 3 (32).

(27) p^{-1} (26) $p \Rightarrow$ (27).

(28)

$$\left(tb^{-1}a^{-1}vabcu\right)^2 = \left(p^2xpxpqp^{-1}xp^2xq^{-1}p\right)^2$$
$$= p\left(xq^{-1}p^2\right)^4 p^{-1} = 1$$

by Theorem 3 (23).

(29) p (24) $p^{-1} \Rightarrow$ (29).

(30) p (7) $p^{-1} \Rightarrow$ (30).

(31) $(zr)^2 = p^{-1}\left(q^2pxpx\right)^4 p = 1$, by Theorem 3 (26).

(32)

$$(zur)^2 = p^{-1}\left(q^2pxpx \cdot q^{-1}pxpxq^{-1}pxpx\right)^2 p$$
$$= p^{-1}q^{-1}(px)^4 qp = 1$$

by Theorem 3 (22), (17).

(33)

$$(dbuc^{-1}b^{-1}a^{-1}vabts)^2 = \left(q^{-1}p^{-1}q^2\,pxpxp^2\,xpxp^2q^{-1}xp^2x\right)^2$$
$$= \left(q^{-1}p^{-1}q^2xp^2\cdot xpq^{-1}x\cdot p^2x\right)^2$$
$$= \left(q^{-1}p^{-1}q^{-1}p^{-1}q^{-1}xpq^{-1}qpqpx\right)^2$$
$$= (pqpqxqpqpx)^2 = 1 .$$

By Theorem 3 (17), (18), (33) follows.

(34) p (33) $p^{-1} \Rightarrow$ (34).

(35) Lemma 2 (6) yields $\left(tb^{-1}wa^{-1}vabcuzc^{-1}ec\right)^2 = (xq)^4 = 1$, by Theorem 3 (17).

(36) By Lemma 2 (3) we have $\left(dwbtsuzb^{-1}d^{-1}\right)^2 = q^2(px)^4q^{-2} = 1$ by Theorem 3 (17).

(37) By Lemma 2 (8) we have $\left(zc^{-1}ecvat\right)^2 = p^2(xq)^4p^{-2} = 1$ by Theorem 3 (17).

(38) By Lemma 2 (8) we have

$$\left(zc^{-1}ecvate^{-1}a^{-1}\right)^2 = \left(p^2xqxpqp\right)^2$$
$$= p^{-1}q^{-1}(xp)^4qp = 1 .$$

This follows by Theorem 3 (18), (17).

(39) By Lemma 2 (9) we have $(dbruz)^2 = p^{-1}(xq)^4p = 1$.

(40) By Lemma 2 (4) we have $\left(zc^{-1}ecvatrvae\right)^2 = p^2q^2(xp)^4q^2p^2 = 1$.

(41) By Lemma 2 (4) we have

$$\left(zc^{-1}ecvatrvaet\right)^2 = p^2\left(q^2xpxp\right)^4p^{-2} = 1 .$$

This last equation follows by Theorem 3 (19).

(42) By Lemma 2 (5) we have

$$\left(dbruzswbe^{-1}a^{-1}\right)^2 = p^{-1}q^{-1}p^{-1}\left(pq^{-1}xpxq\right)^2pqp$$
$$= p^{-1}q^{-1}p^{-1}(xq)^4pqp$$
$$= 1 .$$

We use Theorem 3 (18), (17) for this computation.

(43) By Lemma 2 (5) we have

$$\left(dbruzswbb^{-1}db\right)^2 = p^{-1}q^2(xp)^4q^{-2}p$$
$$= 1$$

which is implied using Theorem 3 (17).

(44) By Lemma 2 (4) we have

$$\left(zc^{-1}ecvatrvad^{-1}abc\right)^2 = p^2q^{-1}p^{-1}\left(pq^{-1}xpxq\right)^2pqp^2$$
$$= p^2q^{-1}p^{-1}(xq)^4pqp^2$$
$$= 1 \, .$$

We have used Theorem 3 (17), (18) for this computation.

(45) By Lemma 2 (6) we have

$$\left(tb^{-1}wa^{-1}vabcuxc^{-1}ecc^{-1}e^{-1}\right)^2 = pq^{-1}\left(qp^{-1}xqxp\right)^2qp^{-1}$$
$$= pq^{-1}(xp)^4qp^{-1}$$
$$= 1$$

using Theorem 3 (17), (18).

(46) By Lemma 2 (5) we have

$$c^{-1}e^{-1}cz\ dbruzswb = q^2xp^2xq^2p^2 \, .$$

The 39th formula of this lemma yields

$$c^{-1}e^{-1}curb^{-1}d^{-1}swb = q^2xp^2xq^2p^2 \, .$$

We apply p^2 to this formula obtaining

$$e^{-1}wtd^{-1}abcrstuc^{-1}b^{-1}a^{-1} = p^2q^2xp^2xq^2 \, .$$

Multiplying both formulae yields

$$c^{-1}e^{-1}curb^{-1}d^{-1}swbe^{-1}wtd^{-1}abcrstuc^{-1}b^{-1}a^{-1} = 1 \, .$$

Use Lemma 5 (4), (2) together with the formula $ecae^{-1}a^{-1}c^{-1} = 1$ which is contained in Chapter 1, Lemma 16, to obtain the formula (46).

(47) We apply p^{-1} to (46) obtaining

$$rste^{-1}a^{-1}rda^{-1}sc^{-1}e^{-1}tc^{-1}b^{-1}a^{-1}dab = 1 \, ,$$

and by Chapter 1, Lemma 16,

$$c^{-1}b^{-1}a^{-1}dabece^{-1}d^{-1} = 1 \, .$$

(48) By Theorem 3 (34) we have

$$1 = \left(qpq^{-1}p^{-1}xqxpxq^{-1}x\right)^4 = q^{-1}\left(q^2pq^2p^{-1}\cdot pxpx\cdot pqxqx\right)^4q$$

which yields $\left(sa^{-1}dsp\right)^4 = 1$. Hence

$$sa^{-1}dspsa^{-1}dsp^{-1}p^2sa^{-1}dsp^{-2}p^{-1}sa^{-1}dsp = 1 \, .$$

We use the formulae for p to conclude (48).

(49) By definition we have

$$dbruc^{-1}ecdbc^{-1}ecabcurabc = q^{-1}p^{-1}\ qp x p x q p x p x\ qpqp^{-1}q^{-1}\ pq^{-1}x\ p x q^{-1}p\ x p x p\ qp^{-1}q^{-1}p$$

$$= q^{-1}\ p^{-1}x p^{-1}x\ p^{-1}q^{-1}x p^{-1}\ x q p^{-1}\ q^{-1}x q x q^{-1}\ x q^{-1}x q^{-1}p$$

$$= q^{-1}x p x q^{-1}x q^{-1}x q^{-1}x q^2 x p$$

$$= \left(q^{-1}x p\right)^2 = 1\ .$$

In this computation Theorem 3 (17), (18), (22) are used.

(50) By definition we have

$$aeb^{-1}a^{-1}vabcub^{-1}d^{-1}becabcuc^{-1}b^{-1}a^{-1}da^{-1}v$$

$$= p^{-1}q^{-1}p^{-1}q^2p^{-1}x p^2 x q^2 pqpqpq^{-1}p^{-1}q^{-1}p^{-1}q^2 x p x p q x q x q^{-1}$$

$$= qpqpq^{-1}p^{-1}x p^2 x qp^{-1}q^2 p^{-1}q^{-1}p^{-1}q^{-1}p^{-1}x q x p q x q x q^{-1}$$

$$= qpqx qp^{-1}x p^{-1}x p^2 x qp^{-1}q^{-1}pq x q x p q x q x q^{-1}$$

$$= q(pqxqx)^4 q^{-1} = 1\ .$$

In this computation Theorem 3 (17), (18), (22) are used.

(51)

$$deb^{-1}wbts = q^2p^2q^{-1}x p x p q^{-1}x p^2 x$$

$$= q^2p^2q^{-1}x p q p x$$

$$= \left(q^2 p x\right)^2\ .$$

Hence (51) follows by Theorem 3 (23).

Lemma 6 contains some immediate consequences of Lemma 5.

LEMMA 6.

(1) $taet = b^{-1}a^{-1}vaeb^{-1}va$;

(2) $secs = a^{-1}vad^{-1}c^{-1}e^{-1}d^{-1}v$;

(3) $rar = b^{-1}sc^{-1}tabcts$;

(4) $rbr = abcbtc^{-1}b^{-1}a^{-1}t$;

(5) $rcr = stc^{-1}ts$;

(6) $td^{-1}abct = wa^{-1}b^{-1}d^{-1}c^{-1}b^{-1}wb$;

(7) $zaez = c^{-1}ecbe^{-1}a^{-1}vaba^{-1}vc^{-1}e^{-1}c$;

(8) $sDs = vDv$;

(9) $svawbs = a^{-1}b^{-1}wa^{-1}va$;

(10) $\quad ra^{-1}vabcur = abcuc^{-1}b^{-1}a^{-1}vac^{-1}b^{-1}a^{-1}$;

(11) $\quad \left(wbcuc^{-1}b^{-1}\right)^2 = 1$;

(12) $\quad \left(vawa^{-1}\right)^4 = 1$;

(13) $\quad \left(vabcuc^{-1}b^{-1}a^{-1}\right)^4 = 1$;

(14) $\quad (st)^4 = 1$;

(15) $\quad (rs)^4 = 1$.

Proof. (1) By Lemma 5 (7), (26)

$$tb^{-1}a^{-1}vaeta^{-1}vaeb^{-1} = 1 .$$

Lemma 5 (24) yields (1).

(2) We apply α on Lemma 5 (21) obtaining (2).

(3) is Lemma 5 (23).

(4) is Lemma 5 (18) together with Lemma 5 (5).

(5) is Lemma 5 (30).

(6) We apply p to (2) obtaining

$$td^{-1}abct = wec^{-1}b^{-1}a^{-1}deb^{-1}wb ,$$

and by Chapter 1 we have $c^{-1}e^{-1}d^{-1}abce^{-1}a^{-1}b^{-1}d^{-1} = 1$.

(7) By Lemma 5 (37), (38)

$$ze^{-1}a^{-1}z = c^{-1}ecvataeta^{-1}vc^{-1}e^{-1}c .$$

Then (1) implies (7).

(8) By (2) we have

$$sa^{-1}ecs = vad^{-1}c^{-1}e^{-1}d^{-1}v \Rightarrow s\left(a^{-1}ec\right)^2 s = v\left(ad^{-1}c^{-1}e^{-1}d^{-1}\right)^2 v .$$

By Chapter 1 we then have

$$\left(a^{-1}ec\right)^2 = D ,$$

$$\left(ad^{-1}c^{-1}e^{-1}d^{-1}\right)^2 = D .$$

(9) We apply p^{-1} to Lemma 5 (28) obtaining

$$(asvawb)^2 = 1$$

which proves (9).

(10) We apply p^2 to Lemma 5 (28) obtaining

$$\left(rabcuc^{-1}b^{-1}a^{-1}va\right)^2 = 1 \ .$$

(11) is Lemma 5 (15).

(14) By Lemma 5 (9), (10), (11)

$$trst = sr \Rightarrow t(rs)^2t = (sr)^2 \ .$$

On the other hand we have

$$tsrst = srs \Rightarrow t(rs)^2t = (rs)^2 \Rightarrow (rs)^4 = 1 \ .$$

We apply p . Then (14) follows.

(15) is also proved by this procedure.

(12) q (14) $q^{-1} \Rightarrow$ (12).

(13) p (12) $p^{-1} \Rightarrow$ (13).

Next we show the formulae which are necessary to prove Theorems 5 and 6.

LEMMA 7.

(1) $rar = b^{-1}sc^{-1}tabcts$;

(2) $rbr = abcbtc^{-1}b^{-1}a^{-1}t$;

(3) $rcr = abcasbs$;

(4) $rdr = stc^{-1}b^{-1}a^{-1}tcsbc^{-1}b^{-1}a^{-1}se^{-1}btsd^{-1}ab^{-1}$
$$sa^{-1}c^{-1}b^{-1}a^{-1}stece^{-1}d^{-1}tecsad^{-1}b^{-1}sc^{-1}$$
$$tabctstabctb^{-1}c^{-1}b^{-1}a^{-1}stc^{-1}b^{-1}a^{-1}tcsbtstabctb^{-1}c^{-1}b^{-1}a^{-1}\ ;$$

(5) $rer = stc^{-1}b^{-1}a^{-1}tcsbda^{-1}sc^{-1}e^{-1}tdec^{-1}e^{-1}ts$;

(6) $rtr = t$;

(7) $rsr = tst$;

(8) r normalizes $\langle a, b, c, d, e, s, t \rangle$;

(9) u normalizes $\langle a, b, c, d, e, v, w \rangle$.

Proof. (1) is Lemma 6 (3).

(2) is Lemma 6 (4).

(3) follows by Lemma 5 (20), (5), (6).

(5) By Lemma 5 (47)

$$ts\ re^{-1}a^{-1}r\ da^{-1}sc^{-1}e^{-1}tdec^{-1}e^{-1} = 1 \ .$$

Using only the first formula of this lemma we obtain (5).

(6) is Lemma 5 (10).

(7) By Lemma 5 (9)

$$1 = rstrst = rsrtst$$

because of Lemma 5 (10).

(4) By Lemma 5 (48)

$$rc^{-1}b^{-1}dbtsabecr = se^{-1}btsd^{-1}as \ .$$

Using (1), (2), (3), (5), (6), (7) we obtain (4).

(9) Obviously the transformation formula in Lemma 3 yields

$$c^{-1}b^{-1}a^{-1}qrq^{-1}abc = u \ ,$$

$$c^{-1}b^{-1}a^{-1}q\langle a, b, c, d, e, s, t\rangle q^{-1}abc = \langle a, b, c, d, e, v, w\rangle$$

such that by (8) we have also proved (9).

Lemma 7' follows in the same way as Lemma 7.

LEMMA 7'.

(1) t *normalizes* $\langle a, b, c, d, e, r, s\rangle$;

(2) w *normalizes* $\langle a, b, c, d, e, u, v\rangle$.

Proof. (1) It is obvious that we must prove

$$tat, \ tbt, \ tct, \ tdt, \ tet, \ trt, \ tst \ \in \langle a, b, c, d, e, r, s\rangle$$

to show (1):

$tbt \in \langle a, b, c, d, e, r, s\rangle$ is clear by Lemma 5 (7);

$tct \in \langle a, b, c, d, e, r, s\rangle$ follows by Lemma 5 (21);

$tat \in \langle a, b, c, d, e, r, s\rangle$ follows by Lemma 5 (22);

$trt \in \langle a, b, c, d, e, r, s\rangle$ follows by Lemma 5 (10);

$tst \in \langle a, b, c, d, e, r, s\rangle$ is Lemma 7 (7).

By Lemma 5 (46)

$$td^{-1}abct \in \langle a, b, c, d, e, r, s\rangle \Rightarrow tdt \in \langle a, b, c, d, e, r, s\rangle \ .$$

By Lemma 5 (48)

$$tb^{-1}esrc^{-1}b^{-1}dbt \in \langle a, b, c, d, e, r, s\rangle \Rightarrow tet \in \langle a, b, c, d, e, r, s\rangle$$

So we have shown (1).

(2) By Lemma 3 we have

$$a^{-1}qtq^{-1}a = w ,$$

$$a^{-1}q\langle a, b, c, d, e, r, s\rangle q^{-1}a = \langle a, b, c, d, e, u, v\rangle .$$

We recall the definitions

$$\Gamma_3 = \langle a, b, c, d, e, r, s, t, u, v, w\rangle ,$$

$$\Gamma_4 = \langle a, b, c, d, e, r, t, u, v, w\rangle ,$$

$$\Gamma_5 = \langle a, b, c, d, e, r, t, v, w\rangle .$$

LEMMA 8.

(1) $zaz = rc^{-1}ecvate^{-1}a^{-1}vrtb^{-1}a^{-1}vaeb^{-1}c^{-1}e^{-1}c$
$$urb^{-1}d^{-1}ruc^{-1}ecdbrvaete^{-1}a^{-1}vrtq^{-1}vc^{-1}e^{-1}c ;$$

(2) $zbz = urb^{-1}d^{-1}b^{-1}d^{-1}swdbaeb^{-1}wsdbruc^{-1}ecbe^{-1}a^{-1}vaba^{-1}vc^{-1}e^{-1}c ;$

(3) $zcz = urb^{-1}d^{-1}ruc^{-1}ecdbuc^{-1}b^{-1}a^{-1}vawbtc^{-1}e^{-1}tb^{-1}wa^{-1}vabcu ;$

(4) $zdz = urb^{-1}d^{-1}ruc^{-1}ecvab^{-1}a^{-1}vaeb^{-1}c^{-1}e^{-1}cur$
$$b^{-1}d^{-1}swbe^{-1}a^{-1}b^{-1}d^{-1}wsdbdbru ;$$

(5) $zez = c^{-1}ecvtrvaete^{-1}a^{-1}vrb^{-1}d^{-1}c^{-1}e^{-1}curbd$
$$ruc^{-1}ecbe^{-1}a^{-1}vabtrvaeta^{-1}vc^{-1}e^{-1}crc^{-1}ecbe^{-1}a^{-1}vaba^{-1}vc^{-1}e^{-1}c ;$$

(6) $zrz = r ;$

(7) $zsz = urb^{-1}d^{-1}b^{-1}d^{-1}wedbrstb^{-1}wd^{-1}sdw$
$$brstb^{-1}d^{-1}wsrstc^{-1}b^{-1}wdbc^{-1}wbtsrdbru ;$$

(8) $ztz = c^{-1}ecvatrvaete^{-1}a^{-1}vrta^{-1}vc^{-1}e^{-1}c ;$

(9) $zuz = rur ;$

(10) $zvz = rc^{-1}ecvate^{-1}a^{-1}vrtb^{-1}a^{-1}vaeb^{-1}c^{-1}e^{-1}c ;$

(11) $zwz = urb^{-1}d^{-1}rstb^{-1}wd^{-1}sdwbrstb^{-1}d^{-1}wsrstc^{-1}b^{-1}wdbc^{-1}wbtsrdbru ;$

(12) z *normalizes* $\Gamma_3 .$

Proof. The formulae (1)-(11) are implied by the formulae Lemma 5 (35)-(45).

The formulae (6) and (9) are implied by Lemma 5 (31), (32).

Then we have, by Lemma 5 (39),

(**)
$$zdbz = urb^{-1}d^{-1}ru$$

and by Lemma 5 (8),

(***) $zc^{-1}ecz = b^{-1}d^{-1}c^{-1}e^{-1}curdbru$.

By Lemma 5 (35), (45)

$$zc^{-1}c^{-1}e^{-1}cz = uc^{-1}b^{-1}a^{-1}vawbtectb^{-1}wa^{-1}vabcu$$.

The third formula of this lemma is obtained by (***).

By Lemma 5 (43), using (**) and Lemma 5 (39),

(****) $zswz = urb^{-1}d^{-1}b^{-1}d^{-1}wsdbdbru$.

By Lemma 5 (42)

$$zswzzbzze^{-1}a^{-1}zzdbruz = urb^{-1}d^{-1}aeb^{-1}ws$$.

Then (****) and Lemma 5 (39), Lemma 6 (7) imply (2).

We use (2) and (**) obtaining immediately (4).

Use Lemma 5 (40) to obtain

$$zc^{-1}ecvatzzrzzvzzaez = e^{-1}a^{-1}vrta^{-1}vc^{-1}e^{-1}c$$.

Then Lemma 5 (31), (37) and Lemma 6 (7) yield (10).

Obviously (8) follows by Lemma 5 (40), (41).

By Lemma 5 (37)

$$zc^{-1}eczzvzzazztz = ta^{-1}vc^{-1}e^{-1}c$$.

Use (***), (10), (8) to conclude (1).

We apply p to (10) obtaining (11). Then (****) and (11) yield (7).

Hence we have shown all formulae (1)-(11).

We obtain (12) when we keep in mind that Γ_3 is generated by the elements $a, b,$ $c, d, e, r, s, t, u, v, w$.

(5) follows by Lemma 6 (7) and Lemma 8 (1).

LEMMA 9.

(1) $sas \in \Gamma_4$;

(2) $sbs \in \Gamma_4$;

(3) $scs \in \Gamma_4$;

(4) $sds \in \Gamma_4$;

(5) $ses \in \Gamma_4$;

(6) $srs \in \Gamma_4$;

(7) $sts \in \Gamma_4$;

(8) $sus \in \Gamma_4$;

(9) $svs \in \Gamma_4$;

(10) $sws \in \Gamma_4$;

(11) s *normalizes* Γ_4 .

Proof. (1) is Lemma 5 (6).

(2) By Lemma 5 (20),

$$sbs = a^{-1}rabc^2r \ .$$

(3) By Lemma 5 (21),

$$scs = rc^2 b^{-1} a^{-1} ra^2 tc^{-1} tb^{-1} a \ .$$

(8) By Lemma 5 (25),

$$sbcus \in \Gamma_4 \Rightarrow sus \in \Gamma_4$$

using (2), (3).

(4) By Lemma 5 (27),

$$sbcuc^{-1}b^{-1}a^{-1}da^{-1}s \in \Gamma_4 \Rightarrow sds \in \Gamma_4$$

using (1), (2), (3), (8).

(5) follows by Lemma 6 (2) and (3).

(6) By Lemma 6 (9), $svawbs \in \Gamma_4$; by Lemma 5 (34) we have

$$secvawbcrs \in \Gamma_4 \Rightarrow (6)$$

using (3), (5).

(7) By Lemma 5 (9) we have $srts = rt$. Then (7) is implied by (6).

(9) By Lemma 5 (33), $sdbuc^{-1}b^{-1}a^{-1}vabts \in \Gamma_4$. The formulae already proved yield $svs \in \Gamma_4$.

(10) By Lemma 5 (29), $stc^{-1}wbrs \in \Gamma_4$. (1)-(9) yield $sws \in \Gamma_4$.

(11) Γ_4 is generated by $a, b, c, d, e, r, t, u, v, w$.

LEMMA 10.

(1) $uau \in \Gamma_5$;

(2) $ubu \in \Gamma_5$;

(3) $ucu \in \Gamma_5$;

(4) $udu \in \Gamma_5$;

(5) $ueu \in \Gamma_5$;

(6) $uru \in \Gamma_5$;

(7) $utu \in \Gamma_5$;

(8) $uvu \in \Gamma_5$;

(9) $uwu \in \Gamma_5$;

(10) u *normalizes* Γ_5 .

Proof. (1) to (5) follow by Lemma 7 (9). (8) and (9) are also implied by Lemma 7 (9).

(7) By Lemma 5 (28),

$$utb^{-1}a^{-1}vabcu \in \Gamma_5 \Rightarrow utu \in \Gamma_5 .$$

(6) By Lemma 6 (10), $uc^{-1}b^{-1}a^{-1}ra^{-1}vabcu \in \Gamma_5$. The formulae already proved yield $uru \in \Gamma_5$.

(10) By definition Γ_5 is generated by $a, b, c, d, e, r, t, v, w$.

We need another subnormal subgroup of Γ_3 .

DEFINITION.

$$\overline{\Gamma}_4 := \langle a, b, c, d, e, s, t, u, v, w \rangle ,$$

$$\overline{\Gamma}_5 := \langle a, b, c, d, e, s, t, u, v \rangle .$$

Then we obtain

LEMMA 10'.

(1) r *normalizes* $\langle a, b, c, d, e, s, t, u, v, w \rangle = \overline{\Gamma}_4$

(2) w *normalizes* $\langle a, b, c, d, e, s, t, u, v \rangle = \overline{\Gamma}_5$

(3) $\overline{\Gamma}_5$ *is subnormal of finite index in* Γ .

Proof. (1) Lemma 7 shows that

$$rar, \; rbr, \; rcr, \; rdr, \; rer, \; rsr, \; rtr \; \epsilon \; \overline{\Gamma}_4 \; .$$

By Lemma 5 (29)

$$rstc^{-1}wbr \; \epsilon \; \overline{\Gamma}_4 \Rightarrow rwr \; \epsilon \; \overline{\Gamma}_4 \; .$$

By Lemma 5 (34)

$$rsecvawbcr \; \epsilon \; \overline{\Gamma}_4 \Rightarrow rvr \; \epsilon \; \overline{\Gamma}_4 \; .$$

By Lemma 6 (10)

$$ra^{-1}vabcur \; \epsilon \; \overline{\Gamma}_4 \Rightarrow rur \; \epsilon \; \overline{\Gamma}_4 \; .$$

Thus we have shown (1).

(2) By Lemma 7' we have

$$waw, \; wbw, \; wcw, \; wdw, \; wew, \; wuw, \; wvw \; \epsilon \; \overline{\Gamma}_5 \; .$$

By Lemma 6

$$wbsa^{-1}vaw \; \epsilon \; \overline{\Gamma}_5 \Rightarrow wsw \; \epsilon \; \overline{\Gamma}_5 \; .$$

By Lemma 5 (51)

$$wbtsdeb^{-1}w \; \epsilon \; \overline{\Gamma}_5 \Rightarrow wtw \; \epsilon \; \overline{\Gamma}_5 \; .$$

Thus we have shown (2).

(3) $\overline{\Gamma}_5$ has index 4 in $\overline{\Gamma}_3$. Moreover, $\overline{\Gamma}_5$ is subnormal in $\overline{\Gamma}_3$. Hence follows (3) by the theorems of Part A.

Now we shall construct finite normal subgroups in Γ . To do this we shall make frequent use of the following lemma.

LEMMA 11. *Let* $\theta \; \epsilon \; \langle a, \, b, \, c, \, d, \, e \rangle$ *and* $p^2, q^2, qp, r, \alpha \; \epsilon \; C_\Gamma(\theta)$. *Then* $\langle\langle\theta\rangle\rangle_\Gamma$ *is finite. In particular we shall prove step by step:*

(1) $a, \, b, \, c, \, d, \, e \; \epsilon \; C_\Gamma(\theta)$;

(2) $t, \, v \; \epsilon \; C_\Gamma(\theta)$;

(3) $waw, \, wbw, \, wcw, \, wdw, \, wew \; \epsilon \; C_\Gamma(\theta)$;

(4) $uau, \, ubu, \, ucu, \, udu, \, ueu \; \epsilon \; C_\Gamma(\theta)$;

(5) $wvu \; \epsilon \; C_\Gamma(\theta)$;

(6) $wvavw, wvbvw, wvcvw, wvdvw, wvevw \in C_\Gamma(\theta)$;

(7) $uvavu, uvbvu, uvcvu, uvdvu, uvevu \in C_\Gamma(\theta)$;

(8) $urvu, utvu, urtu \in C_\Gamma(\theta)$;

(9) $wrvw, wtvw, wrtw \in C_\Gamma(\theta)$;

(10) $wvwavvw, wvwbvvw, wvwcvvw, wvwdvvw, wvwevvw \in C_\Gamma(\theta)$;

(11) $(wv)^4 \in C_\Gamma(\theta)$;

(12) $(wr)^4 \in C_\Gamma(\theta)$;

(13) $(wt)^4 \in C_\Gamma(\theta)$.

Proof. (1) By definition we have

$$\langle a, b, c, d, e \rangle \leq \langle p^2, q^2, qp \rangle \leq C_\Gamma(\theta) \ .$$

(2) Lemma 3 yields

$$p^2 rp^{-2} = t \ ,$$

$$qprp^{-1}q^{-1} = v \ .$$

(3) By Lemma 5 (6), (20), (21) we have

$$sas = a^{-1} \ ,$$

$$sbs = a^{-1}rabc^2r \ ,$$

$$scs = rc^2b^{-1}a^{-1}ra^2tc^{-1}tb^{-1}a \ .$$

Hence we have obviously

$$sas, \ sbs, \ scs \in C_\Gamma(\theta) \ .$$

We apply α to these formulae obtaining

$$sds, \ ses \in C_\Gamma(\theta) \ .$$

But by assumption we have

$$a^{-1}qp \in C_\Gamma(\theta) \ .$$

On the other hand $a^{-1}qp$ normalizes the group $\langle a, b, c, d, e \rangle$ and we have

$$a^{-1}qpsp^{-1}q^{-1}a = w \ .$$

Hence we have shown (3).

(4) p^2 normalizes the group $\langle a, b, c, d, e \rangle$ and we have

$$p^2 w p^{-2} = u .$$

So we have obviously shown (4).

(5) We have

$$p^2 v p^{-2} = c u c^{-1} b^{-1} a^{-1} v a w b \in C_\Gamma(\theta) .$$

Then (5) follows by (3), (4).

(6), (7) follow obviously by (3), (4) and (5).

(8) By Lemma 6 (10) we have

$$r a^{-1} v a b c u r a b c a^{-1} v a b c u c^{-1} b^{-1} a^{-1} = 1 \in C_\Gamma(\theta) .$$

By (1) to (7)

$$u r v u \in C_\Gamma(\theta) \Rightarrow u r w \cdot w v u \in C_\Gamma(\theta) \Rightarrow u r w \in C_\Gamma(\theta) .$$

We apply p^2 obtaining

$$w t u \in C_\Gamma(\theta) \Rightarrow u v t u \in C_\Gamma(\theta) \Rightarrow u r t u \in C_\Gamma(\theta) .$$

(9) Applying p^2 to $u r t u \in C_\Gamma(\theta)$ yields $w r t w \in C_\Gamma(\theta)$. Moreover we have by (8),

$$u r v u = u v w w v r w w v u \in C_\Gamma(\theta) \Rightarrow w v r w \in C_\Gamma(\theta)$$

which suffices to prove (9).

(10) First we show the formulae

$$w u a w , w u b w , w u c w , w u d w , w u e w \in C_\Gamma(\theta) .$$

By Lemma 5 (2), (4), (12)

$$w u a e b c w , w u e b^{-1} w \in C_\Gamma(\theta) .$$

But by Lemma 5 (15),

$$\left(w b c u c^{-1} b^{-1} \right)^2 = 1 \in C_\Gamma(\theta) \Rightarrow w b c w w u c^{-1} b^{-1} w u u b c u c^{-1} b^{-1} \in C_\Gamma(\theta) \Rightarrow w u b c w \in C_\Gamma(\theta) .$$

By Lemma 6 (6)

$$w a^{-1} b^{-1} d^{-1} c^{-1} b^{-1} w = t d^{-1} a b c b t \Rightarrow w u a^{-1} b^{-1} d^{-1} c^{-1} b^{-1} u w = u w a^{-1} b^{-1} d^{-1} c^{-1} b^{-1} w u$$

$$= u t d^{-1} a b c b t u = u t v u u v d^{-1} a b c b v u u v t u \in C_\Gamma(\theta) \Rightarrow w u a^{-1} b^{-1} d^{-1} w \in C_\Gamma(\theta) .$$

By Lemma 5 (50) we have

$$ub^{-1}d^{-1}becabcu = c^{-1}b^{-1}a^{-1}vabe^{-1}a^{-1}vad^{-1}abc \ .$$

We use the same method as above obtaining by (1) to (6),

$$wub^{-1}d^{-1}becabcuw \in C_\Gamma(\theta) \Rightarrow wub^{-1}d^{-1}uw \in C_\Gamma(\theta)$$

where we use the formula which is implied by Chapter I, Lemma 16,

$$ecae^{-1}a^{-1}c^{-1} = 1 \ .$$

But the elements db, dba, bc, $aebc$, eb^{-1} generate $\langle a, b, c, d, e \rangle$. Now use

$$wwwawww = wwiawuauwuvw$$

to conclude from $wuaww \in C_\Gamma(\theta)$ that $wwwawww \in C_\Gamma(\theta)$. Then the formulae already proved yield that the right hand side is contained in $C_\Gamma(\theta)$.

(11) By Lemma 6 (12) we have

$$1 = \left(vawa^{-1}\right)^4 \in C_\Gamma(\theta) \Rightarrow vawa^{-1}vawa^{-1}vawa^{-1}vawa^{-1} \in C_\Gamma(\theta) \ .$$

Obviously, by using (1) to (10) follows

$$(vw)^4 \in C_\Gamma(\theta) \Rightarrow (wv)^4 \in C_\Gamma(\theta)$$

since $v \in C_\Gamma(\theta)$.

(12) First we shall show the formulae

$$uvuauvu, \ uvubuvu, \ uvucuvu, \ uvuduvu, \ uvueuvu \in C_\Gamma(\theta) \ .$$

For $uvuauvu$ proceed in the following way

$$uvuauvu = uvwwuaiawwvu \in C_\Gamma(\theta) \ .$$

Then Lemma 6 (13) yields obviously $(vu)^4 \in C_\Gamma(\theta)$. By Lemma 5 (24), (2) we have

$$tabcut = b^{-1}a^{-1}vu^{-1}uc^{-1}b^{-1}a^{-1}vab \Rightarrow (utabcut)^2 = \left(ub^{-1}a^{-1}vu^{-1}uc^{-1}b^{-1}a^{-1}vab\right)^2 \ .$$

Hence the proofs given hitherto yield that the right hand side of this formula is contained in $C_\Gamma(\theta)$. Then

$$utabcututabcut \in C_\Gamma(\theta) \Rightarrow uvabcututabcu \in C_\Gamma(\theta) \ .$$

Then by (7), $(ut)^4 \in C_\Gamma(\theta)$. We apply p^2 obtaining (12).

(13) If we apply α to (11), then we obtain

$$\left(aeb^{-1}wa^{-1}bt\right)^4 \in C_\Gamma(\theta) \Rightarrow (wt)^4 \in C_\Gamma(\theta) \ .$$

Now we shall prove that $\langle\langle\theta\rangle\rangle_\Gamma$ is finite. To do this we shall show first that $\langle\langle\theta\rangle\rangle_{\Gamma_5}$ is of finite order. We claim that the elements θ, $w\theta w$, $vw\theta wv$, $wvw\theta wvw$ are a conjugacy class in Γ_5. Obviously, all four elements are centralized by a, b, c, d, e. Moreover, we have by (1) to (13),

$$r\theta r = \theta ,$$
$$rw\theta wr = vw\theta wv ,$$
$$rvw\theta wvr = w\theta w ,$$
$$rwvw\theta wvwr = wvw\theta wvw ,$$

$$t\theta t = \theta ,$$
$$tw\theta wt = vw\theta wv ,$$
$$tvw\theta wvt = w\theta w ,$$
$$twvw\theta wvwt = wvw\theta wvw ,$$

$$v\theta v = \theta ,$$
$$vw\theta wv = vw\theta wv ,$$
$$vvw\theta wvv = w\theta w ,$$
$$vwvw\theta wvwv = wvw\theta wvw .$$

Obviously w leaves the above four elements invariant so that $\langle\langle\theta\rangle\rangle_{\Gamma_5}$ is generated by θ, $w\theta w$, $vw\theta wv$, $wvw\theta wvw$.

Because of $\theta \in \langle a, b, c, d, e\rangle$, θ is of finite order and the formulae (3), (6), (10) show that $\langle\langle\theta\rangle\rangle_{\Gamma_5}$ is abelian. Hence $\langle\langle\theta\rangle\rangle_{\Gamma_5}$ is finite.

Γ_5 is a subnormal subgroup of finite index in Γ. Then Theorem 6 of the introduction yields the proof of the lemma.

Next we study the automorphism of Γ_1 which is induced by the element $q^{-1}Y$. By this automorphism we shall obtain the elements θ for which we can use Lemma 11.

LEMMA 12.

(1) $q^{-1}Y(p) = p^{-1}x$;

(2) $q^{-1}Y(q) = q^{-1}x$;

(3) $q^{-1}Y(x) = p^{-1}xq$;

(4) $q^{-1}Y(pq^{-1}) = p^{-1}q$;

(5) $q^{-1}Y(q^{-1}p) = pq^{-1}$;

(6) $q^{-1}Y(p^2) = r$;

(7) $q^{-1}Y((abc)^2) = abc^{-1}b^{-1}a^{-1}dec$;

(8) $q^{-1}Y(u_1 de^{-1}) = (abc)^2$;

(9) $q^{-1}Y\!\left(u_1v_1v_6\right) = u_1v_1v_6$;

(10) $q^{-1}Y\!\left(u_3v_3\right) = u_1u_2u_3v_1v_2v_3$;

(11) $q^{-1}Y\!\left(u_2v_2u_4v_4v_6\right) = u_2v_2u_4v_4v_6$;

(12) $q^{-1}Y\!\left(u_4v_4\right) = u_4v_4$.

Proof. (1) to (6) follow obviously by the transformation formulae in Lemma 1 and by Theorem 3 (18).

(7) By definition we have $(abc)^2 = \left(qp^{-1}q^{-1}p\right)^2$; hence

$$q^{-1}Y\!\left((abc)^2\right) = q^{-1}p^2q^2p^2q^{-1} = abc^{-1}b^{-1}a^{-1}dec$$

using the definitions in Chapter I.

(8) By Chapter 1, Lemma 16, we have

$$u_1de^{-1} = c^{-1}b^{-1}a^{-1}\!\cdot\!abecd = p^{-1}qpq^{-1}\!\cdot\!pq^{-1}\!\cdot\!pq^{-1}p^{-1}q\!\cdot\!qp^{-1}\ .$$

Hence

$$q^{-1}Y\!\left[u_1de^{-1}\right] = qp^2qp^{-1}qp^{-1}q^2p^{-1}q^{-1}p$$

$$= (abc)^2\ .$$

(9) By Chapter I, Lemma 16, we have

$$u_1v_1v_6 = q^{-1}pecp^{-1}q\!\cdot\!ed^{-1}u_1\!\cdot\!c^{-1}b^{-1}a^{-1}$$

$$= q^{-1}ppq^{-1}p^{-1}qp^{-1}qpq^{-1}q^{-1}pqp^{-1}qp^{-1}\ .$$

Hence

$$q^{-1}Y\!\left(u_1v_1v_6\right) = pq^{-1}p^{-1}q^2p^{-1}qp^{-1}qp^{-1}q^{-1}pqp^{-1}q^{-1}pqp^2q$$

$$= (ec)^2c^2b^{-1}a^{-1}dec^{-1}b^{-1}a^{-1}dec$$

$$= u_1v_1v_6$$

by Chapter I, Lemma 16.

(10) By Chapter I

$$u_3v_3 = q^{-1}p\!\cdot\!u_1v_1v_6p^{-1}q\ ;$$

hence

$$q^{-1} Y(u_3 v_3) = pq^{-1} u_1 v_1 v_6 qp^{-1}$$
$$= u_1 u_2 u_3 v_1 v_2 v_3 .$$

(11) By Chapter I

$$u_2 v_2 u_4 v_4 v_6 = pq^{-1} u_3 v_3 qp^{-1} ;$$

hence

$$q^{-1} Y(u_2 v_2 u_4 v_4 v_6) = p^{-1} q u_1 u_2 u_3 v_1 v_2 v_3 q^{-1} p$$
$$= u_2 v_2 v_6 u_4 v_4 .$$

(12) By Chapter I

$$u_4 v_4 = u_1 v_1 v_6 \cdot u_3 v_3 \cdot u_2 v_2 u_4 v_4 v_6 \cdot pq^{-1} \cdot u_1 v_1 v_6 qp^{-1} .$$

Hence we have

$$q^{-1} Y(u_4 v_4) = u_1 v_1 v_6 u_1 u_2 u_3 v_1 v_2 v_3 u_2 v_2 u_4 v_4 v_6 \cdot p^{-1} q u_1 v_1 v_6 q^{-1} p$$
$$= u_4 v_4 ,$$

by the transformation formulae for the p, q on the u_i, v_i .

The following lemmas will serve to show that

$$N = \left\langle \left\langle u_1, a^2, D, de \right\rangle \right\rangle_\Gamma$$

is a finite group. In our computations we shall be a bit sloppy using the same notations for elements contained in Γ and elements contained in factor groups of Γ , since all the Lemmas 1 to 12 apply for any factor group of Γ . In the same way, all formulae stated in Chapter I apply for any factor group of Γ . First we define

DEFINITION.

$$N_1 := \langle\langle u_1 u_2 v_1 v_2, u_4, v_6, u_1 u_3 v_1 v_3 \rangle\rangle_\Gamma .$$

Now Lemma 12 and Lemma 11 imply

LEMMA 13. N_1 *is a finite group.*

Proof. Let us start with the normal subgroup $M_1 = \langle\langle u_4 v_4 \rangle\rangle_\Gamma$. By Chapter I

$$qp, q^2, p^2, a \in C_\Gamma(u_4 v_4) .$$

Then using Lemma 12 (6), (12) we have also

$$r \in C_\Gamma(u_4 v_4) .$$

Hence Lemma 11 implies that M_1 is finite. For the factor group Γ/M_1 we have by Chapter I,

$$q^2,\ p^2,\ qp,\ \alpha \in C_{\Gamma/M_1}\left(u_1u_2v_1v_2\right).$$

Then by Lemma 12 (6), (9), (11),

$$r \in C_{\Gamma/M_1}\left(u_1u_2v_1v_2\right).$$

Then Lemma 11 yields that

$$M_2 = \langle\langle u_1u_2v_1v_2,\ u_4v_4\rangle\rangle_\Gamma$$

is finite. By Chapter I we have

$$pu_1u_2v_1v_2p^{-1}u_1u_2v_1v_2 = u_4v_6.$$

Hence $u_4v_6 \in M_2$. Then, using the transformation formulae of Chapter I, it is easy to see that

$$qp,\ q^2,\ p^2,\ \alpha \in C_{\Gamma/M_2}\left(u_1u_3v_1v_3\right)$$

and

$$qp,\ q^2,\ p^2,\ \alpha \in C_{\Gamma/M_2}\left(u_1u_3v_1v_3v_6\right).$$

Using Lemma 12 (6), (9), (10) the first formulae yield that

$$r \in C_{\Gamma/M_2}\left(u_1u_3v_1v_3v_6\right).$$

Then Lemma 11 shows that

$$M_3 = \langle\langle u_4v_4,\ u_1u_2v_1v_2,\ u_1u_3v_1v_3v_6\rangle\rangle_\Gamma$$

is finite. Since we have also, by Chapter I,

$$pu_1u_3v_1v_3v_6p^{-1}u_1u_2v_1v_2u_1u_3v_1v_3v_6u_4v_4 = v_6 ;$$

hence the proof of Lemma 13 is complete.

Now we use another method to compute the normal subgroup generated by D.

DEFINITION.

$$N_2 := \langle\langle D,\ u_1u_2v_1v_2,\ u_4,\ v_6,\ u_1u_3v_1v_3\rangle\rangle_\Gamma.$$

Then we obtain by Lemma 13

LEMMA 14. N_2 *is a finite group.*

Proof. Obviously we need only show that $\langle\langle D\rangle\rangle_{\Gamma/N_1} \leq \Gamma/N_1$ is finite. Use Chapter I in Γ/N_1 to obtain $a, b, c, d, e \pitchfork D$. By Lemma 6 (8) we have

$$(*) \qquad\qquad sDs = vDv .$$

By Chapter I, Lemma 16 we have $\alpha(D) = D$ in Γ/N_1. We apply α to the formula (*) obtaining

$$(**) \qquad\qquad sDs = tDt .$$

Applying to this formula $p^{-1}q$, we obtain

$$(***) \qquad\qquad uDu = c^{-1}b^{-1}a^{-1}vDvabc .$$

Now we claim that

$$(****) \qquad\qquad a, b, c, d, e \pitchfork sDs .$$

The formula $a \pitchfork sDs$ follows by Lemma 5 (6), $b \pitchfork sDs$ follows by (**) and Lemma 5 (7).

Apply α to the formula $b \pitchfork sDs$ which yields $d \pitchfork sDs$. Now (***) becomes

$$cuDuc^{-1} = sDs .$$

Then by Lemma 5 (2),

$$caeb \pitchfork sDs \rightarrow cae \pitchfork sDs \rightarrow eca \pitchfork sDs$$

by the formula contained in Chapter I, Lemma 16:

$$ecae^{-1}a^{-1}c^{-1} = 1 \Rightarrow ec \pitchfork sDs .$$

Apply α to this to obtain

$$abc \pitchfork sDs \rightarrow c \pitchfork sDs \rightarrow e \pitchfork sDs .$$

Hence we have proved (****).

Now (***) reduces to

$$(***) \qquad\qquad sDs = uDu .$$

Of course, these computations apply only in Γ/N_1 !

The formulae (*)–(****) yield clearly that $\{D, sDs\}$ is a conjugacy class in $\overline{\Gamma}_5$. Since $D \in \langle a, b, c, d\rangle$, (****) yields that the group generated by D and sDs is abelian. Hence the normal subgroup generated by D in $\overline{\Gamma}_5$ is finite. Then Lemmas 9' and 10' (3) and the corollary contained in Part A show the lemma.

On our way to the normal subgroup N, next we come to study the normal subgroup

$$N_3 := \langle\langle u_3 v_3, D, u_4, v_6, u_1 u_2, u_1 u_3 v_1 v_3\rangle\rangle_\Gamma$$

where we can use the same method as in Lemma 13 for N_1 .

LEMMA 15. N_3 *is a finite group.*

Proof. Clearly it suffices to show that $\langle\langle u_3 v_3 \rangle\rangle_{\Gamma/N_2} \le \Gamma/N_2$ is finite. In fact, by Chapter I, we have

$$u_1 u_2 = Dp Dp^{-1} v_6 \in N_2 .$$

Now we have in Γ/N_2 ,

$$qp,\ p^2,\ q^2,\ \alpha \in C_{\Gamma/N_2}(u_3 v_3) ,$$

such that Lemma 12 (10), (6) force $r \in C_{\Gamma/N_2}(u_3 v_3)$.

Using Lemma 11, $\langle\langle u_3 v_3 \rangle\rangle_{\Gamma/N_2}$ is then a finite group. This completes the proof of Lemma 15.

For our next step the method used in Lemma 14 is again of importance. Let

$$N_4 := \langle\langle u_1,\ N_3 \rangle\rangle_\Gamma .$$

LEMMA 16.

(1) $(abcsbs)^2 \in N_3$;

(2) $(secsd^{-1}a)^2 \in N_3$;

(3) $(vavdec)^2\ N_3$;

(4) N_4 *is a finite group.*

Proof. (1) We consider the automorphism induced by $xY^2 pZ$ on $\langle p, q, x \rangle$, for which we have, by Lemma 1,

$$xY^2 pZ : p \longmapsto xpx$$
$$q \longmapsto q$$
$$x \longmapsto xpq^{-1} .$$

By definition $abcb^{-1} = qp^{-1}q^{-1}p^{-1}qp^{-1}q^{-1}p^{-1}$. We apply again the automorphism using Theorem 3 to compute

$$xY^2pZ\left(abcb^{-1}\right) = qxp^{-1}xq^{-1}xp^{-1}xqxp^{-1}xq^{-1}xp^{-1}x$$

$$= qxp^{-1}xp^{-1}xqp^2xqxp^{-1}xq^{-1}xq^{-1}xpq^{-1}$$

$$= q\cdot pxpqp^2xqxp^{-1}qxqpq^{-1}$$

$$= qp^{-1}q^{-1}x\cdot xqp^2xpqp^2xqxp^{-1}q\cdot xqpq^{-1}$$

$$= qp^{-1}q^{-1}xpq^{-1}\cdot xp^{-1}xpqp^{-1}q^{-1}xpxp^{-1}qxqpq^{-1}$$

$$= qp^{-1}q^{-1}xpq^{-1}\cdot xp^{-1}xp^{-1}\cdot p^2qp^{-1}q^{-1}p^{-1}pxpxp^{-1}qpq^{-1}\cdot qp^{-1}xqpq^{-1} \ .$$

Hence we have

$$qp^{-1}xqpq^{-1}xY^2pZ\left(abcb^{-1}\right) = sb^{-1}sc^{-1}b^{-1}a^{-1} \ .$$

The proof of (1) is complete.

(2) Apply α to (1) to obtain (2). We must keep in mind that α is induced by an inner automorphism xY^2 .

(3) Insert in (2) the formula stated in Lemma 6 (2) to obtain (3).

(4) First take into consideration that

$$pu_3v_3p^{-1} = u_3v_2v_4v_6 \in N_3 \Rightarrow v_2v_3 \in N_3 \ !$$

Then we have in Γ/N_3 by Chapter I, Lemma 16,

(*) $$aa^{-1}eca^{-1}c^{-1}e^{-1}a = u_1 \ .$$

By Lemma 6 (2)

$$sa^{-1}ecs = vad^{-1}c^{-1}e^{-1}d^{-1}v \ .$$

Now apply s to (*), then

$$su_1s = sas\cdot sa^{-1}ecssa^{-1}ssc^{-1}e^{-1}as$$

$$= a^{-1}vad^{-1}c^{-1}e^{-1}d^{-1}vavdecda^{-1}v$$

$$= a^{-1}vad^{-1}c^{-1}e^{-1}d^{-1}c^{-1}e^{-1}d^{-1}va^{-1}vda^{-1}v \ .$$

This holds in Γ/N_3 by (3). Moreover, by Lemma 5 (3),

$$su_1s = a^{-1}vad^{-1}c^{-1}e^{-1}d^{-1}c^{-1}e^{-1}d^{-1}vd^{-1}$$

in Γ/N_3 . Then follows in Γ/N_3 ,

$$su_1s = vad^{-1}c^{-1}e^{-1}d^{-1}c^{-1}e^{-1}d^{-1}vda^{-1}$$

$$= vad^{-1}c^{-1}e^{-1}d^{-1}c^{-1}e^{-1}d^{-1}ad^{-1}v \ .$$

The formulae contained in Chapter I, Lemma 16, imply

(1) $su_1s = vu_1v$ in Γ/N_3 .

Apply α to this formula obtaining

(2) $su_1s = tu_1t$.

p (2) p^{-1} implies

(3) $su_1s = ru_1r$.

Now we claim

(4) $a, b, c, d, e \updownarrow su_1s$.

$a \updownarrow su_1s$ follows immediately by Lemma 5 (6); $b \updownarrow su_1s$ follows by (2) and Lemma 5 (7); $c \updownarrow su_1s$ follows by (3) and Lemma 5 (5).

Apply α to the second and third formula, then $d \updownarrow su_1s$ and $e \updownarrow su_1s$.

Evidently all these computations apply only in Γ/N_3 !

We apply q to (2) keeping in mind (1) and (4). Then we obtain

(5) $su_1s = wu_1w$.

p^2 (5) p^{-2} implies

(6) $su_1s = uu_1u$.

Obviously the formulae (1) to (6) imply that

$$\{u_1, su_1s\}$$

is a conjugacy class in Γ_3 . Since $u_1 \in \langle a, b, c, d, e \rangle$, $\langle\langle u_1 \rangle\rangle_{\Gamma_3} . N_3/N_3$ is an abelian group of order 4 at most.

Since Γ_3 is subnormal of finite index in Γ , our Lemma 16 follows.

Now we have gathered enough formulae together to take one of the last steps showing that

$$N := \left\langle\!\!\left\langle a^2, de, D, u_1 \right\rangle\!\!\right\rangle_\Gamma = \left\langle\!\!\left\langle a^2, de, N_4 \right\rangle\!\!\right\rangle_\Gamma$$

is finite.

LEMMA 17. N *is a finite group.*

Proof. Again it will be sufficient to show that $\langle\langle a^2, de \rangle\rangle_{\Gamma/N_4}$ is finite. But now it suffices to show that $\langle\langle a^2 \rangle\rangle_{\Gamma/N_4}$ is finite, since, by Lemma 12 (8),

$$u_1 de^{-1} = Y^{-1}q(abc)^2 q^{-1} Y \; ;$$

that is $de \in \left\langle\!\!\left\langle a^2, N_4 \right\rangle\!\!\right\rangle_\Gamma$.

By Lemma 5 (6) we have, obviously, $sa^2 s = a^2$. But in Γ/N_4 , a^2 is centralized by p and q (by Chapter I, Lemma 16!). Since r, t, u, v, w are all conjugates of s under p, q , we have

$$a^2 \in center \big(\langle a, b, c, d, e, r, s, t, u, v, w \rangle \cdot N_4/N_4 \big)$$
$$= center \big(\Gamma_3 \cdot N_4/N_4 \big) \; .$$

Hence $\langle\langle a^2 \rangle\rangle_{\Gamma_3 N_4/N_4}$ is finite. Since Γ_3 is subnormal of finite index in Γ , our lemma follows.

In order to show the finiteness of Γ , we must collect some relations in Γ/N , where we shall also use the same notations for the elements in Γ and their images in Γ/N . For these computations remember that $\langle a, b, c, d, e \rangle$ in Γ/N is elementary abelian of exponent 2 and that it is generated by a, b, c, d .

LEMMA 18. *In* Γ/N *we have the relations*

(1) $(vb)^2 = 1 \; ;$

(2) $(tad)^2 = 1 \; ;$

(3) $(asr)^2 = 1 \; ;$

(4) $(vr)^2 = 1 \; ;$

(5) $(ws)^2 = 1 \; ;$

(6) $\left(aeb^{-1}ws\right)^2 = 1 \; ;$

(7) $vav = atata \; ;$

(8) $\left(c^{-1}b^{-1}a^{-1}da^{-1}vr\right)^2 = 1 \; ;$

(9) $(bts)^2 = 1 \; ;$

(10) $(vas)^2 = 1 \; ;$

(11) $\left(c^{-1}b^{-1}a^{-1}st\right)^2 = 1 \; ;$

(12) $(ors)^2 = 1 \; ;$

(13) $\quad tact = csbsacsbcs$.

Proof. (1) In Γ/N we have

$$1 = da^{-1}b^{-1}d^{-1}c^{-1}b^{-1}a^{-1}dec = q^2p^2qp^2qp^2q^{-1}p^2q \ .$$

Moreover note that

$$q^{-1}Y : p \mapsto p^{-1}x$$
$$q \mapsto q^{-1}x$$
$$x \mapsto p^{-1}xq \ .$$

Hence

$$q^{-1}Y\left(q^2p^2qp^2qp^2q^{-1}p^2q\right) = xqx \cdot qxpxpq^{-1}pxpq^{-1}pxpxqxpxpx \cdot xq^{-1}x$$
$$= xqx \cdot q(xp)^2q^{-1}pqp^{-1}q^{-1}q(xp)^2q^{-1}qxqx(px)^2 \cdot xq^{-1}x$$
$$= xqx \cdot abcuc^{-1}b^{-1}a^{-1}a^{-1} \cdot abcuc^{-1}b^{-1}a^{-1} \cdot as \cdot s \cdot xq^{-1}x = 1$$

which then implies $uaua = 1$ in Γ/N .

We apply p obtaining (1).

(2) α (1) $\Rightarrow$ (2).

(3) $\qquad\qquad\qquad\qquad ZaZ^{-1} = Zqpq^{-1}p^{-1}Z^{-1}$
$$= xqxp^{-1}xq^{-1}xp$$
$$= q^{-1} \cdot qxqxp^{-1}xp^{-1}xq$$
$$= q^{-1}asrq \ .$$

Then $a^2 = 1$ yields (3) in Γ/N .

(4) We have

$$b^{-1}a^{-1} = p^2qp^2q^{-1} \Rightarrow ZYb^{-1}a^{-1}Y^{-1}Z^{-1} = pxpxq^{-1}xpxpq$$
$$= q^{-1}qpxpxq^{-1}xpxpq$$
$$= q^{-1}vrq \ .$$

In Γ/N we have $(ab)^2 = 1$; hence (4).

(5) p (4) $p^{-1} \Rightarrow$ (5) .

(6) By definition

$$q^2 p^{-1} q^2 p = c^{-1} e^{-1} c \Rightarrow ZYq^2 p^{-1} q^2 pY^{-1}Z^{-1} = xqxqpqxqxp^{-1}$$

$$= q^{-1} qxqxqpqxqxp^{-1} q^{-1} q$$

$$= q^{-1} asaeb^{-1} a^{-1} awa^{-1} q$$

$$= q^{-1} seb^{-1} wa^{-1} q \; .$$

Since in Γ/N , $\left(c^{-1} e^{-1} c\right)^2 = 1$; we get (6).

(7) In Γ/N we have

$$q^2 p^2 q^{-1} p^2 q^2 p^2 qp^2 = e^{-1} d^{-1} b^{-1} a^{-1} bdea^{-1} = 1 \; .$$

Hence

$$q^{-1} Y\left(q^2 p^2 q^{-1} p^2 q^2 p^2 qp^2\right) = q^{-1} xq^{-1} xp^{-1} xp^{-1} qp^{-1} xp^{-1} xq^{-1} xq^{-1} xp^{-1} xp^{-1} xq^{-1} xp^{-1} xp^{-1} x$$

$$= x\left(qpq^{-1} p^{-1} q(xp)^2 q^{-1} (qx)^2 q(xp)^2 q^{-1} (px)^2\right)x$$

$$= xa^2 bcuc^{-1} b^{-1} sabcuc^{-1} b^{-1} a^{-1} sx$$

$$= 1$$

which implies

$$a^2 bcuc^{-1} b^{-1} sabcuc^{-1} b^{-1} a^{-1} s = 1 \; .$$

We apply p obtaining

$$b^{-1} a^{-1} vab^{-1} ta^{-1} vat = 1 \; .$$

By Lemma 5 (24)

$$ta^{-1} t = ba^{-1} vavab \; .$$

Then (1) and $a^2 = 1$ yield (7).

(8) p^{-1} (6) $p \Rightarrow$ (8).

(9) p (3) $p^{-1} \Rightarrow$ (9).

(10) α (9) $\Rightarrow$ (10).

(11) p^2 (9) $p^{-2} \Rightarrow$ (11).

(12) p (9) $p^{-1} \Rightarrow$ (12).

(13) By (12) follows

$$rsrrcr = c^{-1} s \; .$$

We insert rcr using Lemma 7 (3) obtaining

(*) $$rsr = c^{-1} sbcsbs \; .$$

Lemma 7 (1) yields

$$rasrrsr = b^{-1}sc^{-1}tabcts \ .$$

Use (3) and (*) to conclude (13).

Now we show that it suffices to show the finiteness of the group $\langle a, b, c, d, e, r, s, t \rangle$ in Γ/N .

LEMMA 19. *In* Γ/N *we have*

(1) v *normalizes* $\langle a, b, c, d, e, r, s, t \rangle$;

(2) u, v, w *normalize* $\langle a, b, c, d, e, r, s, t \rangle$;

(3) *in* $\langle a, b, c, d, e, r, s, t, u, v, w \rangle$, $\langle a, b, c, d, e, r, s, t \rangle$ *is of index* 8 *at most.*

Proof. (1) By Lemma 18 (7)

$$vav \in \langle a, b, c, d, e, r, s, t \rangle \ ;$$
$$vbv \in \langle a, b, c, d, e, r, s, t \rangle$$

follows by Lemma 18 (1).

$$vdv \in \langle a, b, c, d, e, r, s, t \rangle$$

follows by Lemma 5 (3),

$$vrv \in \langle a, b, c, d, e, r, s, t \rangle$$

follows by Lemma 18 (4).

Then by Lemma 18 (8),

$$vcv \in \langle a, b, c, d, e, r, s, t \rangle \ .$$

By Lemma 5 (24) we have

$$vabtv = tb^{-1}a^{-1} \ .$$

Hence

$$vtv \in \langle a, b, c, d, e, r, s, t \rangle \ .$$

Moreover, by Lemma 18 (10),

$$vsv \in \langle a, b, c, d, e, r, s, t \rangle \ .$$

Hence we have proved (1) in Γ/N (!) .

(2) We have:

$$u = p^{-1}vp \ ,$$
$$w = pvp^{-1} \ .$$

By Lemma 3, $\langle a, b, c, d, e, r, s, t \rangle$ is normalized by p .

(3) $\langle a, b, c, d, e, r, s, t \rangle$ is a normal subgroup in
$\langle a, b, c, d, e, r, s, t, u, v, w \rangle$. But by Lemma 5 (28), (34), (12) we have

$$\left.\begin{array}{r} (uv)^2 \\ (vw)^2 \\ (uw)^2 \end{array}\right\} \in \langle a, b, c, d, e, r, s, t \rangle .$$

Hence the quotient is elementary abelian. Since u, v, w are involutions, it is of order 8 at most.

Next we show that it is sufficient to show the finiteness of the group $\langle a, b, c, d, e, s \rangle$ in Γ/N in order to prove that Γ is finite.

LEMMA 20. *In* Γ/N *we have*

(1) r *normalizes* $\langle a, b, c, d, e, s, t \rangle$,

(2) t *normalizes* $\langle a, b, c, d, e, s \rangle$,

(3) $\langle a, b, c, d, e, s \rangle$ *is in* $\langle a, b, c, d, e, r, s, t \rangle$ *of index* 4 *at most.*

Proof. (1) is our Lemma 7.

(2) By Lemma 5 (7) we have $tbt \in \langle a, b, c, d, e, s \rangle$. Use Lemma 18 (9) to obtain $tst \in \langle a, b, c, d, e, s \rangle$. By Lemma 5 (21) we have $tct \in \langle a, b, c, d, e, s \rangle$. Use Lemma 18 (13) to conclude $tat \in \langle a, b, c, d, e, s \rangle$. Then Lemma 18 (2) implies $tdt \in \langle a, b, c, d, e, s \rangle$.

Obviously these computations suffice to prove (2).

(3) reformulates (1) and (2).

By the next lemma we shall now prove the finiteness of $\langle a, b, c, d, e, s \rangle$ in Γ/N . Then the preceding lemmas show that the finiteness of $\langle a, b, c, d, e, s \rangle$ implies the finiteness of Γ .

LEMMA 21. *The subgroup* $\langle a, b, c, d, e, s \rangle$ *of* Γ/N *is finite.*

Proof. First we claim that in Γ/N , $a, b, c, d, e \not\equiv sbs$, $a, b, c, d, e \not\equiv scs$. To show these formulae, it suffices to see that

(*) $b, c \not\equiv sbs$

since

(**) $a, bcd \in$ center $\langle a, b, c, d, e, s \rangle$)

and since a, b, c, d , are involutions! $bcd \in$ center$(\langle a, b, c, d, e, s \rangle)$ follows by the formula $sbcds = bcd$, which is obtained by applying p^{-1} to Lemma 18 (2). To prove (*) remember that Lemma 5 (21) yields $tct = bsbcs$ in Γ/N . Obviously this formula implies

(***) $b \not\equiv sbcs$.

By Lemma 18 (11) follows $tacttbst = sabc$ which implies $csbsacsbcs.sb = sabc$ by
Lemma 18 (9) and (13) which then implies $c \not\equiv sbs$ and $b \not\equiv scs$. Then, using (***)
we have proved (*). Hence we know that $\Delta := \langle b, sbs, c, scs \rangle$ is an abelian subgroup
of $\langle a, b, c, d, s \rangle$, which is centralized by a, b, c, d, e . Obviously Δ is also
normalized by s . Hence Δ is normal in $\langle a, b, c, d, e, s \rangle$. Δ is of order 16
at most. By (**) the quotient $\langle a, b, c, d, e, s \rangle / \Delta$ is of order 8 at most.

CHAPTER 3

A DESCRIPTION OF G

In this chapter we describe the application of computer implemented algorithms to investigation of the group G , examined in Chapter 1, and some related groups.

All computer calculations were done on a Univac 1100/42 at the Australian National University in Canberra. The principal programs used were an implementation of the nilpotent quotient algorithm (NQA, see Newman (1976)) and implementations of two different versions of coset enumeration, namely the Haselgrove-Leech-Trotter method (HLT) and the Felsch method (see Cannon, Dimino, Havas and Watson (1973)), with certain auxiliary programs. Henceforth reference to the NQA or coset enumeration will include reference to a computer implementation.

First consider the group G (Chapter 1, p. 56). The NQA applied to the presentation for G yields that G has a maximal nilpotent 2-quotient, G_f , of order 2^{21} and class 9 . (Of course the proof in Chapter 1 shows $G = G_f$, and in this chapter we describe an independent proof.)

The consistent power commutator presentation for G_f given by that program follows. Detailed information about G_f (hence G) can be read from this presentation.

We denote the initial generators p and q of G by 1 and 2 , and the additional generators introduced by 3 to 21 . For brevity we omit those power relations which specify that squares of generators are trivial and those commutator relations which specify that a commutator is trivial. For convenient reference we also list the relations defining the additional generators separately.

The relations which define generators 3 to 21 are:

$3 = [2, 1]$, $4 = 1^2$, $5 = 2^2$, $6 = [3, 1]$, $7 = [3, 2]$, $8 = [6, 1]$,
$9 = [6, 2]$, $10 = [7, 2]$, $11 = [8, 1]$, $12 = [8, 2]$, $13 = [9, 1]$,
$14 = [12, 1]$, $15 = [12, 2]$, $16 = [14, 1]$, $17 = [14, 2]$, $18 = [15, 1]$,
$19 = [16, 2]$, $20 = [18, 1]$, $21 = [19, 2]$.

The nontrivial power relations are:

$1^2 = 4$, $2^2 = 5$, $3^2 = 8.9.10.13.15.18.19.21$, $6^2 = 11.14.16.17.18.19$,

$7^2 = 11.15.18.19.20$, $8^2 = 16.17$, $10^2 = 16.17$, $11^2 = 21$, $13^2 = 21$.

The nontrivial commutator relations are:

[2, 1] = 3 ,   [3, 1] = 6 ,   [3, 2] = 7 ,   [4, 2] = 6.8.9.10.11.13.14.15.16.17 ,
[4, 3] = 8.11.14.18.19.21 ,   [5, 1] = 7.8.9.10.14.17.21 ,
[5, 3] = 10.11.15.16.17.18.19.20.21 ,   [5, 4] = 9.13.17.21 ,   [6, 1] = 8 ,
[6, 2] = 9 ,   [6, 3] = 12.13.15.17.19.20 ,   [6, 4] = 11.16.17 ,
[6, 5] = 12.13.14.15.18.21 ,   [7, 1] = 9.12.14.17.20 ,   [7, 2] = 10 ,
[7, 3] = 13.14.15.17 ,   [7, 4] = 13.14.16.21 ,   [7, 5] = 11.18.19.20.21 ,
[7, 6] = 14.15.16.17.20 ,   [8, 1] = 11 ,   [8, 2] = 12 ,   [8, 3] = 14.16.18.19.20.21 ,
[8, 4] = 16.17 ,   [8, 5] = 15 ,   [8, 6] = 16.20 ,   [8, 7] = 18.19 ,   [9, 1] = 13 ,
[9, 2] = 12.13.14.15.18.21 ,   [9, 3] = 14.15.16.19.21 ,   [9, 5] = 16.17.18.20.21 ,
[9, 6] = 16.21 ,   [9, 7] = 16.17.18.19.21 ,   [9, 8] = 20.21 ,
[10, 1] = 12.14.15.16.19 ,   [10, 2] = 11.16.17.18.19.20.21 ,   [10, 3] = 15.20 ,
[10, 4] = 14.16.18.21 ,   [10, 5] = 16.17 ,   [10, 6] = 17.19.20 ,
[10, 7] = 16.17.18.19.20 ,   [10, 8] = 19.20 ,   [10, 9] = 19 ,   [11, 1] = 16.17.21 ,
[11, 2] = 16.17.21 ,   [11, 4] = 21 ,   [11, 5] = 21 ,   [12, 1] = 14 ,   [12, 2] = 15 ,
[12, 3] = 17.18.19.20 ,   [12, 4] = 16 ,   [12, 5] = 16.17.18.19.20 ,
[12, 6] = 19.20.21 ,   [12, 7] = 19.20 ,   [12, 8] = 21 ,   [12, 10] = 21 ,
[13, 1] = 21 ,   [13, 2] = 15.17.19 ,   [13, 3] = 18.20 ,   [13, 5] = 16.17.18.20 ,
[13, 6] = 20.21 ,   [13, 7] = 20 ,   [13, 8] = 21 ,   [13, 9] = 21 ,   [14, 1] = 16 ,
[14, 2] = 17 ,   [14, 3] = 19 ,   [14, 4] = 21 ,   [14, 5] = 19.21 ,   [14, 7] = 21 ,
[15, 1] = 18 ,   [15, 2] = 16.17.18.19.20 ,   [15, 3] = 20 ,   [15, 4] = 20 ,
[15, 6] = 21 ,   [16, 1] = 21 ,   [16, 2] = 19 ,   [16, 5] = 21 ,   [17, 2] = 19.21 ,
[17, 5] = 21 ,   [18, 1] = 20 ,   [18, 4] = 21 ,   [19, 2] = 21 ,   [20, 1] = 21 .

Further investigation of G_f , using the consistent power commutator presentation and computer collection of eighth powers of an appropriate set of words (see Havas and Newman (these proceedings)), reveals that G_f does indeed have exponent 8 .

The NQA does not cast any light on the question whether G is finite or not. For groups given by presentations a major approach to finiteness questions is coset enumeration. However, the order of the maximal nilpotent quotient of G implies that direct coset enumeration is likely to be very difficult, even if the group is finite. The largest subgroups of G which are easily seen (from the presentation for G on p. 56) to be finite have orders at most 8 . This means enumerating at least 2^{18} cosets over these subgroups to prove finiteness directly. Fortunately an indirect approach is available.

Given a finitely presented group K and a subgroup H generated by a set $\{h_j\}$

the following coset enumeration based technique exists for finding a subgroup H_1 , which is a cyclic extension of H (see Havas (1974)). If i is a coset of H such that $i.h = i$ for each element h of the generating set of H then i is called a normalizing (or stabilizing) coset of H . If c_i is a coset representative of the normalizing coset i then $c_i h c_i^{-1} \in H$; so $H_1 = \langle H, c_i \rangle$ is a cyclic extension of H . For our purposes we wish to start with a finite subgroup H of a group K and construct $H_1 = \langle H, c_i \rangle$ with finite order. If c_i is a coset representative for a normalizing coset and m is the least positive integer such that $c_i^m \in H$ (m is the order of c_i modulo H) then $|\langle H, c_i \rangle| = m|H|$.

The implementations of coset enumeration available in Canberra include facilities for finding normalizing cosets, for finding coset representatives and for determining the orders of group elements modulo the subgroup whose cosets are being enumerated. There is no difficulty in completely determining all normalizing cosets, coset representatives and the orders modulo the subgroup of the coset representatives from a complete table of cosets for a subgroup of a group.

What is vital here is that from an incomplete coset table, under favourable circumstances, normalizing cosets (and coset representatives) can be found. Observe that if we find from an incomplete coset table for a subgroup H of a group K that coset i normalizes H , and that $c_i^m \in H$ then, due to possibly undiscovered coincidences between cosets, the actual order modulo H of c_i is a divisor of m . This is enough to ensure that the cyclic extension $H_1 = \langle H, c_i \rangle$ is finite provided H is finite. However, from an incomplete table it is possible that no power of c_i can be found to be in H . Fortunately, when a normalizing coset (and coset representative) can be found from an incomplete table a multiple of the order of the coset representative can frequently also be found. This suffices for our purposes.

This technique can be applied repeatedly provided a suitable (possibly incomplete) coset table can be found, which yields normalizing cosets with representatives of finite order. Under these circumstances we can construct a subnormal chain of subgroups

$$H = H_0 \trianglelefteq H_1 \trianglelefteq H_2 \trianglelefteq \cdots .$$

Moreover, when H is finite and multiples of the orders (modulo the subgroup being extended) of the extending elements are calculable, then upper bounds for the orders of the subgroups in the chain are calculable.

Let us look at the application of the normalizing coset technique to G . For

brevity we include details of only one chain of subgroups constructed by the method.
Consider the subgroup H of G generated by $pqpq^{-1}$. From the initial presentation
for G, $|H|$ divides 8. An HLT coset enumeration for $G\,H$ with a limit of 8000
cosets does not of course complete. However four normalizing cosets are found, cosets
with representatives q^2, pq^2p^{-1}, $pq^{-1}pq$ and $q^{-1}p^{-1}qp$ with orders modulo H
dividing 2, 2, 8 and 8, respectively. Let $H_1 = \langle H,\, pq^{-1}pq \rangle$. Then $|H_1|$
divides 2^6. With a limit of 8000 cosets the HLT enumeration $G|H_1$ does not
complete, but three normalizing cosets are found, with representatives pq^{-1}, $q^{-1}p^{-1}$
and q^{-2} which have orders modulo H_1 dividing 4, 4 and 2 respectively. Let
$H_2 = \left\langle H_1,\, q^{-1}p^{-1} \right\rangle = \langle pq^{-1},\, pq \rangle$. Then $|H_2|$ divides 2^8, so $\left[G : H_2 \right]$ is a
multiple of $2^{13} = 8192$. Using the Felsch method of coset enumeration the index
$\left[G : H_2 \right]$ was determined to be 8192 (with a maximum of 8192 cosets). It follows
that $|G|$ divides 2^{21}. This information combined with that given by the NQA proves
that $|G| = 2^{21}$.

A modification to the NQA yields that the multiplicator of G has rank 7.
This implies that 9 is a lower bound for the number of relations required for a
presentation of G (see Huppert (1967), p. 631). With this motivation consider the
presentation

$$\langle p, q \mid p^4 = q^4 = (pq)^4 = \left(pq^{-1}\right)^4 = \left(p^2q^2\right)^4 = \left(pqp^{-1}q^{-1}\right)^4$$
$$= \left(p^2qpq^2pq\right)^2 = \left(p^2q\right)^8 = \left(pq^2\right)^8 = 1\rangle$$

(that is, a presentation with the 9 "simplest" relations of G). Clearly the group
with this presentation, say G_m, has G as a homomorphic image. The NQA reveals G
is the maximal nilpotent quotient of G_m.

If G_m and G are isomorphic then we have found a minimal presentation for G.
Since their nilpotent quotients are identical, to prove that G_m and G are
isomorphic it suffices to show that G_m is a finite 2-group.

To investigate the finiteness of G_m we cannot follow the same subnormal chain
we used in G because we cannot read from the presentation for G_m that the
starting point, $pqpq^{-1}$, has finite order. Consider the subgroup H of G_m
generated by p^2q. Clearly $|H|$ divides 8. With a limit of 20 000 cosets the
HLT enumeration $G_m|H$ does not complete, but yields one normalizing coset with

representative $pqp^{-1}q^{-1}p^{-2}q^{-1}p^{-2}q^{-1}p^{-1}qp^{-1}$ with order modulo H dividing 2 . (An HLT enumeration with an 8000 coset limit does not yield any normalizing cosets.)

Let $H_1 = \langle H,\ pqp^{-1}q^{-1}p^{-2}q^{-1}p^{-2}q^{-1}p^{-1}qp^{-1}\rangle$. Then $|H_1|$ divides 2^4 . With a limit

of 8000 cosets, the HLT enumeration $G_m|H$ does not complete, but yields one

normalizing coset with representative $pqp^{-1}q^{-1}p^{-1}qp^{-1}$ with order modulo H_1

dividing 2 . Let $H_2 = \left\langle H_1,\ pqp^{-1}q^{-1}p^{-1}qp^{-1}\right\rangle$. Then $|H_2|$ divides 2^5 . With a

limit of 8000 cosets the HLT enumeration $G_m|H_2$ does not complete, but yields one

normalizing coset with representative $pq^{-2}p^{-1}$ with order modulo H_2 dividing 2 .

Let $H_3 = \left\langle H_2,\ pq^{-2}p^{-1}\right\rangle$. Then $|H_3|$ divides 2^6 . Using a limit of 8000 cosets,

the HLT enumeration $G_m|H_3$ does not complete, but yields one normalizing coset with

representative $pq^{-1}p^{-1}$ with order modulo H_3 dividing 2 . Let

$H_4 = \left\langle H_3,\ pq^{-1}p^{-1}\right\rangle = \langle p^2q,\ pq^{-1}p^{-1}\rangle$. Then $|H_4|$ divides 2^7 , so $\left[G_m : H_4\right]$ is a

multiple of $2^{14} = 16\ 384$ if $G_m = G$. Using the Felsch method of coset enumeration

the index $\left[G_m : H_4\right]$ was determined to be 16 384 (with a maximum of 16 384

cosets!).

It follows that the nine relator presentation for G_m is a minimal presentation

for G .

Observe the effectiveness of the normalizing coset extension method in this case. Each coset representative used actually had as order modulo subgroup the number which was given as a multiple of its order from the incomplete coset table.

CHAPTER 4

A DESCRIPTION OF Γ

In this chapter we describe the application of computer implemented algorithms to the investigation of the group Γ examined in Chapter 2 and some related groups.

The NQA applied to the presentation for Γ (Chapter 2, p. 124) yields that Γ has a maximal nilpotent 2-quotient, Γ_f, of order 2^{33} and class 21 . The consistent power commutator presentation for Γ_f , given by the NQA follows. (The proof in Chapter 2 implies that $\Gamma = \Gamma_f$, so the presentation below gives detailed information about Γ .)

We denote the initial generators Y and Z of Γ by 1 and 2 , and the additional generators introduced by 3 to 33 . For brevity we omit those power relations which specify that squares of generators are trivial and those commutator relations which specify that a commutator is trivial. For convenient reference we also list the relations defining the additional generators separately.

The relations which define generators 3 to 33 are:

$3 = [2, 1]$, $4 = 1^2$, $5 = [3, 1]$, $6 = [3, 2]$, $7 = [5, 1]$, $8 = [7, 2]$,
$9 = [8, 1]$, $10 = [9, 1]$, $11 = [9, 2]$, $12 = [11, 1]$, $13 = [12, 1]$,
$14 = [12, 2]$, $15 = [13, 1]$, $16 = [14, 1]$, $17 = [15, 2]$, $18 = [16, 2]$,
$19 = [17, 1]$, $20 = [19, 1]$, $21 = [19, 2]$, $22 = [20, 1]$, $23 = [21, 1]$,
$24 = [22, 2]$, $25 = [23, 1]$, $26 = [24, 1]$, $27 = [25, 1]$, $28 = [26, 2]$,
$29 = [28, 1]$, $30 = [29, 1]$, $31 = [29, 2]$, $32 = [30, 1]$, $33 = [32, 2]$.

The nontrivial power relations are:

$1^2 = 4$, $2^2 = 3.4.5.6.7.10.11.13.15.17.19.21.23.26.27.29.30.31.32$,

$3^2 = 5.6.10.12.15.17.18.21.22.23.24.25.26.33$,

$5^2 = 11.12.13.17.18.25.26.28.31.32.33$,

$7^2 = 10.15.16.17.18.19.21.22.26.27$,

$8^2 = 14.18.20.23.25.28.30.31.33$,

$9^2 = 16.19.20.23.24.25.27.28.30.33$,

$11^2 = 18.26.27.28.29.30$, $12^2 = 26.27.30.31$, $17^2 = 33$.

The nontrivial commutator relations are:

[2, 1] = 3 ,   [3, 1] = 5 ,   [3, 2] = 6 ,   [4, 2] = 6 ,

[4, 3] = 7.10.11.12.13.15.16.18.19.21.22.25.27.30.32.33 ,   [5, 1] = 7 ,

[5, 2] = 7.8.9.10.11.16.21.24.25.27.28.29.32.33 ,

[5, 3] = 8.12.13.14.16.17.19.24.25.26.27.28.31.32.33 ,

[5. 4] = 9.12.13.14.16.19.20.24.25.27.29.33 ,

[6, 1] = 7.8.11.13.14.15.18.21.24.25.26.28.30.33 ,

[6, 2] = 7.9.10.11.12.13.14.17.24.27.30.32.33 ,

[6, 3] = 8.12.13.14.16.17.19.24.25.26.27.28.31.32.33 ,

[6, 5] = 10.11.13.15.21.22.23.24.26.27.28.29.30.31.32 ,

[7, 1] = 9.10.12.13.14.15.17.18.20.21.24.28.33 ,   [7, 2] = 8 ,

[7, 3] = 9.10.12.13.16.17.18.19.20.22.31.33 ,

[7, 4] = 10.13.16.19.20.23.25.26.27.28.32.33 ,

[7, 5] = 10.12.13.16.17.18.19.22.23.29.33 ,   [7, 6] = 10.13.18.21.23.25.28.30.31.33 ,

[8, 1] = 9 ,   [8, 2] = 9.14.16.20.21.22.23.25.27.28.29.31.32.33 ,

[8, 3] = 10.11.13.14.16.18.19.21.22.23.24.25.27.29.30.32 ,

[8, 4] = 10.16.19.20.23.24.25.27.28.30.32.33 ,

[8, 5] = 13.15.16.19.21.22.23.24.27.28.29.30.32.33 ,

[8, 6] = 14.17.18.20.21.22.24.26.27.30.31.33 ,

[8, 7] = 13.15.16.17.20.21.25.27.28.33 ,   [9, 1] = 10 ,   [9, 2] = 11 ,

[9, 3] = 12.17.19.20.21.23.27.30.31 ,   [9, 4] = 15.16.19.25.26.27.32 ,

[9, 5] = 13.15.16.17.23.25.28.30.32.33 ,   [9, 6] = 13.17.21.24.27.29.30.32.33 ,

[9, 7] = 15.20.21.24.26.27.29.31.33 ,   [9, 8] = 19.20.22.23.24.26.27.28.32 ,

[10, 1] = 15.16.19.25.26.27.32 ,   [10, 2] = 13.14.15.16.19.20.21.25.26.27.32.33 ,

[10, 3] = 15.16.19.22.24.27.28.29.31.33 ,   [10, 4] = 27.33 ,

[10, 5] = 19.20.21.22.26.29.31.32 ,   [10, 6] = 28.30.31.33 ,   [10, 7] = 22.24.30.32 ,

[10, 8] = 20.27.29.33 ,   [10, 9] = 22.25.26.27.29.30 ,   [11, 1] = 12 ,

[11, 2] = 12.17.18.31.32 ,   [11, 3] = 13.14.18.19.21.22.23.24.26.29.30.32.33 ,

[11, 4] = 13.26.27.30.31 ,   [11, 5] = 17.19.23.25.27.29.30.32 ,

[11, 6] = 18.20.24.25.26.27.28.29.30 ,   [11, 7] = 17.19.20.24.26.30 ,

[11, 8] = 24.26.29.31.32.33 ,   [11, 9] = 21.22.24.26.27.29.30 ,

[11, 10] = 24.25.32.33 ,   [12, 1] = 13 ,   [12, 2] = 14 ,

[12, 3] = 15.16.22.24.25.28.29.32.33 ,   [12, 4] = 15 ,

[12, 5] = 17.19.20.21.23.25.29.30.31.33 ,   [12, 6] = 17.32.33 ,

[12, 7] = 19.20.21.23.24.26.27.28.32.33 ,   [12, 8] = 21.22.23.25.29.32 ,

[12, 9] = 23.24.25.27.30.31.33 ,   [12, 10] = 26.27.30.31 ,

[12, 11] = 29.30.31.32.33 ,   [13, 1] = 15 ,

[13, 2] = 15.17.18.20.24.25.26.27.28.29.31.32.33 ,

[13, 3] = 17.19.22.24.26.29.30.31 ,   [13, 4] = 32 ,

[13, 5] = 20.21.24.25.26.28.29.30.31.32 , [13, 6] = 30.33 ,
[13, 7] = 20.22.24.25.31.32.33 , [13, 8] = 24.25.31 , [13, 9] = 25.31.32 ,
[13, 10] = 30.33 , [13, 11] = 33 , [13, 12] = 30.32 , [14, 1] = 16 ,
[14, 2] = 16.24.25.26.27.28.29.30.31.32 , [14, 3] = 18.20.26.27.29.31.32.33 ,
[14, 4] = 20.30.32 , [14, 5] = 24.26 , [14, 6] = 30 , [14, 7] = 24.26.31.32 ,
[14, 8] = 30.31.32.33 , [14, 9] = 28.32 , [14, 10] = 30.33 , [14, 11] = 33 ,
[14, 12] = 31.32 , [15, 1] = 32 , [15, 2] = 17 ,
[15, 3] = 19.20.21.22.23.24.26.33 , [15, 5] = 20.22.23.24.27.28.29.33 ,
[15, 6] = 20.33 , [15, 7] = 22.25.26.31.32 , [15, 8] = 25.26.30.31.33 ,
[15, 9] = 27.30.32 , [15, 10] = 32 , [15, 11] = 30 , [15, 12] = 32 ,
[15, 14] = 33 , [16, 1] = 20.30.32 , [16, 2] = 18 , [16, 3] = 22.26.27.30.31.33 ,
[16, 4] = 22.32 , [16, 5] = 24.26.28.29.32 , [16, 6] = 24.33 ,
[16, 7] = 26.28.29.30.31 , [16, 8] = 28.29.30.32.33 , [16, 9] = 29.30.32 ,
[16, 10] = 32.33 , [16, 11] = 31.32 , [16, 13] = 33 , [17, 1] = 19 ,
[17, 2] = 19.20.21.33 , [17, 3] = 20.21.22.23.24.26 , [17, 4] = 20 ,
[17, 5] = 24.25.26.27.28.29.30.31.32.33 , [17, 6] = 33 ,
[17, 7] = 24.25.26.27.28.32 , [17, 8] = 28.32.33 , [17, 9] = 28.30.31.32 ,
[17, 10] = 33 , [17, 15] = 33 , [18, 1] = 26.27 , [18, 2] = 26.27.30.31.33 ,
[18, 3] = 30.31.32.33 , [18, 4] = 30.33 , [18, 5] = 33 , [18, 7] = 33 ,
[19, 1] = 20 , [19, 2] = 21 , [19, 3] = 22.23.25.26.28.32.33 , [19, 4] = 22 ,
[19, 5] = 24.25.26.27.29.31 , [19, 6] = 24.25.30.33 , [19, 7] = 26.27.28.31.33 ,
[19, 8] = 28.30.31.33 , [19, 9] = 29.32.33 , [19, 10] = 33 , [19, 11] = 33 ,
[19, 12] = 33 , [19, 13] = 33 , [20, 1] = 22 , [20, 2] = 22.24.33 ,
[20, 3] = 24.26 , [20, 5] = 28.31.32.33 , [20, 7] = 32 , [20, 8] = 30.33 ,
[20, 9] = 30.33 , [20, 12] = 33 , [21, 1] = 23 , [21, 2] = 23.25.26.27.29.31.32 ,
[21, 3] = 25.26.28.32 , [21, 4] = 25 , [21, 5] = 28.31.32 , [21, 7] = 28.29.30.31 ,
[21, 8] = 30 , [21, 9] = 31.33 , [21, 10] = 33 , [21, 11] = 33 , [22, 2] = 24 ,
[22, 3] = 26.28.29.30.32 , [22, 5] = 29.30 , [22, 6] = 30.33 , [22, 7] = 30.32 ,
[22, 8] = 30.33 , [22, 9] = 32 , [22, 11] = 33 , [23, 1] = 25 ,
[23, 2] = 26.27.29.31.32.33 , [23, 3] = 27 , [23, 4] = 27 , [23, 5] = 28.29.32 ,
[23, 6] = 28.30.31 , [23, 7] = 29.30.31.32.33 , [23, 8] = 31.32.33 , [24, 1] = 26 ,
[24, 2] = 26.28.30.33 , [24, 3] = 28.29.30.32 , [24, 4] = 30.33 , [24, 5] = 30.31 ,
[24, 7] = 30.32 , [24, 8] = 33 , [25, 1] = 27 , [25, 2] = 27.28.31 ,
[25, 3] = 28.29.31.33 , [25, 5] = 30.31.33 , [25, 7] = 30.32.33 , [25, 8] = 33 ,
[26, 1] = 30.33 , [26, 2] = 28 , [26, 3] = 29.30.31 , [26, 4] = 32 ,
[26, 5] = 30.32.33 , [26, 6] = 30.33 , [26, 7] = 32 , [27, 2] = 28.31 ,
[27, 3] = 29.30.31.32.33 , [27, 5] = 30.32 , [27, 6] = 30 , [27, 7] = 32 ,
[28, 1] = 29 , [28, 2] = 29.30.31 , [28, 3] = 30.31.32 , [28, 4] = 30 ,
[28, 5] = 33 , [28, 7] = 33 , [29, 1] = 30 , [29, 2] = 31 , [29, 3] = 32.33 ,
[29, 4] = 32 , [29, 5] = 33 , [29, 6] = 33 , [30, 1] = 32 , [30, 2] = 32.33 ,
[30, 3] = 33 , [31, 1] = 33 , [31, 2] = 33 , [32, 2] = 33 .

Computer collection of eighth powers of an appropriate set of words reveals that Γ_f does not have exponent 8 , but that

$$\Gamma_f^8 = \langle\langle \left(yz^4y^2z^2\right)^8\rangle\rangle_{\Gamma_f} = \langle\, 32,\ 33\,\rangle\ ,$$

so $\Gamma^* = \Gamma/\langle\langle \left(yz^4y^2z^2\right)^8\rangle\rangle_\Gamma$. It follows that Γ^* has a maximal nilpotent quotient, Γ_f^* , of order 2^{31} and class 19 . (Of course $\Gamma = \Gamma_f$ implies that $\Gamma^* = \Gamma_f^*$.) A consistent power commutator presentation for Γ_f^* can be read from the consistent power commutator presentation given above for Γ_f by setting generators 32 and 33 trivial.

A coset enumeration based proof for the finiteness of Γ is not yet available. Attempts at proofs like those given in Chapter 3 have not succeeded because of an inability to define sufficient cosets. Attempts have been made with presentations for Γ^* and with presentations for low index subgroups of Γ^* , with a limit of 40000 cosets. (It is appropriate to note that the checking of the proof of Chapter 2 done by R.I. Yager involved use of some special purpose programs that he wrote to manipulate words.)

The rank of the multiplicator of Γ is 9 as is the rank of the multiplicator of Γ^* . By investigation of groups with 11 relators selected from the presentations for Γ and Γ^* the following presentations were found.

The presentation

$$\langle Y,\ Z\ |\ Y^4 = Z^8 = \left(YZ^{-1}\right)^2 = (YZ)^4 = \left(Y^2Z^2\right)^8 = \left(YZ^2\right)^8 = \left(Z^4Y^2\right)^8 = \left(Z^3Y\right)^8$$
$$= \left(Z^3Y^2\right)^8 = \left(Z^2Y^2ZY\right)^8 = \left(Z^4YZ^2Y^2ZY\right)^8 = 1\rangle$$

of a group Γ_m , say, has Γ as its maximal nilpotent quotient and a minimal number of relations. Replacing the relation $\left(Z^4Y^2\right)^8 = 1$ by $\left(YZ^4Y^2Z^2\right)^8 = 1$ gives a presentation for a group Γ_m^* which has Γ^* as its maximal nilpotent quotient and has a minimal number of defining relations. Questions as to whether $\Gamma_m = \Gamma$ and $\Gamma_m^* = \Gamma^*$ are as yet unsolved.

The fact that a minimal presentation for G was found in Chapter 3 enables us to simplify the presentations for Γ and Γ^* somewhat. A study of the proof that Γ is finite reveals that the relations (ix), (xi), (xxi), (xxii), (xxiii), (xxiv), (xxv) given in the presentation for Γ on p. 124 are used only to prove that the relations (8), (10), (12), (13), (14) and (15) (Chapter 2, Theorem 3) hold in Γ . In view of the minimal presentation for G , this second set of relations is not required to prove the finiteness of G , so it follows that the first set is not required in the presentation for Γ . Further relation (xvi) is redundant because it is used only to prove that relation (27) holds in Γ . This is used only to prove Lemma 5 (19) which is not used elsewhere, so is not essential to the proof.

The above information gives us an 18 relator presentation for Γ (relations (i)-(viii), (x), (xii)-(xv), (xvii)-(xx)) and a 19 relator presentation for Γ^* (namely the 18 relations for Γ plus $\left(yz^4y^2z^2\right)^8 = 1$).

Recall that G was selected as a preimage of a subgroup of Γ^*, namely $\langle (zy)^2z^{-2}, (yz)^2z^{-2}\rangle$, say S . A program auxiliary to the NQA enables computation of a subgroup of a group given generators for the subgroup and a consistent power commutator presentation for the group. The subgroup S in Γ^* was computed and found to have order 2^{14} .

It follows that the mapping of G onto S has kernel of order 2^7 . The availability of the consistent power commutator presentation for Γ^* enables us to compute generators for the kernel of this mapping. By mapping elements of the kernel into B we find words in the generators of B which are trivial in B .

References

С.И. Адян [S.I. Adjan] (1971), "О подгруппах свободных периодических групп нечетного
показателя" [On subgroups of free periodic groups of odd exponent], *Trudy Mat.
Inst. Steklov.* 112, 64-72, 386; *Proc. Steklov Inst. Math.* 112, 61-69 (1973).
MR48#2266; Zbl.259.20028; RZ [1972], 1A331.

S.I. Adyan (1974), "Periodic groups of odd exponent", *Proc. Second Internat. Conf.
Theory of Groups* (Canberra, 1973), pp. 8-12 (Lecture Notes in Mathematics, 372.
Springer-Verlag, Berlin, Heidelberg, New York). MR51#3318; Zbl.306.20044.

С.И. Адян [S.I. Adjan] (1975), *Проблема Бернсайда и тождества в группах* (Izdat
"Nauka", Moscow). English translation: S.I. Adian (1979), *The Burnside problem
and identities in groups* (translated by J. Lennox, J. Wiegold. Ergebnisse der
Mathematik und ihrer Grenzgebiete, 95. Springer-Verlag, Berlin, Heidelberg, New
York). MR55#5753; Zbl.306.20045; RZ [1975], 12A241 .

W. Burnside (1902), "On an unsettled question in the theory of discontinuous groups",
Quart. J. Pure Appl. Math. 33, 230-238. FdM33,149.

John J. Cannon, Lucien A. Dimino, George Havas and Jane M. Watson (1973),
"Implementation and analysis of the Todd-Coxeter algorithm", *Math. Comp.* 27,
463-490. MR49#390; Zbl.314.20028; RZ [1974], 5A243.

H.S.M. Coxeter (1940), "A method for proving certain abstract groups to be infinite",
Bull. Amer. Math. Soc. 46, 246-251. MR1,258; FdM66,73; Zbl.24,254.

H.S.M. Coxeter and W.O.J. Moser (1957), *Generators and relations for discrete groups*
(Ergebnisse der Mathematik und ihrer Grenzgebiete, 14. Springer-Verlag, Berlin,
Göttingen, Heidelberg. Third edition: 1972). MR19,527; Zbl.77,28;
RZ [1961], 11A222 .

Fritz Grunewald - Jens Mennicke (1973), "Über eine Gruppe vom Exponenten acht"
(Dissertation zur Erlangung des Doktorgrades der Fakultät für Mathematik der
Universität Bielefeld, Bielefeld).

George Havas (1974), "Computational approaches to combinatorial group theory" (PhD
thesis, University of Sydney, Sydney). See also: Abstract, *Bull. Austral. Math.
Soc.* 11, 475-476. RZ [1975], 8A224.

Franz-Josef Hermanns (1976), "Eine metabelsche Gruppe vom Exponenten 8 "
(Diplomarbeit, Fakultät für Mathematik, Universität Bielefeld, Bielefeld).

B. Huppert (1967), *Endliche Gruppen I* (Die Grundlehren der Mathematischen
    Wissenschaften, **134**. Springer-Verlag, Berlin, Heidelberg, New York). MR37#302;
    RZ [1968], 6A256н.

Eugene F. Krause (1964), "Groups of exponent 8 satisfy the 14th Engel congruence",
    Proc. Amer. Math. Soc. **15**, 491-496. MR29#2298; Zbl.122,30; RZ [1965], 1A171.

I.D. Macdonald (1973), "Computer results on Burnside groups", *Bull. Austral. Math.
    Soc.* **9**, 433-438. MR49#5165; Zbl.267.20011; RZ [1974], 10A201.

Wilhelm Magnus, Abraham Karrass, Donald Solitar (1966), *Combinatorial group theory:
    Presentations of groups in terms of generators and relations* (Pure and Applied
    Mathematics, **13**. Interscience [John Wiley & Sons], New York, London, Sydney.
    Second, revised edition: Dover, New York, 1976). MR34#7617; Zbl.138,256;
    RZ [1967], 9A143.

M.F. Newman (1976), "Calculating presentations for certain kinds of quotient groups",
    SYMSAC '76, Proc. ACM Symposium on Symbolic and Algebraic Computation (New
    York, 1976), pp. 2-8 (Association for Computing Machinery, New York).

Derek J.S. Robinson (1972), *Finiteness conditions and generalized soluble groups*, Part
    I (Ergebnisse der Mathematik und ihrer Grenzgebiete, **62**. Springer-Verlag,
    Berlin, Heidelberg, New York). MR48#11314; Zbl.243.20032; RZ [1973], 4A325н.

И.Н. Санов [I.N. Sanov] (1947), "О проблеме Бернсайда" [On Burnside's problem], *Dokl.
    Akad. Nauk SSSR (N.S.)* **57**, 759-761. MR9,224; Zbl.29,102.

И.Н. Санов [I.N. Sanov] (1951), "О некоторой системе соотношений в периодических
    группах с периодом степиньн простого числа" [On a certain system of relations in
    periodic groups with period a power of a prime number], *Izv. Akad. Nauk SSSR Ser.
    Mat.* **15**, 477-502. MR14,722; Zbl.45,302.

David Shield (1977), The class of a nilpotent wreath product", *Bull. Austral. Math.
    Soc.* **17**, 53-89. MR57#3266; Zbl.396.20015; RZ [1978], 9A232.

Sonderforschungsbereich,   Universität Bielefeld,
Theoretische Mathematik,   Bielefeld,
Universität Bonn, Bonn,   Federal Republic of Germany;
Federal Republic of Germany;

Department of Mathematics,   Department of Mathematics,
Institute of Advanced Studies,   Institute of Advanced Studies,
Australian National University,   Australian National University,
Canberra, ACT, Australia;   Canberra, ACT, Australia.

FINITENESS PROOFS FOR GROUPS OF EXPONENT 8

F.J. GRUNEWALD AND J. MENNICKE

A. Introduction

In this paper we want to demonstrate some of the techniques used in the
finiteness proofs in the preceding paper - which will be referred to as [A]. The
situation here will be somewhat easier. Let

$$\Delta_2 = \langle X, Y \mid X^2 = Y^4 = \left(XYXY^2\right)^4 = 1; \text{ exponent 8} \rangle .$$

The addition of exponent n to the presentation of a group G means that the law
$g^n = 1$ is required to be satisfied in G.

We prove

THEOREM. Δ_2 *is finite.*

To establish this theorem we shall apply some of the results of [A].

Before proceeding with the proof we collect some more information on Δ_2 and on
similar groups.

DEFINITION.

$$\Delta = \langle X, Y \mid X^2 = Y^4 = 1; \text{ exponent 8} \rangle ,$$

$$\Delta_1 = \langle X, Y \mid X^2 = Y^4 = \left(XY^2\right)^4 = 1; \text{ exponent 8} \rangle ,$$

$$\Delta_3 = \langle X, Y \mid X^2 = Y^4 = (XY)^4 = 1; \text{ exponent 8} \rangle ,$$

$$\Delta_4 = \langle X, Y \mid X^2 = Y^4 = \left(XYXY^{-1}\right)^4 = 1; \text{ exponent 8} \rangle .$$

If G is a group we put

$$G^f = \bigcap_{\substack{N \leq G \\ |G:N| \text{ finite}}} N \ ,$$

$$\overline{G} = G/G^f \ ,$$

$\#G$ = the order of G ,

$c(G)$ = the nilpotency of class of G ,

$r(G)$ = rank of the 2-multiplicator of G .

We give some information on the above groups in the following table.

G		$\#\overline{G}$	$c(\overline{G})$	$r(\overline{G})$
Δ	?	$\leq 2^{205}$	≤ 26	?
Δ_1	finite	2^{31}	19	11
Δ_2	finite	2^{26}	17	9
Δ_3	finite	2^8	6	4
Δ_4	?	2^{48}	21	14

Of course if G is finite then $G = \overline{G}$. The data on Δ contained in this table are contained in the following paper [B]. The group Δ_1 is a quotient group of the group Γ in [A, Chapters 2, 4]. The group Δ_2 is the subject of the present paper. To obtain the results on Δ_3 is an easy exercise.

All data besides the finiteness results were obtained by using the nilpotent quotient algorithm. See [B] for more details on this computer program. Such a computation (if completed) shows that the groups in question have a maximal finite quotient. It also gives a multiplication table for this quotient. We produce such a table for the group $\overline{\Delta}_2$ in part D of this paper.

It will be apparent from part B of this paper that for our finiteness proofs a detailed knowledge of the groups in question is helpful. To get this information tables like the one in part D are of great use.

Part B of this paper describes the major steps in the proof of our theorem. Part C contains the detailed computations necessary for part B. Part D gives a multiplication table for $\overline{\Delta}_2$.

B. The main theorems

Let J be a subgroup of finite index in Δ_2 . The number of generators one gets by the Reidemeister-Schreier method for J grows drastically with the index $|\Delta_2 : J|$. Since Δ_2 is finite many of these generators will clearly be superfluous. But eliminating generators we usually get relations between the remaining ones which are difficult to use. An important first step in our proofs is to find a subgroup J of Δ_2 which has as high as possible index, but which can easily be seen to have not too many generators. To describe such a J we introduce some elements of Δ_2 .

DEFINITION.

$$R = Y^2 ,$$

$$S = XY^2X ,$$

$$T = YXY^2XY^{-1} ,$$

$$U = XYXY^2XY^{-1}X ,$$

$$V = YXYXY^2XY^{-1}XY^{-1} ,$$

$$W = XYXYXY^2XY^{-1}XY^{-1}X ,$$

$$P = YXYXYXY^2XY^{-1}XY^{-1}XY^{-1} ,$$

$$J_0 = \langle R, S, T, U, V, W, P \rangle , \quad J = \langle R, S, T, V, W \rangle ,$$

$$J_1 = \langle R, T, W \rangle .$$

We then prove

THEOREM 1. J *is of finite index in* Δ_2 .

The next step is to accumulate relations in J and to look out for finite subgroups in J . In this case we were lucky and could deduce from a result in [A]

THEOREM 2. $J_1 = \langle R, T, W \rangle$ *is finite.*

There are other subgroups in J which can more or less easily be seen to be finite. We work with the group $\langle R, T, W \rangle$ since the relations one gets in J easily show that $\langle R, T, W \rangle$ contains a large subgroup which is normal in J . For this let

DEFINITION.

$$f_1 = RT ,$$

$$f_2 = WRTW ,$$

$$f_3 = RWRTWR ,$$

$$N = \langle f_1, f_2, f_3 \rangle .$$

THEOREM 3. N *is normal in* J .

A quick glance at Lemma 6 proves Theorem 3. So we have found a finite normal subgroup of the group J which is itself of finite index in Δ_2 . We apply the following trivial but useful lemma. A proof of this lemma is contained in [A]. We denote by $\langle\langle H_1 \rangle\rangle_G$ the normal subgroup in G generated by H_1 .

LEMMA. *Let* G *be a group,* $H \leq G$ *a subnormal subgroup of finite index,* H_1 *a finite normal subgroup of* H . *Then* $\langle\langle H_1 \rangle\rangle_G$ *is finite.*

Since Δ_2 is a 2-group it is clear from the lemma that $\langle\langle N \rangle\rangle_{\Delta_2}$ is finite. Now our main theorem is proved by:

THEOREM 4. $\langle\langle N \rangle\rangle_{\Delta_2}$ *is of finite index in* Δ_2 .

Proof. We have $RT \in \langle\langle N \rangle\rangle_{\Delta_2}$. We repeatedly conjugate this element by X, Y , use Lemma 1 and get

$$SU, \ SR, \ TV, \ UW, \ VP \in \langle\langle N \rangle\rangle_{\Delta_2} \ .$$

Hence Y^2 is central in $\Delta_2 / \langle\langle N \rangle\rangle_{\Delta_2}$. We get

$$|\Delta_2 : \langle\langle N \rangle\rangle_{\Delta_2}| \leq 2^5 \ .$$

We shall now describe the steps for proving Theorems 1 to 4 in more detail. We first observe that

$$J_0 = \langle\langle Y^2 \rangle\rangle_{\Delta_2} \ .$$

This is established by the formulae of Lemma 1. Δ_2 / J_0 is obviously the dihedral group of order 2^4 .

DEFINITION.

$$N_1 = \langle RS, \ VT, \ UW, \ TRT, \ URU, \ TST, \ USU, \ RTU, \ RUT, \ RVU, \ RWT \rangle \ ,$$

$$N_2 = \langle RS, \ RT, \ RV, \ RW, \ TV, \ TW, \ VW \rangle \ .$$

In order to establish Theorem 1 we prove

THEOREM 5. (1) N_1 *is of finite index in* $\langle R, \ S, \ T, \ U, \ V, \ W \rangle$.

(2) N_2 *is of finite index in* $\langle R, \ S, \ T, \ V, \ W \rangle$.

(3) N_1 *is normalized by* P .

(4) N_2 *is normalized by* U .

(5) J *is of finite index in* J_0 .

Proof. (1) We claim that $\{1, R, T, U\}$ contains a set of representatives for the right cosets of N_1 in $\langle R, S, T, U, V, W \rangle$. Let $\theta \in \langle R, S, T, U, V, W \rangle$ be an arbitrary word. By multiplying θ with one of the first three generators of N_1 we make sure that θ starts with R, T, U or is equal to the identity. If θ involves more than two generators we have the following possibilities for the initial two generators

$$RS, \; RT, \; RU, \; RV, \; RW, \; TR, \; TS, \; TU, \; TV, \; TW, \; US, \; UR, \; UT, \; UV, \; UW \; .$$

An easy inspection shows that θ can then be shortened by multiplying with an element from N_1 . So the set $\langle 1, R, S, T, U, V, W \rangle$ contains a set of representatives for N_1 in $\langle R, S, T, U, V, W \rangle$. Because of the first three generators the elements S, V, W can be left off.

(2) N_2 is the subgroup in $\langle R, S, T, V, W \rangle$ which consists of the words of even length.

(3) is proved by Lemma 3.

(4) is proved by Lemma 4.

(5) is obvious from (3), (4).

A more detailed analysis of the situation shows that we have constructed the following chain of subgroups in Δ_2 :

$$
\begin{array}{l}
\Delta_2 \\[4pt]
\quad\Big|\;2^4 \\[4pt]
\langle R, S, T, U, V, W, P \rangle = \langle\langle Y^2 \rangle\rangle_{\Delta_2} \\[4pt]
\quad\Big|\;2^3 \\[4pt]
\langle R, S, T, U, V, W \rangle \\[4pt]
\quad\Big|\;2^2 \\[4pt]
\langle R, S, T, V, W \rangle = J \; .
\end{array}
$$

We shall now describe a proof of Theorem 2. To do this we use results of [A]. Let Γ be the group given in the beginning of [A, Chapter 2], and let

$$a = y^2$$
$$b = y^2 z^{-1} y^{-1} \left.\right\} \in \Gamma \ .$$
$$c = yz^{-1}$$

We hope it is not confusing that we have used a generator y both for Δ_2 and Γ .
Let

$$\Gamma_1 = \langle a, b, c \rangle \ .$$

We shall need

THEOREM 6. (1) Γ_1 *has index* 2 *in* Γ .

(2) *The group*

$$\overline{\Gamma}_1 = \langle a, b, c \mid$$

(i) $a^2 = b^2 = c^2 = 1$

(ii) $(ab)^4 = (ac)^4 = 1$

(iii) $(abc)^4 = 1$

(iv) $(babcac)^4 = 1$

(v) $\left(a(bc)^4\right)^4 = 1$

(vi) $\left(a(baca)^4\right)^4 = 1$

(vii) $(bcbacabac)^4 = 1$

(viii) $(bcacbcacbabacba)^4 = 1$

(ix) $(bacabacabacbcbc)^4 = 1$

(x) $(cbcabcbacbabcbacacb)^4 = 1$

(xi) $(bacabcabacbcacbacbabaca)^4 = 1$

(xii) exponent 8$\rangle$

is an image of Γ_1 .

Proof. (1) is clear.

(2) can easily be proved by using the Reidemeister-Schreier method. One clearly gets only finitely many defining eighth powers for Γ_1 . But we shall not need these here.

For our approach it is very useful to have a lot of information on groups of exponent 8 which are generated by three involutions. Some groups of this type are described in the paper by Hermanns in these proceedings. We shall need a variation of the group in Theorem 6. For this we define

DEFINITION.

$$\Gamma_2 = \langle x,\, y,\, z \mid \quad \text{(i)} \quad x^2 = y^2 = z^2 = 1\,,$$
$$\text{(ii)} \quad (xy)^4 = (xz)^4 = (yz)^4 = 1\,,$$
$$\text{(iii)} \quad (xyz)^4 = 1\,,$$
$$\text{(iv)} \quad (yxzx)^4 = 1\,,$$
$$\text{(v)} \quad \text{exponent 8)}\,,$$

$$L_2 = \langle\langle (yxyzxz)^4 \rangle\rangle_{\Gamma_2}\,,$$

$$\overline{\Gamma}_2 = \Gamma_2/L_2\,.$$

We denote by $\overline{x},\, \overline{y},\, \overline{z}$ the images of $x,\, y,\, z$ in $\overline{\Gamma}_2$.

THEOREM 7. (1) L_2 *is finite.*

(2) *The map*

$$a \mapsto \overline{x}\,,$$
$$b \mapsto \overline{y}\,,$$
$$c \mapsto \overline{z}\,,$$

defines a surjective homomorphism

$$\overline{\Gamma}_1 \to \overline{\Gamma}_2\,.$$

Proof. (1) Lemma 7 (7), (8) shows that the centralizer of $(yxyzxz)^4$ is of finite index in Γ_2. We then use our lemma to establish (1).

(2) The formulae in Lemma 8 show that the relevant relations are satisfied in $\overline{\Gamma}_2$.

We now apply our considerations to a finiteness proof for J_1, that is, we prove Theorem 2.

DEFINITION.

$$L_1 = \langle\langle (WRT)^4 \rangle\rangle_{J_1}\,,$$

$$\overline{J}_1 = J_1/L_1\,.$$

We denote by $\overline{R},\, \overline{T},\, \overline{W}$ the images of $R,\, T,\, W$ in $\overline{J}_1$.

THEOREM 8. (1) L_1 *is finite.*

(2) *The map*

$$x \longmapsto \overline{R} ,$$
$$y \longmapsto \overline{W} ,$$
$$z \longmapsto \overline{T} ,$$

defines a surjective homomorphism

$$\Gamma_2 \to \overline{J}_1 .$$

Proof. (1) follows from Lemma 5 (12).

(2) The necessary relations are contained in Lemma 2 (1), (25), (33) and Lemma 5 (7), (8).

By Theorem 7 we know that Γ_2 is finite. Hence Theorem 8 implies that $J_1 = \langle R, T, W \rangle$ is finite. This was the content of Theorem 2. The following diagram shows the various groups used in the finiteness proof for $J_1 = \langle R, T, W \rangle$.

$$\langle 1 \rangle \leftarrow \overline{\Gamma}_1 \leftarrow \Gamma_1 \subseteq \Gamma$$
$$\downarrow$$
$$\langle 1 \rangle \to L_2 \to \Gamma_2 \to \overline{\Gamma}_2 \to \langle 1 \rangle$$
$$\downarrow \qquad \downarrow$$
$$\langle 1 \rangle \to L_1 \to J_1 \to \overline{J}_1 \to \langle 1 \rangle$$
$$\downarrow$$
$$\langle 1 \rangle$$

C. Proofs

We shall give now the details of the computations needed to complete the proofs of the theorems in Parts A, B. We first give a list of the conjugation laws on J_0 .

LEMMA 1. (1) $XRX^{-1} = S$, $XSX^{-1} = R$, $XTX^{-1} = U$, $XUX^{-1} = T$, $XVX^{-1} = W$, $XWX^{-1} = V$, $XPX^{-1} = RTVPSUW$.

(2) $YRY^{-1} = R$, $YSY^{-1} = T$, $YTY^{-1} = RSR$, $YUY^{-1} = V$, $YVY^{-1} = RUR$, $YWY^{-1} = P$, $YPY^{-1} = RWR$.

Proof. The only non-obvious relation is

$$XPX^{-1} = RTVPSUW .$$

We use the definitions of R, S, T, U, V, W, P and find

$$XPXWUSPVTR = (XY)^4 (YX)^8 (XY)^4 .$$

From $(YX)^8 = 1$ the relation then follows.

The following lemma contains some immediate relations in J_0 .

LEMMA 2.

(1) $R^2 = S^2 = T^2 = U^2 = V^2 = W^2 = P^2 = 1$,

(2) $(RST)^2 = 1$,

(3) $(RSU)^2 = 1$,

(4) $(RTV)^2 = 1$,

(5) $(USW)^2 = 1$,

(6) $(VTP)^2 = 1$,

(7) $(RTVPS)^2 = 1$,

(8) $WUTRUWRT = 1$,

(9) $VTUSTVSU = 1$,

(10) $(ST)^4 = 1$,

(11) $(RU)^4 = 1$,

(12) $(RV)^4 = 1$,

(13) $(SW)^4 = 1$,

(14) $(TP)^4 = 1$,

(15) $(URTVPSUW)^4 = 1$,

(16) $(VSUWRTVP)^4 = 1$,

(17) $(WRTVPVTR)^4 = 1$,

(18) $(PSUWUS)^4 = 1$,

(19) $(RTVPSUW)^2 = 1$,

(20) $(PTUWUS)^2 = 1$,

(21) $(PVTRWSRV)^2 = 1$,

(22) $PVRURTVP = SUWSRTVTRWTV$,

(23) $PSUSTP = RSUWUSVUWRTV$,

(24) $(RS)^4 = 1$,

(25) $(RT)^4 = 1$,

(26) $(US)^4 = 1$,

(27) $(VT)^4 = 1$,

(28) $(UW)^4 = 1$,

(29) $(PV)^4 = 1$,

(30) $(UT)^4 = 1$,

(31) $(RTVP)^4 = 1$,

(32) $(RSUW)^4 = 1$,

(33) $(WRTR)^4 = 1$,

(34) $(VRU)^4 = 1$.

Proof. (1) follows from $Y^4 = 1$.

(2) $RST = Y\left(YXY^2X\right)^2 Y^{-1}$. Hence (2) follows.

(3) $X(2)X^{-1} \Rightarrow (3)$. By this we mean that a conjugation of the relation (2) by X using Lemma 1 yields (3).

(4) $Y(3)Y^{-1} \Rightarrow (4)$.

(5) $X(4)X^{-1} \Rightarrow (5)$.

(6) $Y(5)Y^{-1} \Rightarrow (6)$.

(7) $X(6)X^{-1} \Rightarrow (7)$.

(8) $Y(7)Y^{-1} \Rightarrow (SUWRT)^2 = 1$. Hence $WUSRTSUWRT = 1$ by (5) implies (8) by using (2).

(9) $X(8)X^{-1} \Rightarrow (9)$.

(10) Using the definitions of R, S, T we find

$$TRS = YXY^2XYXY^2X ,$$

$$SRSTS = XY^2XY^2XY^2XYXY^2XY^{-1}XY^2X .$$

We compute

$$TRS.SRSTS.SRT.STSRS = Y^2\left(Y^{-1}XY^2X\right)^2 Y^{-1}\left(Y^2X\right)^4 Y^2\left(YXY^2X\right)^2 Y\left(Y^2X\right)^4$$
$$= Y^2XY^2XY^{-1}.\left(XY^2XY\right)^4.YXY^2XY^2$$
$$= 1 .$$

Using this relation we deduce (10) with the help of (2).

(11) $X(10)X^{-1} \Rightarrow (11)$.

(12) $Y(11)Y^{-1} \Rightarrow (12)$.

(13) $X(12)X^{-1} \Rightarrow (13)$.

(14) $Y(13)Y^{-1} \Rightarrow (14)$.

(15) $X(14)X^{-1} \Rightarrow (15)$.

(16) $Y(15)Y^{-1} \Rightarrow (16)$.

(17) $X(16)X^{-1} \Rightarrow (17)$.

(18) $Y(17)Y^{-1} \Rightarrow (18)$.

(19) follows from Lemma 1 (1) and $P^2 = 1$.

(20) Using the definitions we get

$$PTUWUS = (YX)^4\left(XYXY^{-1}\right)^4(YX)^4 \ ;$$

hence (20).

(21) $Y(20)Y^{-1} \Rightarrow (21)$.

(23) Computing from the definitions we find

$$VSUWUSPPVPPVTVPYXPXY^{-1}SUWUSSUSSPSUSTYXPXY^{-1}S = \left(YXYXY^{-1}X\right)^8 = 1 \ ;$$

using Lemma 1 and (19) we find

$$VSUWUSRPSUSTPVTRWU = 1 \ ;$$

hence (23).

(22) $Y(23)Y^{-1} \Rightarrow (22)$ using (19).

(24) $RS = \left(Y^2X\right)^2$; hence (24) follows from $\left(Y^2X\right)^8 = 1$.

(25) $Y(24)Y^{-1} \Rightarrow (25)$.

(26) $X(25)X^{-1} \Rightarrow (26)$.

(27) $Y(26)Y^{-1} \Rightarrow (27)$.

(28) $X(27)X^{-1} \Rightarrow (28)$.

(29) $X(28)X^{-1} \Rightarrow (29)$.

(30) From the definitions of U, T we find $UT = \left(XYXY^2XY^{-1}\right)^2$; hence (30).

(31) From the definitions we find $RTVP = Y\left(Y^2XY^{-1}XY^{-1}X\right)^2Y^{-1}$; hence (31).

(32) $Y(31)Y^{-1} \Rightarrow (32)$.

(33) From the definitions we find $WRTR = Y^{-1}XY^{-1}(YX)^4Y^2\left(XY^{-1}\right)^4Y^2YXY$; hence (33).

(34) From definitions we find $VRU = \left(YXYXY^2XY^{-1}X\right)^2$; hence (34).

LEMMA 3.

(1) $PRSP = VRSV$,

(2) $PVTP = TV$,

(3) $PUWP = VUWV$,

(4) $PTRTP = VTWUVTRTRWTVSUWRTVTUWUS$,

(5) $PURUP = VUWTVTRURWUVTVSUWUTVTRWUSVTWRTRTVUWVUWRURTVTWUV$,

(6) $PTSTP = SUWUTVTRWUSVTWRTRTVUWTSRV$,

(7) $PUSUP = VUWTVTRURWUVTVSUWUTVTRWUSVTWRTRTVUWVUWRURTVTWUSRV$,

(8) $PRTUP = SUWUTVTRWUSVTWRTRTVUWVUWRURTVTWUV$,

(9) $PRUTP = VRSUWVTUSUW$,

(10) $PRVUP = VRSWSURUVRSUWUWTVTRURWUVTVSUWUTVTRWUSVTWRTRTVUWVUWRURTVTWUSRV$,

(11) $PRWTP = VTVRSWVTSTV$.

Proof. (1) follows from Lemma 2 (6), (7).

(2) follows from Lemma 2 (6).

(3) By Lemma 2 (19), (5) we have $PWUSRTVP = VTRSUW$. From (1), (2) we get $PWUP = VTRWUSTRSV$. Lemma 2 (2), (8) completes the proof of (3).

(9) From Lemma 2 (20) we get $PTUWUSP = SUWUT$. Lemma 2 (5) yields $PTUSUWP = WUSUT$. (1), (3) complete the proof.

(4) From Lemma 2 (22), (5) we get $PVRURP = WURTVTRWTVTV$. (2), (9) complete the proof.

(11) From Lemma 2 (22), (2), we get $PVTPPRWTPPRSPPTVP = VRSWRTV$. From (1), (2) we deduce (11).

(8) and (5) From Lemma 2 (22), (2), (4), (5) we get $PWUVTRTRWTVP = VRURTV$. Using (3), (2), $PRTRWP = VTVUWRURTVTV$. From Lemma 2 (2) and (4) we deduce (8); furthermore we get $PRTRUP = VTVUWRURTVTWUV$.

(8) now implies (5).

(6) From Lemma 2 (2) we get $TRTTST = SR$, and (1), (4) imply (6).

(7) From Lemma 2 (3) we get $URUUSU = SR$ and (1), (5) imply (7).

(10) From Lemma 2 (23) we get $PRSUWUSVSWUSRTVP = SUST$. (1), (2), (3) yield $PUSVRP = VWUSRVUSURWSRV$. (7) implies then (10).

LEMMA 4.

(1) $URSU = SR$,

(2) $URTU = WRTW$,

(3) $URVU = WRTWSTVS$,

(4) $URWU = SRWS$,

(5) $UTVU = STVS$,

(6) $UTWU = WTRWSRWS$,

(7) $UVWU = SVTRWTRS$.

Proof. (1) follows from Lemma 2 (3).

(2) follows from Lemma 2 (8).

(3) From Lemma 2 (9) we get $UVTU = SVTS$. (2) implies then (3).

(4) follows from Lemma 2 (5) and (1).

(5) follows from Lemma 2 (8) and (3).

(6) follows from Lemma 2 (5) and (1), (2).

(7) follows from Lemma 2 (8) and (2), (5).

We now collect some relations in $\langle R, S, T, V, W \rangle$. Note that we write

$$g \not\leftrightarrow h$$

if two elements g, h of a group commute.

LEMMA 5.

(1) $SR \not\leftrightarrow WTRW$,

(2) $SVTS \not\leftrightarrow WTRW$,

(3) $VT \not\leftrightarrow WTRW$,

(4) $(WTRS)^4 = 1$,

(5) $(TW)^4 = (RW)^4$,

(6) $RWTRWR = WRWRTWRWWRTWTR$,

(7) $(RW)^4 = 1$

(8) $(TW)^4 = 1$,

(9) $VT \not\downarrow WRWTRWRW$,

(10) $(VRS)^4 = (WRT)^4$,

(11) $(WSR)^4 = (WRT)^4$,

(12) $R, T, W \not\downarrow (WRT)^4$.

Proof. (1) From Lemma 2 (2) we get $RS \not\downarrow TR$. If we conjugate this relation with U and use Lemma 4 we get (1).

(2) From Lemma 2 (4) we get $VT \not\downarrow TR$. We conjugate this relation by U and we use Lemma 4 to get (2).

(4) We start out from Lemma 2 (30), (26) and make the following computation using Lemma 4 and (1):

$$1 = (UT)^4 = UTSUUSUTUTSUUSUT$$
$$= WTRWSRUSUSSTWTRWSRUSUT$$
$$= WTRWSRSUSRSWRTWTUT$$
$$= RSRS(WTRS)^4 SRSR .$$

We have also used Lemma 2 (13).

(5) From (1), (4) we get

$$1 = WTRSWTRWWSWSWWRTWSRT$$
$$= WTWTRWRSWSWSWSRWRTWT$$

From this formula (5) follows with the use of Lemma 2 (13).

(6) is an obvious consequence of (5).

(7) From Lemma 2 (32), (28) and Lemma 4 we get $RSWUWUWSRWRSUWRSUW = 1$. If one repeatedly applies the formulae of Lemma 4 together with Lemma 2 (13) one gets (7).

(8) follows from (5) and (7).

(3) From (1), (2) we deduce $RVTS \not\downarrow WTRW$. Lemma 2 (4) yields $TVRS \not\downarrow WTRW$. Hence (9) follows from (1).

(9) follows from (6), (1), (2), (3).

(10) From Lemma 2. (34) and Lemma 4 we get $(VRSVTSWTRW)^2 = 1$. Using (2), (3) we deduce $SVTSVRSVTSVTRT = (WRT)^4$. With Lemma 2 (2), (4) we find $T(RSV)^4 T = (WRT)^4$. Using $VT \not\downarrow (WRT)^4$ which follows from Lemma 2 (4) and (3) we get (10).

(11) We conjugate (10) by X ; $(WSR)^4 = (VSU)^4$. Using Lemma 4 and (1) we get (11).

(12) From (1) and Lemma 2 (2) we get $RS \not< (WRT)^4$; from (3) and Lemma 2 (4) we get $VT \not< (WRT)^4$; from (10) and (11) we have $VRS \not< (WRT)^4$, $WSR \not< (WRT)^4$. Trivially we have $WRT \not< (WRT)^4$. These relations are enough to prove (12).

LEMMA 6.

(1) $Rf_1R = f_1^{-1}$,

(2) $Rf_2R = f_3$,

(3) $Rf_3R = f_2$,

(4) $Sf_1S = f_1^{-1}$,

(5) $Sf_2S = f_3$,

(6) $Sf_3S = f_2$,

(7) $Tf_1T = f_1^{-1}$,

(8) $Tf_2T = f_1^{-1}f_3f_1$,

(9) $Tf_3T = f_1^{-1}f_2f_1$,

(10) $Vf_1V = f_1^{-1}$,

(11) $Vf_2V = f_1^{-1}f_3f_1$,

(12) $Vf_3V = f_1^{-1}f_2f_1$,

(13) $Wf_1W = f_2$,

(14) $Wf_2W = f_1$,

(15) $Wf_3W = f_3f_1f_2^{-1}$.

Proof. (1) to (3) are obvious.

(4) follows from Lemma 2 (2).

(5) and (6) follow from Lemma 5 (1).

(7) to (9) are obvious.

(10) follows from Lemma 2 (4).

(11) and (12) follow from Lemma 5 (3).

(13) and (14) are obvious.

(15) follows from Lemma 5 (6).

Lemma 7 will contain some relations in Γ_2 .

LEMMA 7.

(1) $\left((yxz)^2(zyzxz)^2\right)^4 = 1$,

(2) $\left(yxz(zyzxz)^2yxz\right)^4 = 1$,

(3) $\left(yxz(zyzxz)^2yxz\right) \not\nsim \left((yxz)^2(zyzxz)^2\right)$,

(4) $xz(zyzxz)^2yxz(yxz)^2(zyzxz)^2 = (yxyzxz)^2$,

(5) $yxz\left(yxz(zyzxz)^2yxz\right)^2(yxz)^{-1} = \left((yxz)^2(zyzxz)^2\right)^2$,

(6) $yxz\left((yxz)^2(zyzxz)^2\right)^2(yxz)^{-1} = \left(yxz(zyzxz)^2yxz\right)^2$,

(7) $x,\ zxz,\ yz \not\nsim (yxyzxz)^4$,

(8) $\langle x,\ zxz,\ yz\rangle$ is of index 2 in Γ_2 .

Proof. (1) Using relation (iii) we find $(yxz)^2(zyzxz)^2 = zx(yzxyz)^2xz$ and hence relation (v) implies (1).

(2) is an obvious consequence of (1).

(3) Using the relations $(xyz)^4 = (yz)^4 = 1$ we find

$$\left(yxz(zyzxz)^2yxz\right)\left((yxz)^2(zyzxz)^2\right)\left(zxy(zyzxz)^2zxy\right)\left((zyzxz)^2(yxz)^2\right) = yxz(zy)^4zxy = 1$$

and hence (3).

(4) follows using $(xyz)^4 = 1$.

(5) is obvious.

(6) is a consequence of (2).

(7) $x,\ zxz \not\nsim (yxyzxz)^4$ are obvious. From (3), (4), (5), (6) we deduce $yxz \not\nsim (yxyzxz)^4$. So (7) is proved.

(8) $\langle x,\ zxz,\ yz\rangle$ clearly is the normal subgroup generated by x and yz .

The following lemma will contain some relations in $\overline{\Gamma}_2$. From the definition of $\overline{\Gamma}_2$ we contend that

$$\Gamma_2 = \langle \bar{x}, \bar{y}, \bar{z} \mid \quad \text{(i)} \quad \bar{x}^2 = \bar{y}^2 = \bar{z}^2 = 1 \,,$$

$$\text{(ii)} \quad (\overline{xy})^4 = (\overline{xz})^4 = (\overline{yz})^4 = 1 \,,$$

$$\text{(iii)} \quad (\overline{xzy})^4 = 1 \,,$$

$$\text{(iv)} \quad (\overline{yxzx})^4 = 1 \,,$$

$$\text{(v)} \quad (\overline{yxyzxz})^4 = 1 \,,$$

$$\text{(vi)} \quad \text{exponent } 8 \rangle \,.$$

LEMMA 8. *In* $\bar{\Gamma}^2$ *hold*

(1) $(\overline{yzyxzxyxz})^4 = 1 \,,$

(2) $(\overline{yzxzyzxzyxyxzyx})^4 = 1 \,,$

(3) $(\overline{yxzxyxzxyxzyzyz})^4 = 1 \,,$

(4) $(\overline{zyzxyzyxzyxyzyxzxzy})^4 = 1 \,,$

(5) $(\overline{yxzxyzxyxzyzxzyxzyzyxyxzx})^4 = 1 \,.$

Proof. (1)

$$\overline{yzyxzxyxz} = \overline{yzxzxyxzxyx} \quad \text{by (iv)}$$
$$= \overline{yxzyzxyzyxy} \quad \text{by (ii), (iii)}$$
$$= \overline{yxzyzyyxzyzyzxy} \quad \text{by (ii);}$$

hence (iii) implies (1).

(2)

$$\overline{yzxzyzxzyxyxzyx} = \overline{xy}.\overline{yxyxzyyzxzxyxzxzyxyz}.\overline{yx} \quad \text{by (ii), (iii)}$$
$$= \overline{xy}.\overline{yxzxzyzxzxyxz}.\overline{yx} \quad \text{by (iv), (ii), (iii)}$$
$$= \overline{xyzxyxz}.\overline{yzxyxzyxz}\cdot\overline{zxyxzyx} \quad \text{by (ii), (iii)}$$
$$= \overline{xyzxyxzyxz}.\overline{xzy}.\overline{zxyzxyxzyx} \quad \text{by (ii), (iii);}$$

hence (2) follows from (iii).

(3)

$$\overline{yxzxyxzxyxzyzyz} = \overline{yzxzyz}.\overline{xzy}.\overline{zyzxzy}$$

by (ii), (iii); hence (3) follows from (iii).

(4)

$$\overline{zyzxyzyxzyxyzyxzxzy} = \overline{yz}.\overline{yzxyzxzxyzxzx}.\overline{zy} \quad \text{by (ii), (iii)}$$
$$= \overline{yzxzyzx}.\overline{zxy}.\overline{xzyzxzy} \quad \text{by (ii), (iii);}$$

hence (4) follows from (iii).

(5)

$$\overline{yxzxyzxyxzyzxzxyzzyxzyxyzxz} = \overline{xyx}.\overline{yzyxzxyzyxyxyzxyz}.\overline{xyx} \quad \text{by (ii), (iii)}$$

$$= \overline{xyxzyzyzxyzx}.\overline{zxy}.\overline{xzyxzyzyzxyx} \quad \text{by (ii), (iii);}$$

hence (5) follows from (iii).

D. A power commutator presentation for Δ_2

We give here a power commutator presentation for Δ_2 . The table was computed using the Canberra nilpotent quotient algorithm in Bielefeld. For more detailed information see [B].

The generators 1 to 26 are defined as

$$1 = X , \qquad 2 = Y , \qquad 3 = [2, 1] ,$$
$$4 = Y^2 , \qquad 5 = [3, 1] , \qquad 6 = [3, 2] ,$$
$$7 = [6, 2] , \qquad 8 = [7, 1] , \qquad 9 = [7, 2] ,$$
$$10 = [8, 2] , \quad 11 = [9, 2] , \quad 12 = [10, 1] ,$$
$$13 = [12, 2] , \quad 14 = [13, 1] , \quad 15 = [13, 2] ,$$
$$16 = [14, 2] , \quad 17 = [15, 2] , \quad 18 = [16, 1] ,$$
$$19 = [17, 1] , \quad 20 = [18, 2] , \quad 21 = [20, 1] ,$$
$$22 = [20, 2] , \quad 23 = [21, 2] , \quad 24 = [23, 1] ,$$
$$25 = [24, 2] , \quad 26 = [25, 1] .$$

The nontrivial squares of the generators are computed as

$$2^2 = 4$$
$$3^2 = 5 \quad 11 \quad 13 \quad 14 \quad 15 \quad 16 \quad 17 \quad 19 \quad 20 \quad 26$$
$$5^2 = 11 \quad 13 \quad 14 \quad 15 \quad 16 \quad 17 \quad 19 \quad 20 \quad 26$$
$$6^2 = 9 \quad 10 \quad 11 \quad 12 \quad 14 \quad 15 \quad 16 \quad 17 \quad 19 \quad 21 \quad 23 \quad 25 \quad 26$$
$$7^2 = 11 \quad 15 \quad 16 \quad 18 \quad 21 \quad 23 \quad 25$$
$$8^2 = 14 \quad 15 \quad 17 \quad 18 \quad 21 \quad 22 \quad 23 \quad 24 \quad 25$$
$$9^2 = 16 \quad 17 \quad 20 \quad 22 \quad 23 \quad 25 \quad 26$$
$$10^2 = 16 \quad 17 \quad 20 \quad 23$$
$$12^2 = 18 \quad 19 \quad 22 \quad 23 \quad 25$$
$$13^2 = 25$$
$$14^2 = 26$$
$$15^2 = 26$$

The nontrivial commutators of the generators are computed as

$$[2, 1] = 3$$
$$[3, 1] = 5$$
$$[3, 2] = 6$$
$$[4, 1] = 5 \quad 6 \quad 8 \quad 10 \quad 13 \quad 14 \quad 16 \quad 17 \quad 18 \quad 19 \quad 20 \quad 22 \quad 25 \quad 26$$

```
[4, 3] =  7   9  10  12  14  17  22  23  24
[5, 1] = 11  13  14  15  16  17  19  20  26
[5, 2] =  8   9  12  16  21  22  26
[5, 4] = 10  11  13  14  15  16  19  20  23
[6, 1] =  9  10  11  12  14  15  16  17  18  21  22  24  26
[6, 2] =  7
[6, 3] =  8  10  11  15  18  19  24  25
[6, 4] =  9  11  15  16  18  21
[6, 5] = 12  14  15  16  17  20  21  22  23
[7, 1] =  8
[7, 2] =  9
[7, 3] = 10  13  14  19  20  21  26
[7, 4] = 11  16  17  20  22  23  25  26
[7, 5] = 13  14  17  19  24
[7, 6] = 13  14  15  17  18  19  22  23  26
[8, 1] = 14  15  17  18  21  22  23  24  25
[8, 2] = 10
[8, 3] = 12  13  14  16  17  18  21  22  24  25  26
[8, 4] = 14  19  22  23  24
[8, 5] = 14  19  20  21  22  23  26
[8, 6] = 14  19  20  21  23  26
[8, 7] = 15  17  18  19  22  23  25  26
[9, 1] = 12  14  18  19  24
[9, 2] = 11
[9, 3] = 13  19  20  21  25
[9, 4] = 16  17  20  23  25  26
[9, 5] = 16  18  21  23  24
[9, 6] = 15  16  21  26
[9, 7] = 17  20  23
[9, 8] = 21  22  24
[10, 1] = 12
[10, 2] = 14  16  17  19  20  22  24
[10, 3] = 13  14  19  20  21  23  25  26
[10, 4] = 16  20  26
[10, 5] = 16  20  23
[10, 6] = 15  16  18  19  20  21  23  25  26
[10, 7] = 17  21  25
[10, 8] = 23  26
[10, 9] = 22  24
[11, 1] = 14  18  21  26
[11, 2] = 16  17  20  23  25  26
[11, 3] = 16  19  20  23  25
```

```
   [11, 5] = 20   24   25
   [11, 6] = 20   22   24   25
   [11, 7] = 22   23   25   26
   [11, 8] = 22   23   24   26
   [11, 9] = 23   25
  [11, 10] = 23   26
   [12, 1] = 18   19   22   23   25
   [12, 2] = 13
   [12, 3] = 14   16   18   25
   [12, 4] = 15   25
   [12, 5] = 18   21   23   26
   [12, 6] = 18   19   24   25   26
   [12, 7] = 19   23   24
   [12, 8] = 24
   [12, 9] = 21   22   23   25   26
  [12, 10] = 21   22   23   24   25
  [12, 11] = 24
   [13, 1] = 14
   [13, 2] = 15
   [13, 3] = 16   18   26
   [13, 4] = 17   26
   [13, 5] = 20   23   24   25
   [13, 6] = 19   20   21   22   24
   [13, 7] = 20   21   22   24
   [13, 8] = 21   22   23   24
   [13, 9] = 24
  [13, 11] = 25
   [14, 1] = 26
   [14, 2] = 16
   [14, 3] = 18   20   21
   [14, 4] = 22   26
   [14, 5] = 21   24   25
   [14, 6] = 21   25
   [14, 7] = 22   23
   [14, 8] = 26
  [14, 11] = 26
   [15, 1] = 19   26
   [15, 2] = 17
   [15, 3] = 19
   [15, 4] = 26
   [15, 5] = 21   23   25
   [15, 6] = 21   22   23   24   25   26
```

```
   [15, 7] = 22   24   26
   [15, 8] = 26
   [15, 9] = 26
  [15, 11] = 26
   [16, 1] = 18
   [16, 2] = 22   26
   [16, 3] = 20   21
   [16, 4] = 23   25
   [16, 5] = 23   25   26
   [16, 6] = 22   23   25   26
   [16, 7] = 23   25
   [16, 9] = 26
   [17, 1] = 19
   [17, 2] = 26
   [17, 3] = 20   21   23   24
   [17, 5] = 23   26
   [17, 6] = 22   23   24
   [17, 7] = 23
   [17, 9] = 26
   [18, 2] = 20
   [18, 3] = 21   23   24
   [18, 4] = 22
   [18, 5] = 24   26
   [18, 6] = 26
   [18, 7] = 24   26
   [19, 2] = 20   23
   [19, 3] = 21   24   25   26
   [19, 4] = 22   26
   [19, 5] = 24
   [19, 7] = 24
   [20, 1] = 21
   [20, 2] = 22
   [20, 3] = 23   24
   [20, 4] = 23   25
   [20, 5] = 25
   [20, 6] = 24   25   26
   [20, 7] = 25
   [21, 2] = 23
   [21, 3] = 24   25   26
   [21, 4] = 26
   [21, 5] = 26
   [21, 6] = 26
```

$$[21, \ 7] = 26$$
$$[22, \ 1] = 24 \quad 26$$
$$[22, \ 2] = 23 \quad 25$$
$$[22, \ 3] = 24 \quad 26$$
$$[22, \ 5] = 26$$
$$[22, \ 7] = 26$$
$$[23, \ 1] = 24$$
$$[23, \ 2] = 26$$
$$[23, \ 3] = 25 \quad 26$$
$$[23, \ 6] = 26$$
$$[24, \ 2] = 25$$
$$[24, \ 3] = 26$$
$$[24, \ 4] = 26$$
$$[25, \ 1] = 26$$
$$[25, \ 2] = 26$$

The machine has found that the group $\tilde{\Delta}_2$ defined below satisfies

$$\tilde{\Delta}_2 = \overline{\tilde{\Delta}}_2 = \Delta_2 \ .$$

We believe that $\tilde{\Delta}_2$ is isomorphic to Δ_2, but this would need a more detailed analysis in our proof.

$$\tilde{\Delta}_2 = \langle X, \ Y \mid X^2 = Y^4 = \left(XYXY^2\right)^4 = (XY)^8 = \left(XYXY^{-1}\right)^8 = \left(XY^2\right)^8 = \left(XYXY^{-1}XYXY^{-1}XY\right)^8$$
$$= \left(XYXY^{-1}XY^{-1}XY^2\right)^8 = \left(XYXY^{-1}XY^2\right)^8 = 1 \rangle \ .$$

References

[A] Fritz J. Grunewald, George Havas, J.L. Mennicke, M.F. Newman, "Groups of exponent eight", these proceedings, pp. 49-188.

[B] George Havas and M.F. Newman, "Application of computers to questions like those of Burnside", these proceedings, pp. 211-230.

Sonderforschungsbereich,   Universitat Bielefeld,
Theoretische Mathematik,   Bielefeld,
Universitat Bonn, Bonn,   Federal Republic of Germany.
Federal Republic of Germany;

PROCEEDINGS OF THE BURNSIDE WORKSHOP
BIELEFELD, June-July 1977, 211-230.

20-04, 20F50

APPLICATION OF COMPUTERS
TO QUESTIONS LIKE THOSE OF BURNSIDE

George Havas and M.F. Newman

Computers have been used in seeking answers to questions related to those
about periodic groups asked by Burnside in his influential paper of 1902.
A survey is given of results obtained with the aid of computers and a key
program which manipulates presentations for groups of prime-power order is
described.

1. Introduction

Computers are becoming increasingly important in studying questions about groups
like those about periodic groups raised by Burnside in his influential paper of 1902.
The actual questions asked by Burnside are discussed in the paper Problems in these
proceedings. He was concerned with what are now called Burnside groups. We consider
here similar questions about a wider class of groups whose description may include
individual relations and residual finiteness in addition to an exponent law. For
simplicity we refer to all these questions as Burnside questions. The main results
are described in section three below.

This report has its origins in a lecture given by one of us at the workshop.
Since then both of us have lectured a number of times on this theme. There is a brief
report on this subject in the lecture "Computers in Group Theory" given by Marshall

Hall to the Galway Summer School in 1973 (see M. Hall 1977, pp. 34-35).

The growing impact of computers derives from their capacity for fast accurate symbol manipulation on a grand scale. The symbol manipulation procedure relating to groups for which computers have been most extensively used is coset enumeration. A detailed account can be found in a paper by Cannon, Dimino, Havas and Watson (1973). It has been widely used in the context of Burnside questions. For instance, given a finite set $X = \{a, b\}$ and the finite set

$$R = \{a^4, b^4, (ab)^4, (a^{-1}b)^4, (a^2b)^4, (ab^2)^4, (a^2b^2)^4, (a^{-1}b^{-1}ab)^4, (a^{-1}bab)^4\}$$

of group words on X, coset enumeration yields that the group G generated by X and having R as a defining set of relators has order $4096 = 2^{12}$ and a pair of permutations on 4096 letters which generate a group isomorphic to G. On a Univac 1100/42 using the coset enumeration program, Todd-Coxeter V2.2, this result can be obtained in about 30 seconds. From the output it can be deduced that G has exponent 4. Hence $B(2, 4)$ has order 2^{12} and has a presentation on two generators which has a defining set of only nine fourth powers. This example shows, in a simplified form, the role of coset enumeration in the study of groups. This is to find, where possible, the set $G \backslash S$ of cosets of a subgroup S in a group G and a representation of G by permutations of $G \backslash S$ given a finite presentation for G and a finite generating set for S. Further examples of results obtained using coset enumeration are given in section three.

In the context of the Burnside questions good progress has been made using computers to perform manipulations with descriptions of groups of prime-power order. For example, $B(2, 4)$ cannot be defined by fewer than nine fourth powers (Macdonald 1974), $\bar{B}(2, 5)$, the largest residually finite quotient of $B(2, 5)$, has order 5^{34} (Havas, Wall and Wamsley 1974), $B(4, 4)$ has order 2^{422} and $\bar{\Delta}$, the largest residually finite quotient of

$$\Delta := \langle x, y \mid x^2 = y^4 = 1, \text{ exponent } 8 \rangle ,$$

has order dividing 2^{205}. In section three we will give an outline of how the latter results were obtained. Recently the same method has been used to obtain a related result; see the note by Alford and Pietsch in these proceedings. The programs used are based on algorithms for calculating information about nilpotent quotients of groups given by finite presentations and finite sets of identical relations or laws. There have been a few rather brief accounts of such nilpotent quotient programs published (Macdonald 1973, 1974, Wamsley 1974, Newman 1976). While these describe applications to Burnside questions, none of them goes into sufficient detail to discuss a key point in such applications which is how to make law enforcement practical. This point is taken up in section two of this account.

Computers have also been programmed to perform other symbol manipulation procedures related to groups. Two such procedures which are playing an important role in some current investigations are the Reidemeister-Schreier algorithm for calculating presentations for subgroups of finite index in finitely presented groups (Havas 1974) and manipulations such as Nielsen and Tietze transformations with presentations.

The programs we use are written in a readily portable superset of 1966 Standard FORTRAN. Further information on both programs and results can be obtained from the authors.

We are indebted to many colleagues, too many to name individually, for helpful discussions and access to programs. We thank them all collectively.

2. A nilpotent quotient program

In this section we describe a computer program, the Canberra nilpotent quotient program, which, with its forerunners, played an important role in many of the applications to be discussed in the next section. The program provides the capacity to manipulate presentations for groups of prime-power order efficiently. This makes it possible to study many groups of order p^n for moderate values of p and n, say $p < 1000$, $n < 20$, and selected groups with n well into the hundreds and occasionally even higher. There have been essentially two previous programs of this kind; one produced by Macdonald (1974) and one by Bayes, Kautsky and Wamsley (1974). The Canberra program has its origins in the latter and incorporates some important features from it.

The core of the program is a set of routines for manipulating what we call power-commutator presentations. These are finite presentations which present groups of prime-power order. Moreover every group of prime-power order has such presentations. A *power-commutator presentation* consists of a finite set $A = \{a_1, \ldots, a_n\}$ of generators and a finite set of $n(n+1)/2$ relations

$$a_i^p = \prod_{l=i+1}^{n} a_l^{\alpha(i,l)} \, ,$$

(2.1)

$$a_j a_i = a_i a_j \prod_{l=j+1}^{n} a_l^{\alpha(i,j,l)} \, , \quad i < j \, ,$$

where p is a prime and $\alpha(i, l)$, $\alpha(i, j, l)$ belong to $\{0, 1, \ldots, p-1\}$. It will be convenient to denote this presentation $(n; p, \alpha)$ and the group it presents $\langle n; p, \alpha \rangle$. The way in which such presentations may be thought of as occurring will be explained later. The word "commutator" is used because in most contexts the second type of relation is written with a commutator on the left; it is convenient here to defer this change a little.

It is easy to see that every element of $\langle n; p, \alpha \rangle$ can be represented by a word

of the form $\prod_{l=1}^{n} a_l^{\xi(l)}$ with $\xi(l)$ in $\{0, 1, \ldots, p-1\}$; we call such a word *normal*
and abbreviate it to ξ , and use $\emptyset$ for the identity element. At the heart of this
whole approach is the simple fact that the given set of relations makes it straight-
forward to calculate a normal word equivalent to the product of two normal words. It
will be convenient to discuss this (collection) process in the wider context of all
semigroup words on A ; inverses play little role in the present discussion. Notice
that the set of left sides of the relations in $(n; p, \alpha)$ is precisely the set of
minimal non-normal semigroup words on A . A *collection process* relative to
$(n; p, \alpha)$ is a procedure of the following kind:

(2.2) Given a semigroup word w on A ,

    1 if w is normal, stop;

    2 replace a minimal non-normal subword u of w by the right side of
       the relation whose left side is u , and return to 1 .

It is routine to prove that collection processes terminate, in a normal word. There
are many collection processes because of the choice in step 2. A detailed discussion
of collection in the Canberra program is given in a paper by Havas and Nicholson
(1976). The rest of this account is based on the collection process in the Canberra
program; it would apply equally well to any other (fixed) collection process. The
normal word resulting from applying this collection to a word w will be denoted
(w) .

The first question which arises is: what is the order of $\langle n; p, \alpha \rangle$? Since the
number of normal words is p^n , the group $\langle n; p, \alpha \rangle$ has order at most p^n .
Unfortunately it often happens that $\langle n; p, \alpha \rangle$ has order less than p^n . In this
case it is easy to see that there are normal words ξ, η, ζ such that
$((\xi\eta)\zeta) \neq (\xi(\eta\zeta))$. This leads to the following straight-forward test for determining
whether the order of $\langle n; p, \alpha \rangle$ is p^n . Let T_n be the set of words consisting of
all

$$a_k a_j a_i , \qquad i < j < k ,$$

$$a_j^p a_i, \; a_j a_i^p , \quad i < j ,$$

$$a_i^{p+1} .$$

Observe that each word in T_n has exactly two minimal non-normal subwords and that
these overlap. Hence each word in T_n can be uniquely written $\xi\eta\zeta$ with ξ, η, ζ
non-empty normal words and $\xi\eta, \eta\zeta$ minimal non-normal. The group $\langle n; p, \alpha \rangle$ has
order p^n if and only if for each word $\xi\eta\zeta$ in T_n , $((\xi\eta)\zeta) = (\xi(\eta\zeta))$. If this

is satisfied $(n, p; \alpha)$ is said to be *consistent*. If $(n; p, \alpha)$ is not consistent, then some test word will collect to two different normal words.

This leads to another basic procedure, reduction, which, given a power-commutator presentation $(n; p, \alpha)$ and a pair μ, ν of distinct normal words, produces on a subset of $(n-1$ elements of$)$ A a power-commutator presentation for the group defined by $(n; p, \alpha)$ and $\mu = \nu$. It is not difficult to describe reduction in general; however, it is easier to describe a special case which suffices for the present discussion. Before doing so, we complete our consideration of the order question. When the words μ, ν come from collecting a word in T_n the group defined by the resulting power-commutator presentation is clearly isomorphic to $\langle n; p, \alpha \rangle$. Repeating this process leads, eventually, to a consistent power-commutator presentation for $\langle n; p, \alpha \rangle$; the order of $\langle n; p, \alpha \rangle$ can be read off from this.

From now on the relations in a power-commutator presentation will be taken in power and commutator form. Let us say that a normal word ξ is *obviously central and of order* p if, for every generator a_l which occurs in ξ $\bigl($that is, $\xi(l) \neq 0\bigr)$, all the relations with a_l on the left have $\emptyset$ on the right. We only need a *reduction* procedure when the words μ, ν are obviously central and of order p. Let G be the group defined by $(n; p, \alpha)$ and $\mu = \nu$. Since μ, ν are distinct there is a greatest positive integer k such that $\mu(k) \neq \nu(k)$. Let $\eta(l)$ be the least non-negative integer such that

$$\eta(l)\bigl(\mu(k)-\nu(k)\bigr) \equiv \nu(l) - \mu(l) \text{ modulo } p .$$

The relation $\mu = \nu$ is equivalent to

(2.3) $$a_k = \prod_{l=1}^{k-1} a_l^{\eta(l)}$$

$\bigl($with a_l obviously central and of order p whenever $\eta(l) \neq 0\bigr)$. Let α' be the function defined by

$$\alpha'(i, l) \equiv \alpha(i, l) + \eta(l)\alpha(i, k) \text{ modulo } p ,$$
$$\alpha'(i, j, l) \equiv \alpha(i, j, l) + \eta(l)\alpha(i, j, k) \text{ modulo } p ,$$

and

$$0 \leq \alpha'(i, l) , \quad \alpha'(i, j, l) < p .$$

Then G is defined by $(n; p, \alpha')$ and 2.3. Since $\alpha'(i, k) = \alpha'(i, j, k) = 0$ for all i, j, the generator a_k only occurs on the left of relations in $(n; p, \alpha')$; and when it does the relation is a consequence of 2.3 and relations not involving a_k. Hence the presentation obtained from $(n; p, \alpha')$ by deleting all relations involving a_k and a_k from the generating set is a power-commutator presentation for G on $n - 1$ generators; this is what was required.

Clearly it would be desirable from the point of view of computational efficiency to make the set of test words for consistency smaller. There are some obvious redundancies; for example, one never needs to test $a_n a_j a_i$. We now introduce some further structure into power-commutator presentations which leads to a useful reduction in the test set for consistency. This additional structure also leads to a criterion which makes it practical to test whether a group given by a power-commutator presentation has exponent p^f .

(2.4)   A power-commutator presentation will be called *weighted* if

    (i)   some of its relations are called *definitions*; and

    (ii) it admits a weight function ω from $\{1, \ldots, n\}$ to the positive integers such that

        (a) $\omega(1) = 1$ and $\omega(i) \leq \omega(i+1)$;

        (b) for each k with $\omega(k) > 1$ there is exactly one definition whose right side is a_k ;

        (c) if the definition of a_k is $a_i^p = a_k$, then $\omega(k) = \omega(i) + 1$;

        (d) if the definition of a_k is $[a_j, a_i] = a_k$, then
$$\omega(k) = \omega(i) + \omega(j) ;$$

        (e) if $\alpha(i, l) \neq 0$, then $\omega(l) \geq \omega(i) + 1$,
if $\alpha(i, j, l) \neq 0$, then $\omega(l) \geq \omega(i) + \omega(j)$.

Note there may be more than one relation which has a_k as right side but only one of these is the definition of a_k . It is easy to see that there are power-commutator presentations which admit no such weight function; indeed presentations of this kind occur in the course of the calculations to be described. A weighted power-commutator presentation $(n; p, \alpha)$ with weight ω will be denoted $(n; p, \alpha; \omega)$. The positive integer $\omega(n)$ is the *class* of $(n; p, \alpha; \omega)$ and of $\langle n; p, \alpha; \omega \rangle$, the group it defines. This class is, see 2.6, the exponent-p-central class which is usually defined in terms of the lower exponent-p-central series of a group G ; that is, the chain

$$G = G_0 \geq \ldots \geq G_h \geq G_{h+1} \geq \ldots$$

of subgroups where $G_{h+1} = [G_h, G]G_h^p$. If $G_{c-1} > G_c = E$, then G has *exponent-p-central* class c . Note that a group with exponent-p-central class c is nilpotent and has nilpotency class at most c . In the study of groups of prime-power order the exponent-p-central class often seems the more significant. An important observation about the lower exponent-p-central series is

(2.5) $[G_j, G_h] \leq G_{j+h+1}$.

This is proved in essentially the same way as the corresponding result for the lower
central series.

(2.6)   The exponent-p-central class of $\langle n; p, \alpha; \omega \rangle$ is $\omega(n)$.

Let $G = \langle n; p, \alpha; \omega \rangle$. We prove that G_h is generated by all the a_k with
$\omega(k) \geq h + 1$. This is clear for $h = 0$. For $h \geq 1$, suppose G_{h-1} is generated
by all the a_n with $\omega(l) \geq h$. It follows that G_h is generated by all the a_l^p
and all the $[a_l, a_i]$ with $\omega(l) \geq h$ and a_i arbitrary, and so, by clause (a) of
the definition of a weighted power-commutator presentation, G_h is contained in the
subgroup generated by all a_k with $\omega(k) \geq h + 1$. Conversely, if $\omega(k) \geq h + 1 > 1$,
then either $a_k = a_i^p$ with $\omega(k) = \omega(i) + 1$ or $a_k = [a_j, a_i]$ with
$\omega(k) = \omega(i) + \omega(j)$. Hence, by the inductive assumption and 2.5, a_k belongs to
G_h .

A first gain which comes from using weighted power-commutator presentations is
that the set T_n of test words for consistency can be reduced to T_n^* :

(2.7) T_n^* is the set of all words

$$a_k a_j a_i , \qquad i < j < k ,$$

with $\omega(i) + \omega(j) + \omega(k) \leq \omega(n)$,

$$a_j^p a_i, \ a_j a_i^p , \quad i < j ,$$

with $1 + \omega(i) + \omega(j) \leq \omega(n)$, and

$$a_i^{p+1} , \qquad \text{with } 1 + 2\omega(i) \leq \omega(n) .$$

This is so because it is straight-forward to check that for every other test word
collection yields the same result; for example, if $1 + 2\omega(i) > \omega(n)$, then
$\left(a_i^p\right) a_i = a_i \left(a_i^p\right)$ since $\left(a_i^p\right)$ only involves a_l with $\omega(l) \geq \omega(i) + 1$. While this
is a useful reduction, in practice redundancy is usually observed once the class
exceeds 3 ; moreover this redundancy increases with class. It would be desirable to
have this observation reflected in still smaller test sets.

A second gain is a criterion which makes testing whether an exponent law holds
reasonably practical. For a normal word ξ , let us call $\sum\limits_{l=1}^{n} \omega(l)\xi(l)$ the *weight* of
ξ .

(2.8)   The group $\langle n; p, \alpha; \omega \rangle$ has exponent p^f provided $\xi^{p^f} = \emptyset$ for all normal words of weight at most $\omega(n)$.

The criterion can be proved by a simple varietal argument. The crux is covered by an argument of Higman (1959, p. 169; or see §3.3 of Hanna Neumann's book (1967) where, however, it is not as conveniently put). It is applied to the word

$(x_1 \ldots x_s)^{p^f}$ in the free group F freely generated by $X = \{x_1, \ldots, x_s\}$ to give

$$(x_1 \ldots x_s)^{p^f} = \prod_J \left(x_J^{p^f}\right)^{\pm 1} . u$$

where x_J is the product of some proper (ordered) subset of X and u is a product of commutators in X each of which involves each element of X . From this the criterion is proved by induction. Let ξ be a normal word. If the weight of ξ is at most $\omega(n)$, then clearly $\xi^{p^f} = \emptyset$. Otherwise assume $\eta^{p^f} = \emptyset$ whenever the weight of η is less than that of ξ . Let $s = \sum_{l=1}^{n} \xi(l)$. Let θ be the homomorphism from F to $\langle n; p, \alpha; \omega \rangle$ which maps x_k to a_i where

$$\sum_{l=1}^{i-1} \xi(l) < k \le \sum_{l=1}^{i} \xi(l) ,$$

then $(x_1 \ldots x_s)\theta = \xi$. Hence by the above

$$\xi^{p^f} = \prod_J \left((x_J\theta)^{p^f}\right)^{\pm 1} . u\theta .$$

Clearly $x_J\theta$ is a normal word and its weight is less than that of ξ , so

$(x_J\theta)^{p^f} = \emptyset$. From the structure of u it follows that $u\theta$ is a product of commutators in A each with at least $\xi(l)$ entries a_l for each l . Since the weight of ξ exceeds $\omega(n)$ and $\langle n; p, \alpha; \omega \rangle$ has class $\omega(n)$, it follows readily that $u\theta = \emptyset$. Hence $\xi^{p^f} = \emptyset$ as required.

This criterion is incorporated into the program of Bayes, Kautsky and Wamsley (though their paper (1974) does not mention it). Macdonald's (1974) criterion seems to be of a similar kind. While the criterion makes testing whether exponent laws hold quite practical, the calculation of even the powers specified by it is the most severe limitation to these explorations of groups of Burnside type. As with consistency testing much redundancy is observed, so it would be desirable to find further restrictions on the powers that need to be calculated.

In practice we have used other considerations to extend the scope of what can be
done. In our initial work on $B(4, 4)$ we used related commutator laws and
congruences. The relevant calculations are more complicated to program, but are then
quicker to execute. The simplest example of this comes from the law

$$[y, x]^2[y, x, x, x][y, x, y, x][y, x, y, y] = \emptyset$$

which holds in $B(2, 4)$ and therefore in all groups of exponent 4 . This yields the
criterion that $\langle n; 2, \alpha; \omega \rangle$ has exponent 4 provided $a_k^4 = \emptyset$ for all k and

$$[\eta, \xi]^2[\eta, \xi, \xi, \xi][\eta, \xi, \eta, \xi][\eta, \xi, \eta, \eta] = \emptyset$$

for all normal words ξ, η such that $\xi\eta$ is a normal word of weight at most $\omega(n)$.
More complicated criteria come from allowing more variables. The present nilpotent
quotient program can handle $B(4, 4)$ without such special considerations. We hope,
if resources permit, to implement these more refined methods in such a way that they
will allow the routine testing of other laws such as commutator laws. In studying $\overline{\Delta}$
our main goal was to prove finiteness. For this it suffices to show that some
preimage of $\overline{\Delta}$ is finite; or, equivalently, to find a set E of eighth powers such
that the residually finite group generated by $\{x, y\}$ with $\{x^2, y^4\} \cup E$ as a
defining set of relators is finite. Therefore we used heuristic considerations to
limit the eighth power calculations done; more details are given in the next section.

Just as reduction is used to make a power-commutator presentation consistent, so
it can be used to enforce exponent laws; that is, given p^f , to yield from $(n; p, \alpha)$
a power-commutator presentation for the largest exponent p^f quotient of $\langle n; p, \alpha \rangle$.
If testing shows that $\xi^{p^f} \neq \emptyset$ for some ξ , then the largest exponent p^f quotient
of $\langle n; p, \alpha \rangle$ is the same as that of the group defined by $(n; p, \alpha)$ and
$\xi^{p^f} = \emptyset$. Applying reduction yields a power-commutator presentation for this group
which has fewer generators. Repeating this process of testing and reduction clearly
leads to the required power-commutator presentation.

We introduce next the p-covering presentation for a consistent weighted power-
commutator presentation $(n; p, \alpha; \omega)$. Let $P = \langle n; p, \alpha; \omega \rangle$. Suppose
$\omega(d) = 1 < \omega(d+1)$, then P can be generated by d , and no fewer, elements.

The *p-covering presentation* for $(n; p, \alpha; \omega)$ has a generating set
$a_1, \ldots, a_{d+n(n+1)/2}$ and relations of three kinds:

(1) the $n - d$ definitions in $(n; p, \alpha; \omega)$,

(2) $d + n(n-1)/2$ of the form $u_k = v_k a_k$ for $k \in \{n+1, \ldots, d+n(n+1)/2\}$

   where $u_k = v_k$ is a relation in $(n; p, \alpha; \omega)$ which is not a definition,

(3) those which specify that $a_{n+1}, \ldots, a_{d+n(n+1)/2}$ are central and of

order p .

Relations of the first two kinds are definitions. Because $(n; p, \alpha; \omega)$ is
consistent, if the consistency test fails for a test word relative to the p-covering
presentation, the resulting normal words are obviously central and of order p , and
so reduction can be used to yield a consistent power-commutator presentation
$(n^*; p, \alpha^*)$ with $n^* - d$ definitions. This presentation can not in general be made
into a weighted presentation. The group $P^* = \langle n^*; p, \alpha^* \rangle$ is the p-*covering group*
of P . Clearly P^* has class at most $\omega(n) + 1$. If Q is a group of class
$\omega(n) + 1$ whose largest class $\omega(n)$ quotient is P , it can be shown that Q is a
homomorphic image of P^* . If the class of P^* is $\omega(n)$, there is no such Q and
P is said to be *terminal*; otherwise, there is such a Q and P is said to be
capable. If the relations $u_k = v_k$ which are not definitions are suitably ordered

then the class of P^* can be determined from the definition of a_{n+1} in

$(n^*; p, \alpha^*)$. If this is $a_i^p = a_{n+1}$ with $\omega(i) = \omega(n)$ or $[a_j, a_i] = a_{n+1}$ with
$\omega(i) + \omega(j) = \omega(n) + 1$, then the class of P^* is $\omega(n) + 1$ and P is capable;
otherwise the class is $\omega(n)$ and P is terminal. The kernel of the mapping from P^*
to P is isomorphic to the second homology group $H_2(P, Z_p)$ of P over the field of
p elements; we call it the p-*multiplicator* of P . The rank $n^* - n$ of the
p-multiplicator is the least number of relations needed to define P as a group of
p-power order on d generators; in saying a presentation $\{X|R\}$ defines a group G
as a group of p-power order, we mean that the group $\langle X|R \rangle$ has a largest quotient of
p-power order and that this quotient is isomorphic to G . It follows that $n^* - n$
is a lower bound on the number of relations needed to define P as a d generator
group.   It remains an open question whether this lower bound is always attained.

Since the class of P^* is at most $\omega(n) + 1$, it is possible to introduce fewer
generators in writing down a power-commutator presentation for P^* unless
$\omega(n) = 1$. It suffices to add a new generator for each relation in $(n; p, \alpha; \omega)$
which neither is a definition nor has its left side of the form $[a_j, a_i]$ with
$\omega(i) + \omega(j) \geq \omega(n) + 2$. This is of value because it means fewer reduction steps are
needed to reach a consistent power-commutator presentation. Further cuts are
possible. For example, if $\omega(k) = 2$ and $[a_j, a_i] = a_k$ is the definition of a_k
and if $\omega(l) = \omega(n) - 1$, then

$$[a_l, a_k] = [a_l, a_j, a_i][a_l, a_i, a_j]^{p-1} .$$

To tell much more of this story requires technical details that it is not appropriate
to go into here. Note though that it implies that $\omega(i) = 1$ in all definitions of

the form $[a_j, a_i] = a_k$.

We have now developed enough machinery to outline how to determine a consistent power-commutator presentation for the largest class c quotient of the Burnside group $B(d, p^f)$ which we denote $B(d, p^f; c)$. Begin with a consistent weighted power-commutator presentation for $B(d, p^f; 1)$: it has a generating set $\{a_1, \ldots, a_d\}$ with relations $a_i^p = \emptyset$ and $[a_j, a_i] = \emptyset$ for all i, j and weight function the constant function with value 1 . For $c > 1$, let $(n; p, \alpha; \omega)$ be a consistent weighted power-commutator presentation for $B(d, p^f; c-1)$ with $\omega(n) = c - 1$. Write down a power-commutator presentation for the p-covering group of $\langle n; p, \alpha; \omega \rangle$ and reduce it to a consistent presentation $(n^*; p, \alpha^*)$. If this shows $\langle n; p, \alpha; \omega \rangle$ is terminal, then the largest residually finite quotient, $\overline{B}(d, p^f)$, of $B(d, p^f)$ is presented by $(n; p, \alpha; \omega)$. It follows that $\overline{B}(d, p^f)$ has order p^n , class $c - 1$ and can be defined as a group of p-power order by d generators and $n^* - n$ p^fth powers. If $\langle n; p, \alpha; \omega \rangle$ is capable, enforce the exponent p^f on $(n^*; p, \alpha^*)$; note that in this context the normal word collected from a p^fth power will be obviously central and of order p , so the condition needed for reduction holds. Let the resulting (consistent) presentation be $(n'; p, \alpha')$. If $n' = n$, then again $\overline{B}(d, p^f)$ is presented by $(n; p, \alpha; \omega)$ and the other consequences follow. For example, a power-commutator presentation for $B(2, 4; 5)$ has 12 generators, and that of its 2-covering group has 21 generators; this latter presentation shows $B(2, 4; 5)$ is capable, however enforcing exponent 4 reduces it to the presentation for $B(2, 4; 5)$; hence $\overline{B}(2, 4)$ has order 2^{12} , class 5 and needs 9 relations, which can be 4th powers, to define it as a group of 2-power order; since $B(2, 4)$ is well-known to be finite, these conclusions apply to it as well. The function ω can be extended to a weight function ω' on $(n'; p, \alpha')$ by: $\omega'(i) = \omega(i)$ for $i \leq n$ and $\omega'(i) = c$ for $i > n$. Thus this procedure can be iterated. If $\overline{B}(d, p^f)$ is finite, as it is known to be when $f = 1$ by a result of Kostrikin, the procedure will eventually stop. In practice (for $p \geq 5$) this has happened so far only for $\overline{B}(2, 5)$.

In the study of groups, such as Δ , whose description includes individual relations, we work with presentations which, in effect, combine the given finite presentation and an appropriate power-commutator presentation. Thus in studying a group given by the finite presentation

$$\{b_1, \ldots, b_m; u_1 = v_1, \ldots, u_r = v_r\} ,$$

we work with augmented power-commutator presentations which have

$\{b_1, \ldots, b_m, a_1, \ldots, a_n\}$ as set of generators, and relations

$$u_1 = v_1, \ldots, u_r = v_r, \quad b_h = \prod_{l=1}^{n} a_l^{\rho(h,l)}, \quad 1 \le h \le m,$$

with $\rho(h, l)$ in $\{0, 1, \ldots, p-1\}$, and all the relations 2.1.

Where relevant, there is a weight function ω as before on $\{1, \ldots, n\}$;
however, now there is a definition for each a_k , the first d definitions, where
$\omega(d) = 1 < \omega(d+1)$, have left side b_h for some h .

In this context every semigroup word on the generators can be collected to a
normal word on $\{a_1, \ldots, a_n\}$ by adding the steps: replace b_h by $\prod a_l^{\rho(h,l)}$.
Where the inverse of a generator b_i is involved we deal with it by introducing a new
generator b_{m+i} and including among the defining relations $b_i b_{m+i} = \emptyset$. We say the
presentation is *consistent* if, in addition to the previous conditions, it satisfies
$(u_k) = (v_k)$, where ( ) denotes the result of collection, for all k in
$\{1, \ldots, r\}$. Thus, if an augmented power-commutator presentation is consistent, the
group it defines has order p^n and is a quotient of the group given by the input
presentation. We next modify the procedure for writing down p-covering
presentations. A new generator is added to each relation which is neither given nor a
definition (nor has large weight left side). Note this includes the case when
$n = 0$; then there are no definitions and the p-covering presentation has generating
set $\{b_1, \ldots, b_m, a_1, \ldots, a_m\}$ and relations

$$u_1 = v_1, \ldots, u_r = v_r,$$

$$b_k = a_k, \quad 1 \le k \le m,$$

and each a_i is central and has order p .

We can finally outline how to determine a consistent augmented power-commutator
presentation for the largest class c quotient of

$$\left\langle b_1, \ldots, b_m \mid u_1 = v_1, \ldots, u_r = v_r, \text{ exponent } p^f \right\rangle .$$

Begin with the largest class 0 quotient: it has generating set $\{b_1, \ldots, b_m\}$ and
relations

$$u_1 = v_1, \ldots, u_r = v_r, \quad b_1 = \emptyset, \ldots, b_m = \emptyset .$$

For $c \ge 1$, let there be given a consistent augmented power-commutator presentation
P for the largest class $c - 1$ quotient. Write down the p-covering presentation of
it and reduce to a consistent presentation. If this is P , the given group has a
largest quotient of p-power order which is presented by P . Otherwise enforce the

exponent p^f . The same conclusion follows if this reduces the presentation to P .
If not, extend the weight function as before.

3. Results

3.1 EXPONENT FOUR

Leech (1963) was the first to publish results on Burnside groups obtained with
the aid of a computer. He produced presentations for some groups of exponent three
and four. For example he found, using computer coset enumeration, various sets of
nine fourth powers which suffice as relators for $B(2, 4)$. He also reported on a
group of Burnside type suggested by Philip Hall, namely

$$I3 = \langle a, b, c \mid a^2 = b^2 = c^2 = \emptyset, \text{ exponent } 4 \rangle .$$

Leech computed the order of $I3$ and found ten relator presentations for this group.
With a view towards $B(3, 4)$ he investigated the group

$$I2 = \langle a, b, c \mid a^2 = b^2 = \emptyset, \text{ exponent } 4 \rangle .$$

Macdonald (1973, 1974) used a nilpotent quotient program to prove various results
about Burnside groups. He showed that minimal presentations for $B(2, 4)$ have nine
relations and for $I3$ have ten relations, as achieved by Leech. He also showed that
$I2$ has order 2^{19} . Bayes, Kautsky and Wamsley (1974) computed the order of
$B(3, 4)$, 2^{69} , and Alford, Havas and Newman (1975) computed the order of $B(4, 4)$,
2^{422} , in both cases using a nilpotent quotient program. More recently Havas (to
appear) has used coset enumerations in $B(2, 4)$ to express the fifth Engel word as a
product of fourth powers.

We present here some new results about various groups of exponent four and in
particular about $I2$ and about $B(4, 4)$. We include a description of the
computations involved.

Using the Canberra nilpotent quotient program we readily confirm Macdonald's
result that $I2$ has order 2^{19} . We determine that the rank of its 2-multiplicator
is 20 , which shows that at least 18 fourth power relations in addition to the 2
involutory relations are required to provide a presentation for $I2$. The Canberra
nilpotent quotient program also produces a list of 18 fourth powers which (together
with the two involutory relations) is sufficient to define $I2$ as a group of 2-power
order. This list is routinely obtained as the lexicographically shortest words in
terms of the power-commutator presentation generators which suffice. This kind of
list is quite suitable for nilpotent quotient calculations but less satisfactory for
other purposes, such as coset enumeration, because the free group word length is not
usually particularly short. In the case of $I2$ one of the longest such words is

$(acac^{-1}acbc^{-1}acab)^4$, a free group word of length 12 , raised to the fourth power. However other sets of fourth powers can be found readily. One approach is to calculate the 2-covering group of $I2$ and then to collect fourth powers in that covering group. Collection of short powers (short in the free group sense) reveals that the following relations suffice to define $I2$ as a group of 2-power order:

$$a^2 = b^2 = c^4 = (ab)^4 = (ac)^4 = (bc)^4 = (abc)^4 = (acb)^4 = (acc)^4 = (bcc)^4$$
$$= (abac)^4 = (abcb)^4 = (abcc)^4 = (acbc)^4 = (acbc^{-1})^4 = (ababc)^4 = (abcbc)^4$$
$$= (bacac)^4 = (bc^{-1}aca)^4 = (bcacc)^4 = \emptyset .$$

Such short relations are much more suitable for coset enumeration. It is easy to enumerate the 8192 cosets of $\langle a, c \rangle$ in the group presented this way. (The corresponding coset enumeration in the group presented with fourth powers routinely given by the Canberra nilpotent quotient program has not been successfully performed.) Since $\langle a, c \mid a^2 = c^4 = (ac)^4 = (acc)^4 = \emptyset \rangle$ presents a group of order 64 (very easy by coset enumeration) it follows that the above 20 relations provide a minimal presentation for $I2$.

Note that some of the claims in the literature about $I2$ are wrong. Thus Leech (1963) gives the incorrect order for the group. This arises from the misapprehension that $\langle ab, c \rangle$ is isomorphic to $B(2, 4)$ of order 2^{12} . In fact $\langle ab, c \rangle$ is a group of order 2^{10} , which follows from the above calculations. This may also be shown by explicit calculation with the power-commutator presentation for $I2$. E. Oppelt of RWTH Aachen has implemented routines which augment the Canberra nilpotent quotient program and provide facilities for subgroup construction and investigation by computer, and which can be used for this purpose.

Leech (1967) endeavours to improve on the presentation for $I2$ and also refers to an attempt by Sinkov at obtaining a presentation for that group (which is the presentation given by Cannon (1974)). The presentations of Leech and Sinkov define different groups of order 2^{21} which have $I2$ as a quotient.

Using the Canberra nilpotent quotient program we have computed the rank of the 2-multiplicator of $B(3, 4)$. This rank is 105 and we have found a set of 105 fourth powers which suffice to define $B(3, 4)$ as a group of 2-power order.

A very significant computational effort was put into determining a consistent power-commutator presentation for $B(4, 4)$. The initial motivation was to gain information on the solubility question for groups of exponent four. At the time that the computations were first performed it seemed possible that the determination of $B(4, 4)$ could perhaps be combined with results of Gupta and Newman (1974) to prove the solubility of all groups of exponent four, if that were in fact the case. Not surprisingly this did not succeed because insoluble groups of exponent four do exist, as was proved by Razmyslov (1978). However Vaughan-Lee (1979) has used the detailed information available on $B(4, 4)$, via the consistent power-commutator presentation computed for it, as a basis for determining the solubility length of $B(d, 4)$ explicitly in terms of d . As mentioned in section two, a number of technical innovations were made for the initial computation of $B(4, 4)$. A significant saving in computer time was made by the use of formulas instead of direct collection. Whereas the Canberra nilpotent quotient program uses direct collection both to process the consistency equations and to perform exponent testing, in the initial run for $B(4, 4)$ commutator formulas were calculated and used instead, where possible. Thus formulas for Jacobi identities, which correspond to consistency equations, and for fourth power expansions were computed. Substitution in this type of formula is superior to direct collection because the bulk of collection required has been done once and for all in computation of the formula. With improvements in computer technology and the development of the latest version of the Canberra nilpotent quotient program, formula substitution methods are no longer needed in order to compute $B(4, 4)$. Indeed we have used the current version of the Canberra nilpotent quotient program to recompute $B(4, 4)$ and also to compute the rank of its 2-multiplicator, namely 1055 . This calculation took about 10 hours on a Univac 1100/42 . Note that the rank of the 2-multiplicator shows that a presentation for $B(4, 4)$ requires at least 1055 relations. We have computed a set of 1055 fourth power relations which suffice to define $B(4, 4)$ as a group of 2-power order.

Other groups of exponent four for which we have computed consistent power-commutator presentations are:

$\langle a, b, c \mid a^2 = \emptyset$, exponent 4$\rangle$, which has order 2^{37} and class 7 ;

$\langle a, b, c, d \mid a^2 = b^2 = c^2 = d^2 = \emptyset$, exponent 4$\rangle$, which has order 2^{38} and class 5 ;

$\langle a, b, c, d \mid a^2 = b^2 = c^2 = \emptyset$, exponent 4$\rangle$, which has order 2^{66} and class 6 ;

$\langle a, b, c, d \mid a^2 = b^2 = \emptyset$, exponent 4$\rangle$, which has order 2^{120} and class 8 ;

$\langle a, b, c, d \mid a^2 = \emptyset$, exponent 4$\rangle$, which has order 2^{224} and class 10 ;

$\langle a, b, c, d, e \mid a^2 = b^2 = c^2 = d^2 = e^2 = \emptyset$, exponent 4$\rangle$, which has order 2^{138} and class 6 ;

$\langle a, b, c, d, e \mid a^2 = b^2 = c^2 = d^2 = \emptyset$, exponent 4$\rangle$, which has order 2^{228} and class 7 .

3.2 EXPONENT FIVE

The order of $\overline{B}(2, 5)$ was shown by Havas, Wall and Wamsley (1974) to be 5^{34} using computer calculations in two different ways, one via nilpotent quotient computations and the other using Lie algebraic methods. More recently we have obtained a consistent power-commutator presentation for $B(3, 5; 9)$ using the Canberra nilpotent quotient program. $B(3, 5; 9)$ has order 5^{916} which is less than the order of the free 3-generator Lie algebra satisfying the 4th Engel condition, namely 5^{926}. This provides an independent confirmation of a result of Wall (1974).

Currently a number of people, including M. Hall Jr., G. Havas, J.S. Richardson and J. Wilkinson, are investigating the unrestricted problem for exponent five with computer assistance. Calculations in search of a finiteness proof for $B(2, 5)$ are being performed using nilpotent quotient, coset enumeration, subgroup presentation and presentation manipulation programs.

3.3 EXPONENT SEVEN

We have constructed a consistent power-commutator presentation for $B(2, 7; 13)$. This is a group of order 7^{668}. If x and y are generators of this group then $[y, {_8x}] \neq \emptyset$, so the 8th Engel identity does not hold.

3.4 EXPONENT EIGHT

Detailed results about a number of groups of exponent eight are presented in other papers in these proceedings. Here we present some results and computational details about one particular group of exponent 8.

In the paper by Grunewald, Havas, Mennicke and Newman in these proceedings the problem as to whether the subgroup $\langle a^4, b^2 \rangle$ of $B(2, 8)$ is finite is posed. We show that it has a largest finite quotient by showing that a residually finite preimage is finite.

Let $\Delta := \langle a, b \mid a^2 = b^4 = \emptyset,$ exponent $8 \rangle$. Then Δ has a preimage D which has a maximal nilpotent quotient of order 2^{205} and class 26. This preimage was found by using the exponent testing procedure to partially enforce the exponent law. The class 10 quotient of Δ was calculated in the normal way. With increasing class the exponent tests were taking more and more time. As pointed out in section two, the exponent law test has a lot of redundancy so that, for example, to compute the class 10 quotient of Δ only two eighth powers are actually required, eighth powers of ab and abb which are words of weight 2 and 3. This suggests that it may suffice to test only words of low weight. From class 10 onwards eighth power testing was performed only for words with weight less than 11. D is the group defined by two generators, one of order 2 and one of order 4, in which all normal words of weight less than 11 have order dividing 8 . To define this as a group of 2-power order 69 eighth powers are needed, with those of highest weight having weight 10 . Subsequently further exponent 8 checking was applied to the consistent power-commutator presentation for the nilpotent quotient of D . It was found that all normal words of weight less than 18 have order dividing 8 . This strongly suggests that the nilpotent quotients of Δ and D are in fact the same. The rank of the 2-multiplicator of the maximal nilpotent quotient of D is 69 .

3.5 EXPONENT NINE

We have computed a consistent power-commutator presentation for $B(2, 9; 12)$ which is a group of order 3^{724} .

Macdonald (1973) points out that "even groups generated by two elements of order 3 lead to depressingly large results", and this is confirmed by Shield (1977). Macdonald showed that the group $\langle a, b \mid a^3 = b^3 = (ab)^3 = \emptyset$, exponent 9$\rangle$ has a largest finite quotient of order 243 . It is easily shown by coset enumeration that $\langle a, b \mid a^3 = b^3 = (ab)^3 = (aab)^9 = \emptyset \rangle$ is a presentation for this finite group, and in fact a minimal presentation.

By applying the Canberra nilpotent quotient program we have shown that the restricted Burnside problem is answered affirmatively for the freest group of exponent 9 generated by two elements of order 3 whose commutator has order 3 . Thus the nilpotent quotient program shows that a preimage of

$$\langle a, b \mid a^3 = b^3 = [a, b]^3 = \emptyset, \text{ exponent } 9 \rangle$$

has a largest finite quotient of class 18 and order 3^{71} .

The unrestricted problem has been solved for a quotient of the free group considered above, namely $\langle a, b \mid a^3 = b^3 = [a, b]^3 = (ababb)^3 = \emptyset$, exponent 9$\rangle$. The nilpotent quotient program shows that this group has a largest finite quotient of class 10 and order 3^{12} . The nilpotent quotient program provides three ninth powers $\left((ab)^9, (aab)^9, (aabab)^9 \right)$ which suffice to define this group as a group of 3-power order when added to the four explicitly given third powers. Enumeration of the cosets of $\langle ab \rangle$ in

$$\langle a, b \mid a^3 = b^3 = [a, b]^3 = (ababb)^3 = (ab)^9 = (aab)^9 = (aabab)^9 = \emptyset \rangle$$

yields index 59049 , completing the solution of the unrestricted problem in this case.

We have applied the nilpotent quotient program to the freest group of exponent 9 generated by two elements of order 3 whose product has order 3 . Computations to date reveal that it has a preimage whose class 18 quotient has order 3^{603}

228

George Havas and M.F. Newman

References

William A. Alford, George Havas and M.F. Newman (1975), "Groups of exponent four",
    Notices Amer. Math. Soc. **22**, A.301.

A.J. Bayes, J. Kautsky and J.W. Wamsley (1974), "Computation in nilpotent groups
    (application)", *Proc. Second Internat. Conf. Theory of Groups* (Canberra, 1973),
    pp. 82-89 (Lecture Notes in Mathematics, **372**. Springer-Verlag, Berlin,
    Heidelberg, New York). MR50#7299; Zbl.288.20032; RZ [1975], 6A324.

John J. Cannon, Lucien A. Dimino, George Havas and Jane M. Watson (1973),
    "Implementation and analysis of the Todd-Coxeter algorithm", *Math. Comp.* **27**,
    463-490. MR49#390; Zbl.314.20028; RZ [1974], 5A243.

John Cannon (1974), "A general purpose group theory program", *Proc. Second Internat.
    Conf. Theory of Groups* (Canberra, 1973), pp. 204-217 (Lecture Notes in
    Mathematics, **372**. Springer-Verlag, Berlin, Heidelberg, New York). MR50#7300.

N.D. Gupta and M.F. Newman (1974), "The nilpotency class of finitely generated groups
    of exponent four", *Proc. Second Internat. Conf. Theory of Groups* (Canberra,
    1973), pp. 330-332 (Lecture Notes in Mathematics, **372**. Springer-Verlag, Berlin,
    Heidelberg, New York). MR50#4752; Zbl.291.20034; RZ [1975], 6A320.

Marshall Hall, Jr. (1977), "Computers in group theory", *Topics in group theory and
    computation* (Proceedings of the Summer School, University College, Galway, 1973),
    pp. 1-37 (Academic Press [Harcourt Brace Jovanovich], London, New York, San
    Francisco). MR56#9809; Zbl.382.20002.

George Havas (1974), "A Reidemeister-Schreier program", *Proc. Second Internat. Conf.
    Theory of Groups* (Canberra, 1973), pp. 347-356 (Lecture Notes in Mathematics,
    372. Springer-Verlag, Berlin, Heidelberg, New York).

George Havas (to appear), "Commutators in groups expressed as products of powers",
    Comm. Algebra.

George Havas and Tim Nicholson (1976), "Collection", *SYMSAC '76* (Proc. ACM Sympos. on Symbolic and Algebraic Computation, Yorktown Heights, New York, 1976), pp. 9-14 (Association for Computing Machinery, New York).

George Havas, G.E. Wall, and J.W. Wamsley (1974), "The two generator restricted Burnside group of exponent five", *Bull. Austral. Math. Soc.* 10, 459-470. MR51#3298; Zbl.277#20025; RZ [1975], 6A255.

Graham Higman (1959), "Some remarks on varieties of groups", *Quart. J. Math. Oxford Ser.* (2) 10, 165-178. MR22#4756; Zbl.89,13; RZ [1961], 8A180.

John Leech (1963 - 1967), "Coset enumeration on digital computers", *Proc. Cambridge Philos. Soc.* 59, 257-267. (Privately reprinted 1967.) MR26#4513; Zbl.117,269; RZ [1963], 12A176.

I.D. Macdonald (1973), "Computer results on Burnside groups", *Bull. Austral. Math. Soc.* 9, 433-438. MR49#5165; Zbl.267#20011; RZ [1974], 10A201.

I.D. Macdonald (1974), "A computer application to finite p-groups", *J. Austral. Math. Soc.* 17, 102-112. MR51#13028; Zbl.277.20024; RZ [1974], 11A228.

Hanna Neumann (1967), *Varieties of groups* (Ergebnisse der Mathematik und ihrer Grenzgebiete, 37. Springer-Verlag, Berlin, Heidelberg, New York). MR35#6734; Zbl.251#20001; RZ [1968], 3A194н.

M.F. Newman (1976), "Calculating presentations for certain kinds of quotient groups", *SYMSAC '76* (Proc. ACM Sympos. on Symbolic and Algebraic Computation, Yorktown Heights, New York, 1976), pp. 2-8 (Association for Computing Machinery, New York). See also: Abstract, *Sigsam Bull.* 10, no. 3, 5.

Ю.П. Размыслов [Ju.P. Razmyslov] (1978), "О проблеме Холла-Хигмена" [On a problem of Hall-Higman], *Izv. Akad. Nauk SSSR Ser. Mat.* 42, 833-847. Zbl.394.20030; RZ [1979], 1A246.

David Shield (1977), "The class of a nilpotent wreath product", *Bull. Austral. Math. Soc.* 17, 53-89. MR56#3266; Zbl.396.20015; RZ [1978], 9A232.

M.R. Vaughan-Lee (1979), "Derived lengths of Burnside groups of exponent 4 ", *Quart. J. Math. Oxford* (2) 30, 495-504.

G.E. Wall (1974), "On the Lie ring of a group of prime exponent", *Proc. Second Internat. Conf. Theory of Groups* (Canberra, 1973), pp. 667-690 (Lecture Notes in Mathematics, **372**. Springer-Verlag, Berlin, Heidelberg, New York). MR50#10098; Zbl.286.20050; RZ [1975], 7A389.

J.W. Wamsley (1974), "Computation in nilpotent groups (theory)", *Proc. Second Internat. Conf. Theory of Groups* (Canberra, 1973), pp. 691-700 (Lecture Notes in Mathematics, **372**. Springer-Verlag, Berlin, Heidelberg, New York). MR50#7350; Zbl.288.20031; RZ [1975], 6A323.

Department of Mathematics,
Institute of Advanced Studies,
Australian National University,
Canberra, ACT, Australia.

ON CERTAIN GROUPS OF EXPONENT EIGHT
GENERATED BY THREE INVOLUTIONS

Franz-Josef Hermanns

A. Preliminaries

In their work Grunewald, Havas, Mennicke, and Newman (1979 and these proceedings) examined certain homomorphic images of the group

$$\Delta := \langle x, y \mid 1 = x^2 = y^4, \text{ exponent } 8 \rangle .$$

Specially the finiteness of the group

$$\Gamma := \langle u, v \mid 1 = u^2 = v^4 = \left(uv^2\right)^4, \text{ exponent } 8 \rangle$$

has been proved. A computer implemented algorithm for calculating nilpotent quotient groups called nilpotent quotient algorithm (see Newman (1976)) was applied to these groups. The result is that the maximal residually finite quotient of Δ has order dividing 2^{205} and class at most 26 , and Γ has order 2^{31} and class 19 .

The finiteness of Γ gives also the finiteness of the subgroup $\langle \left(xy^2\right)^2, y \rangle$ of Δ , as Γ maps onto this subgroup via $u \to xy^2x$ and $v \to y$.

This paper deals with a similar situation. Instead of Δ (the freest group of exponent 8 generated by a cyclic group of order 2 and a cyclic group of order 4), here a first step is done to examine the freest group C of exponent 8 generated by a cyclic group of order 2 and an elementary abelian group of order 4 . Let

$$C := \langle \alpha, \beta, \gamma \mid 1 = \alpha^2 = \beta^2 = \gamma^2 = (\alpha\beta)^2, \text{ exponent } 8 \rangle .$$

It follows from very recently completed machine calculations with the nilpotent quotient algorithm (in Bielefeld and Canberra supervised, respectively, by Pietsch and Alford - for more details see the note by Alford and Pietsch in these proceedings) that the largest residually finite quotient of a group generated by three involutions, two of which commute, and 120 eighth powers as defining relations is finite of order

2^{313} and class 22 ; it seems likely the same holds for C .

Now consider the homomorphic image

$$B := \langle a, b, c \mid 1 = a^2 = b^2 = c^2 = (ab)^2 = (bc)^4, \text{ exponent 8}\rangle$$

of C . Similar to the situation above one has a homomorphism of B onto the subgroup $\langle \alpha, \beta, \gamma\beta\gamma \rangle$ of C via $a \to \alpha$, $b \to \beta$, and $c \to \gamma\beta\gamma$. The nilpotent quotient algorithm shows that the largest residually finite quotient of B has order 2^{33} and class 17 . In Part B of this paper the finiteness of B is proved by doing some hand calculation and using a computer implementation of the Todd-Coxeter algorithm (see Cannon, Dimino, Havas, and Watson (1973)).

To give a better understanding of this proof, here is a short outline how these results were obtained.

Adding enough eight powers to the relations $1 = a^2 = b^2 = c^2 = (ab)^2 = (bc)^4$ one gets a presentation of a group B^* which has the same residually finite quotient as B . The group B^* is a preimage of B , and the finiteness of B^* implies $B = B^*$.

To get more information about B^* , possibly to find a "large" (that is, considerably larger than the dihedral group of order 16) finite subgroup, some subgroups were examined with an index in B^* that makes it not too complicated to get a presentation by the Reidemeister-Schreier method (see Magnus, Karrass, Solitar (1966)).

The subgroup

$$V^* := \langle b, c, acbca, acacacbcacaca \rangle$$

has index 16 in B^* . A presentation of V^* is easy to get and one finds that the subgroup

$$S^* := \langle b, c, acbcacacbca \rangle$$

of V^* is finite of order dividing $8192 = 2^{13}$. The power commutator presentation of the residually finite quotient of B^* shows that the image of S^* in this quotient has order 2^{11} .

The index $[V^* : S^*]$ can not be calculated by machine. If one now looks at subgroups of V^* which contain S^* , one finds that the subgroup

$$U^* := \langle b, c, acbcacacbca, acacacbcacaca \rangle$$

has index 32 in V^* . Though some relations get very complicated it is possible to get a presentation of U^* in terms of those four generators. The index $[U^* : S^*]$ still can not be calculated, but one can make use of the following theorem (see

Grunewald, Havas, Mennicke and Newman (these proceedings), Chapter II).

The normal closure in U^* of a finite subgroup of U^* which is normal in a subnormal subgroup of finite index in U^* is finite.

With this one can show that the normal subgroup in U^* generated by an element w of the center of S^* is finite. It is now sufficient to show the finiteness of $U^*/\langle w \rangle^{U^*}$. In $U^*/\langle w \rangle^{U^*}$ again the normal subgroup generated by a central element of the image of S^* is finite. This argument can be repeated several times and not only with central elements of images of S^* but also with central elements of images of $S^{*acacaca}$ which lies also in U^*.

So after several steps one gets a finite normal subgroup N^* of U^* such that the index $[U^*/N^* : S^*N^*/N^*]$ can be calculated by machine. This proves the finiteness of U^*.

In order to shorten the proof of the finiteness of B^* in Part B not the full presentation of U^* is used but only the finite index $[B^* : U^*]$, and those relations of U^* which were needed to establish its finiteness. So actually the finiteness of a preimage B_0 of B^* will be proved. If $U \leq B_0$ is the preimage of U^* then again all unnecessary relations will be deleted and the finiteness of a preimage K of U will be proved.

B. A group of order 2^{40}

Let B_0 be the group generated by three involutions a, b, c with definining relations

$$(1) \qquad\qquad 1 = (ab)^2$$

$$(2) \qquad\qquad\quad = (bc)^4$$

$$(3) \qquad\qquad\quad = (ac)^8$$

$$(4) \qquad\qquad\quad = (acb)^8$$

$$(5) \qquad\qquad\quad = (cacb)^8$$

$$(6) \qquad\qquad\quad = (acacb)^8$$

$$(7) \qquad\qquad\quad = (cacacb)^8$$

$$(8) \qquad\qquad\quad = (acacacb)^8$$

$$(9) \qquad\qquad\quad = (cacacacb)^8$$

$$(10) \qquad\qquad\quad = (cacabcb)^8$$

$$(11) \qquad\qquad\quad = (acacabcb)^8$$

$$(12) \qquad\qquad\quad = (acacbcacb)^8$$

$$(13) \qquad\qquad = (acacacbcacb)^8$$

$$(14) \qquad\qquad = (acacacbcacacb)^8$$

$$(15) \qquad\qquad = (acacabcacbcacb)^8$$

$$(16) \qquad\qquad = (acbcacacbcacb)^8$$

$$(17) \qquad\qquad = (acacabcacacbcacb)^8 .$$

It will be shown that B_0 is finite by using a computer implementation of the Todd-Coxeter algorithm and doing some hand calculations. All coset enumerations were done using the Felsch method (see Cannon, Dimino, Havas and Watson (1973)).

In B_0 let $f := acacaca$, $g := fbf$, $k := acbcacacbca$, and let U be the subgroup of B_0 generated by b, c, g , and k . The index $\begin{bmatrix} B_0 : U \end{bmatrix}$ was determined to be $2^{12} = 4096$ with a maximum of 4096 cosets to be defined. From the relation (3) one gets

$$(18) \qquad\qquad 1 = (cf)^2 ,$$

and from (4) together with (1) and (2),

$$(19) \qquad\qquad 1 = (bkbf)^2 .$$

So f operates on U by interchanging b and g and leaving c and bkb fixed.

LEMMA 1. *In* U *the following relations hold:*

$$(20) \qquad\qquad 1 = b^2$$

$$(21) \qquad\qquad = c^2$$

$$(22) \qquad\qquad = k^2$$

$$(23) \qquad\qquad = (bc)^4$$

$$(24) \qquad\qquad = (bk)^4$$

$$(25) \qquad\qquad = (ck)^4$$

$$(26) \qquad\qquad = (cbckbk)^2$$

$$(27) \qquad\qquad = (bcbk)^4$$

$$(28) \qquad\qquad = (bck)^8$$

$$(29) \qquad\qquad = (bg)^4$$

$$(30) \qquad\qquad = (cbcg)^4$$

$$(31) \qquad\qquad = (kbkg)^4$$

$$(32) \qquad\qquad = (bgkc)^4$$

(33) $= (bckgcgbkbg)^4$

(34) $= cgbkbgcbgckcgb$

(35) $= kbgcgbkgbkckbg$

(36) $= bckbgbkckcgkcgbgckck$

(37) $= (ckcgbg)^4$

(38) $= (kckgbg)^4$

(39) $= (ckcgbgbkbgbg)^2$

(40) $= (kckgbgbcbgbg)^2$

(41) $= (gbgckckbckck)^2$

(42) $= (gcgbkbkckbkb)^2$

(43) $= (gckckgbkcbkc)^2$.

Proof. First consider the automorphism α of B_0 which leaves b and c fixed and maps a to ab . α is indeed an automorphism of B_0 since it leaves the system of defining relations of B_0 invariant. Let β be the inner automorphism of B_0 induced by $acbca$. As α and β are of order 2 and commute with each other, $\alpha\beta$ is of order 2 .

$\alpha\beta$ induces an automorphism on U as it fixes b and maps g to bgb , c to k , and k to c .

With this automorphism the relations (22), (24), (31), (35), (38), and (40) follow immediately from the relations (21), (23), (30), (34), (37), and (39).

The relations (20), (21), and (23) are relations of B_0 .

The relations (25), (26), (27), (28), (29), (30), and (32) follow directly from the relations (7), (5), (10), (16), (7), (9), and (13) respectively by replacing g and k by words in a, b, c and using the relations (1) and (2).

The relations (33) and (34) follow from (17) and (8) respectively by first replacing g by fbf , then applying (18) and (19), and finally replacing f and k by words in a, b, c and using again (1) and (2).

Conjugating (25) with f gives $1 = (cgbkbgcgbkbg)^2$ from which (39) follows with (34).

Conjugating (38) with f gives $1 = (gbkgcgbkbgbgb)^4$ which can be reduced with (35) and (29) to $1 = (bgbkck)^4$. Applying (35) again to $1 = (bgbkckbgbkck)^2$ gives (42).

Now the relations (36), (37), (41), and (43) remain to be proved.

To do this let

(44) $e := acbca$.

By definition one has

(45) $k = ece$,

and from (1) and (2) follow

(46) $1 = (be)^2$

and

(47) $1 = (ef)^4$.

The relations (36), (37) and (41) will be proved by expressing $egbge$ by words in b, c, g, k in three different ways.

With (6) and (1) one gets $1 = (cefb)^4$ which gives with (18), (47), (45), (46) and (19)

$$
\begin{aligned}
1 &= fcfefbcfefefebcefbcefb \\
&= fcfefbcfefefbkfbcefb \\
&= fcfefbcfefebkbgbcefb .
\end{aligned}
$$

This implies with (45) and (46),

(48) $fefbcfef = cgkbegebcb$.

With (18), (45) and (19) one gets from (48),

(49) $fefbfef = cgkbegebcbgbkbg$

and so also

(50) $1 = (cgkbegebcbgbkbg)^2$.

With (7), (1) and (18) one gets

$$
\begin{aligned}
1 &= (cfefcb)^4 \\
&= (fcecfbfcecfb)^2 \\
&= (fcecgcecgf)^2 ,
\end{aligned}
$$

and so with (45),

(51) $1 = (egekcgck)^2$.

One also has with (18), (45), (48), and (19),

$$
\begin{aligned}
1 &= (cfefcb)^4 \\
&= (cfefcbfeffkfcb)^2 \\
(52) \qquad &= (cbcbegebkgcgbkbgcb)^2
\end{aligned}
$$

which gives with (2) and (45),

(53) $$1 = (egcgbkbgbcegbc)^2 .$$

On the other hand (52) leads with (50) to

$$1 = c(bcbegebkgcgbkbg)bcegebkgcgbkbgcb$$
$$= cgbkbgcgkbegebegebkgcgbkbgcb$$

which gives with (46) and (29)

$$egbge = kgcgbkbgcbcgbkbgcgk .$$

Applying (34) twice to this relation gives

(54) $$egbge = kgbgckcgbgckcgbgk .$$

With (45), (46) and (47) one gets from (19),

$$1 = efe(ebcbef)^2 efe$$
$$= efbcbfefefbcbfef$$
(55) $$= egcgefefbcbfef .$$

This gives with (48) and (49)

(56) $$1 = egcgecgkbegebcbgbkbgbcbegebkgc .$$

On the other hand one gets from (55) with (47), (49), (48), (46), and (45),

$$1 = egcgfefbecbfef$$
$$= egcgcgkbegebcbgbkbgebcbegebkgc$$
$$= egcgcgkbegebcbgbkbgbkbgebkgc .$$

Conjugating (23) and (24) with f one gets $1 = (gc)^4 = (bkbg)^4$ which reduces the last relation with (46) and (45) to

$$1 = ecgckbegebckbgbkekgc$$
$$= ecgckbegebckbgbkcegeke .$$

Now with (51) and (46) follows

(57) $$egbge = bckbgbkckcgck .$$

With (46), (45), (2), (53), and (35) one has

$$(bcbegebkgc)^2 = egcge(egcgbkbgbcegbc)bcegebkgc$$
$$= egcgecbgecbgbkbgcgebcegebkgc$$
$$= egcgecbgecbgb(kbgcgbkg)ebkgc$$
$$= egcgecbgecbgbgbkckbebkgc$$

which reduces with (29), (46), and (45) to

$$(bcbegebkgc)^2 = egcgecbgkegbgeckckgc .$$

With (50), (56), and (34) one gets

(59) $\qquad (bcbegeckgc)^2 = gbkbgcgkbegebcbgbkbgbcbegebkgc$
$$= gbkbgegcge$$
$$= cbgckcgbcegcge \ .$$

Now one has with (2), (48), (59), and (58)

$$1 = (bcbegebkgc)^4$$
$$= cbgckckegbgeckckgc$$

which gives with (25),

(60) $\qquad\qquad\qquad\qquad egbge = kckcgbgckck \ .$

Now (36) and (37) follow from (60) together with (57) and (54) respectively. Because of (46) and (29), $egbge$ commutes with b and so (60) and (25) imply (41).

From (14) one gets $1 = (bgcecg)^4$ and so with (45) and (57) one has

$$1 = (bgckegbgekcg)^2$$
$$= (bgckbckbgbkckc)^2 \ .$$

Applying the automorphism $\alpha\beta$ to this relation one gets (43). This proves the lemma. $\square$

Now let K be the group generated by four involutions b, c, g, k with the relations (20) to (43) and their conjugates by f as defining relations. K is a preimage of U . All further calculations will take place in K and so it should not confuse if we use the same notation for the elements of U and K . The automorphism of K which interchanges b and g and leaves c and bkb fixed will also be denoted by f .

Let S be the subgroup of K generated by b, c , and k . Using relations (20) to (28) one can show:

LEMMA 2. *In* S *the following relations hold:*

(61) $\qquad (bcbkck)^2(ckcbkb)^2 = b(cbk)^4b \ ,$

(62) $\qquad (ckcbkb)^2(bcbkck)^2 = ckckb(kbc)^4bckck \ ,$

(63) $\qquad bkckcb(kbc)^4bkckcb = (cbk)^4 \ ,$

(64) $\qquad\qquad\qquad (kckbcb)^4 = (ckcbkb)^4 \ ,$

(65) $\qquad\qquad\qquad (kckbcb)^8 = 1 \ ,$

(66) $\qquad\qquad (bkckbc)^2 = k(bcbkck)^2k \ ,$

(67) $\qquad (bckck)^2 = (bck)^2(bkc)^2 = (bkc)^2(bck)^2 \ ,$

(68) $\qquad\qquad (bckck)^8 = 1$,

(69) $\qquad\qquad (bckck)^2(kckbc)^2 = (bkckbc)^2$,

(70) $\qquad\qquad (bckck)^4(kckbc)^4 = (bkckbc)^4$,

(71) $\qquad\qquad (bkckc)^2(ckcbk)^2 = (bckcbk)^2$,

(72) $\qquad\qquad (bkckc)^4(ckcbk)^4 = (bckcbk)^4$,

(73) $\qquad\qquad (kcbcb)^2(bkbkc)^2 = (kcbkc)^2$,

(74) $\qquad\qquad (kcbcb)^4(bkbkc)^4 = (kcbkc)^4$,

(75) $\qquad\qquad (kcbcb)^8 = (bkbkc)^8 = 1$,

(76) $\qquad\qquad (ckbkb)^2 = cbcb(ckbk)^2$,

(77) $\qquad\qquad (ckbkb)^4 = (ckbk)^4$,

(78) $\qquad (cbk)^4(kbkbc)^4(ckb)^4 = (cbkckb)^4$,

(79) $\qquad (kbc)^4(cbcbk)^4(kcb)^4 = (kbckcb)^4$.

Proof. In this proof a word in brackets is changed in the next line by using the relations (23) to (28). Only the use of the relations (26), (27), or (28) will be mentioned behind the respective line.

To (61):

$$
\begin{aligned}
(bcbkck)^2(ckcbkb)^2 &= bcbkckbcb(kckckc)bkbckcbkb \\
&= bcbkck(bcbckbkb)ckcbkb \\
&= bcbkc(kcbckbkc)kcbkb \quad \text{(with (26))} \\
&= bcbkcbkcbkcbkb \\
&= b(cbk)^4 b \ .
\end{aligned}
$$

To (62):

$$
\begin{aligned}
(ckcbkb)^2(bcbkck)^2 &= (ckcbkbckc)bkcb(kckbcbkck) \\
&= kckcbkbck(ckbkcbck)ckbcbkckc \quad \text{(with (26))} \\
&= kckcbkbckbckbckbcbkckc \\
&= kckcb(kbc)^4 bkckc \ .
\end{aligned}
$$

To (63):

$$
\begin{aligned}
bkckcb(kbc)^4 bkckcb &= bkckcbkbc(kb)ck(bc)kbcbkckcb \\
&= bkck(cbkbcbkb)kbkckckbc(bcbkbcbk)ckcb \quad \text{(with (27))} \\
&= bkckbkbcbk(bckbkckcbckb)cbkbcbckcb \quad \text{(with (26))} \\
&= bkc(kbkb)cb(kck)b(kckc)b(ckc)bk(bcbc)kcb \\
&= bkcb(kbkcbkc)ckcbckckbkck(ckbkcbc)bkcb \quad \text{(with (26))} \\
&= bk(cbcbc)kbckcbckckbkckbc(kbkbk)cb
\end{aligned}
$$

$$= (bkbcbkbc)kcbckckbkc(kbcbkbcb) \quad \text{(with (27))}$$
$$= cbkbcb(kbkcbck)(ckbkcbc)bkbcbk \quad \text{(with (26))}$$
$$= cbk(bcbcbc)kc(kbkbkb)cbk$$
$$= cbkcbkcbkcbk$$
$$= (cbk)^4 \ .$$

Now (64) follows from (61), (62), (63), and (28). $kckbcb$ commutes with c and $ckcbkb$ commutes with k . So (64) implies $(kckbcb)^4 = (bcbkck)^4$ and (65) holds.

To (66):

$$(bkckbc)^2 = bkc(kbcbk)ckbc \quad \text{(with (27))}$$
$$= bkcbc(bkb)c(bkb)cbckbc$$
$$= (bkcbckbk)bkckb(kbkcbckb)c \quad \text{(with (26))}$$
$$= kcbcbkckbcb(ckc)$$
$$= kc(bcbkck)^2ck$$
$$= k(bcbkck)^2k \ .$$

To (67):

$$(bckck)^2 = bckckb(kckc)$$
$$= bck(ckbkc)kc \quad \text{(with (26))}$$
$$= bckbckbkcbkc$$
$$= (bck)^2(bkc)^2 \ .$$

The second equation follows in the same way. (68) follows from (67) and (28).

(69) follows with (23) and (24). From (66) follows that $(bkckbc)^2$ commutes with kck . It also commutes with bcb . And so one has with (69):

$$(bckck)^4(kckbc)^4 = bkckcbckck(bckck)^2(kckbc)^2kckbckckbc$$
$$= bkckbcbcbkck(bkckbc)^2kckbcbbkckbc$$
$$= (bkckbc)^4 \ .$$

This proves (70). The relations (71) and (72) follow similarly to (69) and (70).

To (73):

$$(kcbcb)^2(bkbkc)^2 = kcbc(bkcbckbkcb)kbkc \quad \text{(with (26))}$$
$$= kcb(ckck)bkc$$
$$= (kcbkc)^2 \ .$$

To (74):

$$(kcbcb)^4(bkbkc)^4 = (kcbcb)^2(kcbkc)^2(bkbkc)^2$$
$$= kcbcbkcbc(bk)cbkckcbk(cb)kbkcbkbkc$$
$$= kcbc(bkcbckbk)bkbcbkckcbkbcb(cbckbkcb)kbkc \quad \text{(with (26))}$$
$$= kcbckc(bcbkbcbk)ck(cbkbcbkb)kckbkc \quad \text{(with (27))}$$
$$= kcbckckbcbk(bcbckbkb)cbkbckckbkc$$
$$= kcbckckbc(bkcbckbkcb)kbckckbkc \quad \text{(with (26))}$$
$$= (kcbkc)^4 .$$

To show $(bkbkc)^8 = 1$ it is enough to show that $bkbk$ commutes with
$(bkbkc)^4 = (kcbkc)^4(kcbcb)^4$ (see (74)). But from (23), (24), and (26) follows that
$bkbk$ commutes with $(kcbcb)^4$ and so it is enough to show that $bkbk$ commutes with
$(kcbkc)^4$. With (25) and (26) one has

$$kbkb(kcbkc)^4bkbk = kbkb(kcbckckbkc)^2bkbk$$
$$= kbk(kcbckckbkc)^2kbk$$
$$= kb(cbkck)^4bk .$$

So it is enough to show that b commutes with $(cbkck)^4 = (cbkckb)^4(bkckc)^4$ (see
(69)). b commutes with $(bkckc)^4$ because of (68). From (64), (65), and (66)
follows that b also commutes with $(cbkckb)^4$:

$$b(cbkckb)^4b = (bcbkck)^4 = (bkbckc)^4$$
$$= k(bkbckc)^4k = k(bcbkck)^4k = (bkckbc)^4 = (cbkckb)^4$$

So $(bkbkc)^8 = 1$ holds. $(bcbck)^8 = 1$ is proved in the same way.

To (76):

$$(ckbkb)^2 = cb(kbkcb)kbk \quad \text{(with (26))}$$
$$= cbcbckbkckbk$$
$$= cbcb(ckbk)^2 .$$

(77) follows from (76) since $cbcb$ commutes with $(ckbk)^2$.

To (78):

$$(cbk)^4(kbkbc)^4(ckb)^4 = (cbk)^4(kbkc)^4(ckb)^4 \quad \text{(with (77))}$$
$$= cbkcbkcb(kckck)bkckb(kckck)bckbckb$$
$$= cbkc(bkcbck)cbkckbc(kcbckb)ckb \quad \text{(with 26))}$$
$$= cbkck(cbc)kbcbkckbcbk(cbc)kckb$$
$$= cbkckbc(bcbkbcbk)c(kbcbkbcb)cbkckb \quad \text{(with (27))}$$
$$= cbkckbckbcbk(bcbcbcb)kbcbkcbkckb$$

$$= cbkckbck(bcbkck)^2kbkckb \quad \text{(with (66))}$$

$$= (cbkckb)^4 .$$

(79) follows in the same way, and so the lemma is proved. □

With the relations of Lemma 2 it is easy to prove the finiteness of S. It is too hard to determine the index $[K : S]$. But one can get the following indices:

$$[K : \langle b, c, k, gbg \rangle] = 2^9 ,$$

$$[K : \langle b, c, k, gcg \rangle] = 2^9 ,$$

$$[K : \langle b, c, k, gckckg \rangle] = 2^{12} .$$

These indices were found by a coset enumeration in the split extension of K by f with a maximum of 1024, 1024, and 8192 cosets respectively.

Now let

$$w_0 := (bcbkck)^4 , \quad w_1 := (bckck)^4 ,$$

$$w_2 := (bkbkc)^4 , \quad w_3 := (bkc)^4 ,$$

$$w_4 := (bcbkck)^2 , \quad w_5 := (bckck)^2 ,$$

$$w_6 := (bkbkc)^2 ,$$

and define factor groups of K by $K_0 := K$ and $K_{i+1} := K/\langle w_0, \ldots, w_i \rangle^K$ for $i = 0, \ldots, 6$. Let $\nu_i : K \to K_i$ be the natural homomorphism and $S_i := \nu_i(S)$ for $i = 0, \ldots, 7$. Now one can show:

LEMMA 3. $\nu_i(w_i)$ *is central in* S_i *of order at most two and generates a finite normal subgroup in* K_i *for* $i = 0, \ldots, 6$.

Proof. The first part of the lemma follows by using Lemma 2 and the relations (20) to (28). That $\nu_i(w_i)$ has order at most two follows from the relations (65), (68), (75), and (28) respectively for $i = 0, 1, 2, 3$ and is trivial for $i = 4, 5, 6$.

Because of the relations (23), (24), (25), and (64), w_0 commutes with c and k. Relation (66) implies that w_0 also commutes with b. With the same argument using relations (61) and (66) one has that $\nu_4(w_4)$ is central in S_4.

Since $\nu_1(w_1) [\nu_5(w_5)]$ has order at most two it commutes with the image of b. That it also commutes with the images of c and k follows from the relations (70) and (72) [(69) and (71)].

As $\nu_2(w_2)$ $[\nu_6(w_6)]$ has order at most 2 , it commutes with the image of c . It also commutes with the image of k because of relation (74) [(73)]. $\nu_2(w_2)$ commutes also with the image of kbk and so with the image of b because of relation (77). The same holds for $\nu_6(w_6)$ because of the relations (76), (26) and (24).

As $\nu_3(w_3)$ has order at most 2 and because of the relations (78) and (79) it is central in S_3 .

To show that $\nu_i(w_i)$ generates a finite normal subgroup in K_i it is sufficient to show that it generates a finite normal subgroup in $\nu_i(\langle b, c, k, gbg \rangle)$, $\nu_i(\langle b, c, k, gcg \rangle)$, or $\nu_i(\langle b, c, k, gckckg \rangle)$ as these groups have finite index in K_i .

With the first part of the lemma and the relations (37), (38), (39), (40), and (29) it follows that $\nu_i(gbgw_igbg)$ is in the centralizer of $\nu_i(\langle ckc, kck, b \rangle) = S_i$ in K_i for $i = 0, 4$.

So $\nu_i(\langle w_i, gbgw_igbg \rangle)$ is elementary abelian and normal in $\nu_i(\langle b, c, k, gbg \rangle)$ for $i = 0, 4$.

The relations (29) and (41) imply that $\nu_i(w_i)$ is in the center of $\nu_i(\langle b, c, k, gbg \rangle)$ for $i = 1, 5$.

Conjugating (23) by f one has $1 = (gc)^4$, and so together with relation (42) it follows that $\nu_i(w_i)$ is in the center of $\nu_i(\langle b, c, k, gcg \rangle)$ for $i = 2, 6$.

Finally relation (43) gives that $\nu_3(w_3)$ is in the center of $\nu_3(\langle b, c, k, gckckg \rangle)$.   □

COROLLARY. S *is finite of order dividing* 2^{13} .

Proof. From the first part of Lemma 3 one has $|S \cap \ker \nu_7| \le 2^7$. Because of the relations (23), (24), and (25) and as

$$\nu_7((bckck)^2) = \nu_7((bkbkc)^2) = \nu_7((bcbck)^2) = 1$$

one has $[[S_7, S_7], S_7] = 1$, and so the order of S_7 is at most 2^6 .   □

THEOREM. K *and* B_0 *are finite.*

Proof. By Lemma 3 the elements w_i and w_i^f, $i = 0, \ldots, 6$, generate a finite normal subgroup N in K. f induces an automorphism $\overline{f}$ on K/N. A coset enumeration of SN/N in the split extension of K/N by $\overline{f}$ shows that the index $[K/N : SN/N]$ is 2^{10} with a maximum of 2048 cosets to be defined. So K and B_0 are finite. □

Remark. Applying the nilpotent quotient algorithm (see Newman (1976)) to the groups which are involved in Part B one gets the following results.

(i) B_0 has class 22 and order 2^{40} and exponent 16. Factoring out the normal subgroup generated by $(cacacbcacacabcb)^8$ and $(acabcacbcacacbcb)^8$ gives the group $B = B^*$ defined in Part A.

(ii) $K.\langle f \rangle$ has class 11 and order 2^{35}. K has class 6 and order 2^{34}. $\gamma_i(w_i)$ is a central involution in K_i for $i = 0, 1, 2, 3$. For $i > 3$ the normal subgroup generated by $\nu_i(w_i)$ in K_i is elementary abelian of order 4 if $i = 4, 5$ and of order 16 if $i = 6$. $z := (ckf)^8 = (ckcgbkbg)^4$ is a central involution in $K.\langle f \rangle$. $K.\langle f \rangle/\langle z \rangle$ has class 10, order 2^{34}, and exponent 8. $K/\langle z \rangle$ has class 6 and order 2^{33}.

(iii) The group

$$R := \langle r, s, t \mid 1 = r^2 = s^2 = t^2 = (rs)^4$$
$$= (rt)^4 = (st)^4 = (srstrt)^2 = (rsrt)^4 = (rst)^8 \rangle$$

has class 6, order 2^{13}, and exponent 8. The power commutator presentation of K shows, that R is isomorphic to the subgroup S of K.

References

John J. Cannon, Lucien A. Dimino, George Havas and Jane M. Watson (1973), "Implementation and analysis of the Todd-Coxeter algorithm", *Math. Comp.* **27**, 463–490.

Fritz J. Grunewald, George Havas, J.L. Mennicke, and M.F. Newman (1979), "Groups of exponent eight", *Bull. Austral. Math. Soc.* **20**, 7–16.

Wilhelm Magnus, Abraham Karrass, Donald Solitar (1966), *Combinatorial group theory: presentations of groups in terms of generators and relations* (Pure and Applied Mathematics, 13. Interscience [John Wiley & Sons], New York, London, Sydney; revised 1976).

M.F. Newman (1976), "Calculating presentations for certain kinds of quotient groups",
    SYMSAC '76, 2-8 (Proc. ACM Sympos. on Symbolic and Algebraic Computation,
    Yorktown Heights, New York, 1976. Association for Computing Machinery, New York.
    See also: Abstract, *Sigsam Bull.* 10 (1976), No. 3, 5).

Fakultät für Mathematik,
Universität Bielefeld,
Federal Republic of Germany.

PROCEEDINGS OF THE BURNSIDE WORKSHOP
BIELEFELD, June–July 1977, 246–248.

20E10, 20F50

GENERALIZED POWER LAWS

FRANK LEVIN AND GERHARD ROSENBERGER

The Burnside variety B_k of groups of exponent k can be defined as the class of all groups whose elements satisfy the identity $x^k = 1$. In this note we consider the problem of finding upper bounds for the exponents of groups in two generalizations of B_k , $B(p, q)$ and $B(1, 1, k)$, which can be defined as follows:

   I. $G \in B(p, q)$ if for some fixed $b \in G$ all elements $x \in G$ satisfy
   $$x^p b x^{-q} b^{-1} = 1 ;$$

   II. $G \in B(1, 1, k)$ if for some fixed $a, b, c \in G$ all elements $x \in G$
   satisfy $xaxbx^k c = 1$.

These upper bounds are not finite in all cases. For instance, for any n , $B(n, n)$ contains all abelian groups, and $B(2n, -2n)$ contains the infinite dihedral group $D = \langle u, v \mid (uv)^2 = u^2 = 1 \rangle$, which satisfies $x^{2n} u x^{2n} u^{-1} = 1$ for all $x \in G$. Further, since $B(2, -k) \leq B(1, 1, k)$ for any k , $B(1, 1, -2)$ and $B(1, 1, 2)$ also contain groups with unbounded exponents. Our main result is that finite upper bounds exist for the exponents of groups in all other cases. These are given in Theorem 1 for $B(p, q)$, $(p, q) \neq (n, n)$, $(2n, -2n)$, and in Theorem 2 for $B(1, 1, k)$, $k \neq \pm 2$.

Our notation for group commutators is as follows: $[x, y] = x^{-1} y^{-1} xy$ and $[x, y, z] = [[x, y], z]$. Also, $y^x = x^{-1} yx$ for group elements x, y , and $\exp G$ denotes the exponent of G , that is, the minimum k such that $G \in B_k$.

THEOREM 1. *Let* $G \in B(sd, td)$, *where* $\gcd(|s|, |t|) = 1$, $s \neq t$, $d \geq 1$, *and if* $s = -t = 1$, *then* $(d, 2) = 1$. *Then* $\exp G$ *divides* $d(s-t)\left(s^d - t^d\right)$. *In particular, if* $G \in B(1, -1)$, *then* $\exp G$ *divides* 2 .

Proof. Suppose that for $b \in G$ and all $x \in G$, $x^{sd}bx^{-td}b^{-1} = 1$. Then with $x = b$, $b^{sd-td} = 1$. Further, for any $y \in G$, $\left(x^{sd}\right)^{y^{-1}}b\left(x^{-td}\right)^{y^{-1}}b^{-1} = 1$ so that $x^{sd}b^y x^{-td}b^{-y} = 1$. Thus, $b^y x^{-td}b^{-y} = bx^{-td}b^{-1}$, so $\left[b, y, x^{td}\right] = 1$. Similarly, $\left[b^{-1}, y, x^{sd}\right] = 1$ for any $x, y \in G$, so $\left[b^{-1}, y, \left(x^{sd}\right)^{b^{-1}}\right] = 1$, whence $\left[b, y, x^{sd}\right] = 1$. Since $(s, t) = 1$, the two identities $\left[b, y, x^{sd}\right] = 1$ and $\left[b, y, x^{td}\right] = 1$ imply $\left[b, y, x^d\right] = 1$ which holds for any $x, y \in G$. In particular, replacing x by b and y by yz^{-1}, the latter identity implies that $\left[b^{yz^{-1}}, b^d\right] = 1$, so that $\left[b^y, b^{dz}\right] = 1$. Thus, the normal closure $H = \langle b^d \rangle^G$ of b^d in G is abelian, and, since $b^{sd-td} = 1$, $\exp H$ divides $s - t$.

The group G/H satisfies $x^{sd}bx^{-td}b^{-1} = 1$ with $b^d = 1$. Replacing x by x^t and x^s, in turn, yields $x^{std}bx^{-t^2d}b^{-1} = 1$ and $x^{s^2d}bx^{-std}b^{-1} = 1$. Hence $x^{s^2d}_b{}^2x^{-t^2d}_b{}^{-2} = 1$. More generally, $x^{s^nd}_b{}^nx^{t^nd}_b{}^{-n} = 1$, which follows by a straightforward induction. In particular, $x^{s^dd}_x{}^{-t^dd} = 1$ in G/H, so $\exp G/H$ divides $s^dd - t^dd$. It follows that $\exp G$ divides $\exp H.\exp G/H = (s-t)\left(s^d-t^d\right)d$.

Finally, if G satisfies $xbx^{-1}b^{-1} = 1$, then replacing x by xb^{-1} gives $x^2 = b^2$ for all $x \in G$. Hence $1 = b^2$ so $x^2 = 1$ for all $x \in G$; that is, $\exp G$ divides 2.

THEOREM 2. *Let* $G \in \mathcal{B}(1, 1, k)$, $k \neq \pm 2$. *Then* $\exp G$ *divides* $(k+2)^4$ *if* k *is odd and* $(k+2)^3(k-2)^2/8$ *if* k *is even.*

Proof. Suppose $xaxbx^kc = 1$ for fixed $a, b, c \in G$ and all $x \in G$. Replacing x by 1 gives $abc = 1$, so that $bx^kb^{-1} = x^{-1}a^{-1}x^{-1}a$. Replacing x by x^{-1} now yields $bx^{-k}b^{-1} = xa^{-1}xa$, and it follows that $x^{-1}a^{-1}x^{-1}a = \left(xa^{-1}xa\right)^{-1} = a^{-1}x^{-1}ax^{-1}$. Hence

(1) $[a, x, x] = 1$, for all $x \in G$.

Replacing x by ax in (1) gives

$$1 = [a, ax, ax] = [a, x, ax] = [a, x, x][a, x, a]^x = [a, x, a]^x,$$

so that $[a, x, a] = 1$ for all $x \in G$. Hence $\left[a^x, a\right] = [a, x, a] = 1$, and replacing x by yz^{-1} gives that $\left[a^y, a^z\right] = 1$ for all $y, z \in G$, that is, the normal closure $H = \langle a \rangle^G$ of a in G is abelian.

The identity $xaxbx^kb^{-1}a^{-1} = 1$ can be expressed in the form

(2) $$x^{k+2} = [x^{-1}, a][b^{-1}, x^k] .$$

Replacing x by $x^{b^{-1}}$ in (2) gives

$$(x^{k+2})^{b^{-1}} = [x^{-b^{-1}}, a][b^{-1}, (x^k)^{b^{-1}}] = [x^{-1}, a^b]^{b^{-1}}[b^{-1}, x^k]^{b^{-1}} .$$

Hence $x^{k+2} = [x^{-1}, a^b][b^{-1}, x^k]$, so, by (2), $[a, x] = [a^b, x]$. Thus

$$[a^{-1}a^b, x] = 1 = [a, b, x] ,$$

that is, $[a, b]$ is central in G . Hence, replacing x by $[a, b]$ in (2) gives $[a, b]^{k+2} = 1$. Thus, replacing x by a in (2) gives us

$$a^{k+2} = [b^{-1}, a^k] = [a, b]^k = [b, a]^2 ,$$

since a commutes with its conjugates. It follows that $a^{(k+2)^2} = 1$ if k is odd, while if k is even, then a has order dividing $(k+2)^2/2$. Hence $\exp H$ divides $(k+2)^2$ if k is odd and $(k+2)^2/2$ if k is even. Finally, $\exp G$ divides $(\exp H)(\exp G/H)$, and since $G/H \in \mathcal{B}(2, -k)$, $\exp G/H$ divides $(k+2)^2$ if k is odd and divides $2(\frac{1}{2}k-1)((\frac{1}{2}k)^2-1)$ if k is even, by Theorem 1. This proves Theorem 2.

Since $\mathcal{B}_k$ is locally finite for $k = 1, 2, 3, 4, 6$ (*cf.* Magnus, Karrass, Solitar, *Combinatorial Group Theory*, John Wiley & Sons, 1966, §5.12, 5.13) the following corollary is an immediate consequence of Theorems 1 and 2.

COROLLARY. 1. *If* $G \in \mathcal{B}(sd, td)$, $(|s|, |t|) = 1$, $s \neq t$, $d \geq 1$, *and* $d \neq 2$ *if* $s = -t$, *then* G *is locally finite if* $|d(s^d-t^d)|$ *is* 1, 2, 3, 4 *or* 6 .

2. *If* $G \in \mathcal{B}(1, 1, k)$, $k \neq \pm 2$, k *even, then* G *is locally finite if* $2((\frac{1}{2}k)^2-1)$ *is* 2, 4 *or* 6 , *that is, if* $k = \pm 4$.

3. *If* $G \in \mathcal{B}(1, 1, k)$, k *odd, then* G *is locally finite if* $|k+2|$ *is* 1, 3 , *that is, if* k *is* -5, -3, -1 *or* +1 . (*However, if* $k = -1$ *Theorem 2 shows that* G *is trivial, so local finiteness is also trivial in this case.*)

Ruhr Universität,   Universität,
Bochum,   Dortmund,
Federal Republic of Germany.   Federal Republic of Germany.

PROCEEDINGS OF THE BURNSIDE WORKSHOP
Bielefeld, June-July 1977, 249-254.

20-02, 20F50

PROBLEMS

COMPILED BY M.F. NEWMAN

1. Introduction

The purpose of the following notes is to present some of the outstanding problems
in that part of the theory of groups which has grown in response to the questions
raised in Burnside's famous paper "ON AN UNSETTLED QUESTION IN THE THEORY OF
DISCONTINUOUS GROUPS" (1902). The emphasis here is on stating problems with only
enough related discussion to set the scene. A brief account of some of the work on
the questions asked in the paper is given in sections 5.12 and 5.13 of the book by
Magnus, Karrass and Solitar (1966). More detailed information can be found by
consulting the bibliography which occurs elsewhere in this volume. I am indebted to
the participants in this Bielefeld meeting and to colleagues in Canberra for helpful
comments.

For the record Burnside's original formulation is:

*A still undecided point in the theory of discontinuous groups is whether
the order of a group may be not finite, while the order of every operation
it contains is finite. A special form of this question may be stated as
follows:*

*Let $A_1, A_2, \ldots, A_m$ be a set of independent operations finite in number,
and suppose that they satisfy the system of relations given by*

$$S^n = 1 \, ,$$

where n is a given finite integer, while S represents in turn any and every operation which can be generated from the m given operations A.

Is the group thus defined one of finite order, and if so what is its order?

2. The Burnside questions

The opening sentence of Burnside's paper is usually interpreted as asking: can a finitely generated periodic group be infinite? That the answer is yes was first demonstrated by Golod (1964) who exhibited for each prime p a finitely generated infinite p-group.

Let $B(m, n)$ denote the group defined in the special form of the question; that is, the freest m-generator group of exponent n. With this notation the question becomes: is $B(m, n)$ finite? The answer is trivially yes when the number of generators is 1 and easily seen to be yes when the exponent is 2. Burnside himself showed that the answer is yes when the exponent is 3. Sanov (1940) showed that the answer is yes when the exponent is 4 and M. Hall (1957-58) did when it is 6. Finally Novikov and Adian (1968) in a sequence of remarkable papers showed that the answer is no for most exponents. They felt (see p. 212) no doubt that this would be the case for almost all exponents even though extension of their proof to demonstrate this presented difficulties they had been unable to resolve. The strongest result in this direction is that $B(2, n)$ is infinite whenever n has an odd divisor which is at least 665 (Adian 1975). Moreover Adian (1974, pp. 8-9) has stated that he believes it should be possible to reduce the bound on the divisor to 101 by making a more detailed analysis of the proof and effecting certain changes of a technical nature.

PROBLEM 1. *Prove that $B(2, n)$ is infinite for all but finitely many n.*
It is, of course, sufficient to prove that $B\left(2, 2^k\right)$ is infinite for some k.

It will then remain to settle the precise conditions on m and n under which $B(m, n)$ is finite. Kostrikin (1962, p. 266; 1974, p. 412) has pointed out that $B(2, 5)$ is infinite if it does not satisfy the 6th Engel condition. Some work has been done on $B(2, 8)$ by Grunewald, Havas, Mennicke and Newman. Because of results on the restricted problem (Problem 3), the case of exponent 12 seems to deserve some attention.

PROBLEM 2 (The order problem). *If $B(m, n)$ is finite, what is its order?*

While it is possible to describe algorithms which in principle answer this question, what is usually wanted is a "reasonable" formula for the order. When n is 2, 3 or 6 such formulas exist (Burnside 1902, p. 38; Levi and van der Waerden 1932; Hall and Higman 1956, p. 38). In addition the orders of $B(2, 4)$, $B(3, 4)$ and

$B(4, 4)$ have been shown to be 2^{12}, 2^{69} and 2^{422} respectively (Tobin 1954, Chapter 4; Bayes, Kautsky and Wamsley 1974, pp. 85-88; Alford, Havas and Newman 1975). Lower and upper bounds for the order of $B(m, 4)$ have been by Sanov (1947a, p. 760) and by Higman (1967, p. 155) respectively. In that paper Higman was discussing the sequence of orders of relatively free groups of locally finite varieties. He made a conjecture about these sequences which for groups of exponent four amounts to: the sequence $(\log \log |B(m, 4)|)/m$ is bounded (Postscript a).

3. Intermezzo

These questions of Burnside have, for a variety of reasons - their nature as finiteness questions, their difficulty and developments in the study of groups, led to the formulation of numerous other questions. Among these questions are some which are concerned with other algebraic systems: for instance, the Kurosh problems (1941; see, for example, Herstein 1968, p. 162) about algebraic algebras or questions of nilpotence and local nilpotence of Engel Lie rings (see Kostrikin 1974). There are questions about groups with additional structure: is a compact periodic group locally finite? (3.41 of the Kourovka notebook, 1969.)

The first variations on the original questions arose in a paper of Burnside (1905) in which he exhibited four finiteness conditions for groups of linear substitutions. He proved, for example, that a group of linear substitutions with only finitely many conjugacy classes of elements is finite. There are infinite groups with just two conjugacy classes (Higman, Neumann and Neumann 1949); are there infinite periodic groups with finitely many conjugacy classes?

The following sections contain some of the questions which have arisen about groups.

4. The restricted problem

It seems that during the 1930s the view developed that the Burnside question might have a negative answer but that there might be a largest one among the finite m-generator groups of exponent n. The question of whether this is so is nowadays called the restricted Burnside problem. A convenient formulation is in terms of the largest residually finite quotient, $\bar{B}(m, n)$, of $B(m, n)$; that is, $\bar{B}(m, n) = B(m, n)/R$ where R is the intersection of the (normal) subgroups of finite index in $B(m, n)$.

PROBLEM 3 (The restricted problem). *Is $\bar{B}(m, n)$ always finite?*

The main result on this question is: yes when the exponent is square-free. This answer was obtained in stages. The first main step was a theorem of Hall and Higman (1956, Theorem 4.4.1) which reduced the problem to the prime-power exponent case and to a question about automorphism groups of non-abelian simple groups. The next main step was Kostrikin's proof (1958) that the answer is yes when the exponent is prime. The third main step consisted in showing that all finite simple groups with square-free exponent are "known"; this was achieved when Walter (1969) showed that all simple groups with abelian Sylow 2-subgroup are known up to the uncertainty residing in the so-called Ree group problem (Postscript b); for in this context there is enough information about these groups of Ree-type to exclude them. The answer is also affirmative for exponent 12 (Hall and Higman 1956, pp. 3-4) because of the affirmative answer for exponent 4 . This latter result together with the reduction theorem leads to the question: is every finite simple group whose exponent divides four times a square-free odd integer known? Kostrikin (1974, p. 412) raises the particular case of exponent 60 and suggests it may be easier than the corresponding question for general $\{2, 3, 5\}$-groups.

The results on the restricted Burnside problem combined with those of Novikov and Adian show that $B(m, n)$ need not be residually finite.

PROBLEM 4. *For what values of* m *and* n *is* $B(m, n)$ *residually finite?*
In this connection it should be pointed out that (in spite of some comments to the contrary) there is no proof, as far as I am aware, that the p-groups constructed by Golod are not of finite exponent; so some of these groups might provide examples to answer the question.

There is also an order problem for the restricted case.

PROBLEM 5. *If* $\overline{B}(m, n)$ *is finite, what is its order?*

The only answer in addition to those to Problem 2 is that $\overline{B}(2, 5)$ has order 5^{34} (Havas, Wall and Wamsley 1976). Some lower bounds are known; see Kostrikin 1957b.

5. Conditions on subgroups

In the same era quite different questions arose. These ask about finiteness when conditions are imposed on subgroups of a group. Two examples, taken from the epilogue of Kurosh's book (1944), are: is a group all of whose ascending and descending chains of subgroups break-off finite?; are there infinite groups, apart from quasicyclic groups, all of whose proper subgroups are finite? (Šmidt's problem). Even much more stringent conditions are not known to guarantee finiteness. The resulting type of finiteness problem is often referred to as a Tarski problem. (Postscript c.)

PROBLEM 6. *Is a group finite if all its proper subgroups have prime order (for the same prime)?*

The original unsettled point may be interpreted as asking for finiteness when all one-generator subgroups are finite. This leads to the question of whether a finitely generated group is finite if for some m all its m-generator subgroups are finite. Golod also provided a negative answer to this question: see his address (1968) to the ICM held in Moscow in 1966. The Burnside version, in the general context of varieties

of groups, has been stated by B.H. Neumann (1967, Problem 7) and called the extended
Burnside problem.

PROBLEM 7. *Is a finitely generated group finite if for some* m *and* n *all its*
m-generator subgroups are finite and have exponent dividing n ?
Neumann also states (see his Problem 9) a restricted extended Burnside problem.

6. Properties of finite Burnside groups

When the groups $B(m, n)$ or $\overline{B}(m, n)$ are finite one can ask for information
about them other than simply their order. Some of this can be expected to be easier
to obtain than the order. For instance for prime-power exponents the groups will be
nilpotent.

PROBLEM 8. *What is the nilpotency class of* $\overline{B}(m, p)$ *for a prime* p ?
In those cases where the order problem has been solved so has this (see the
references cited there). Kostrikin (1957a, Theorem 7) has established for primes p
(exceeding 3) the lower bound $2p$ for the class of $\overline{B}(2, p)$. Razmyslov (1971) has
shown that $\overline{B}(m, p)$ has class at least $2m - 1$ (again for p exceeding 3). For
exponent four one has the curious situation that either the class of $B(m, 4)$ is
$3m - 2$ for all m exceeding 2 (Postscript a) or it is of the form $m + k$ for some
k when m is sufficiently large (see Gupta and Newman 1974). The first case arises
if the free group of countably infinite rank, $B(\infty, 4)$, is insoluble and the second
if $B(\infty, 4)$, and therefore every group of exponent four, is soluble.

PROBLEM 9 (Postscript a). *Is every group of exponent four soluble?*
Recent evidence, including that of Doyle, Mandelberg, Vaughan-Lee (1978), points to
the answer no. The work of Razmyslov shows that $\overline{B}(\infty, p)$ is insoluble for all primes
p greater than 3 . Thus to complete this part of the story it remains to show that
$\overline{B}(\infty, 8)$ and $\overline{B}(\infty, 9)$ are insoluble.

The work of Hall and Higman (1956, Theorem A) shows that finite soluble groups of
exponent dividing n have for each odd non-Fermat prime p dividing n p-length at
most e where p^e is the highest power of p dividing n . This leads to the
question (*cf.* Gross 1967) whether the nilpotent length (Fitting height) of finite
soluble groups of exponent dividing n is at most the number of primes (counted with
repetitions) in the prime decomposition of n . (Postscript d.)

7. Properties of infinite Burnside groups

Novikov and Adian established a number of properties for the $B(m, n)$ with n
odd that they proved infinite. For instance they showed that every commutative
subgroup is finite (1968e, Theorem 3). A little later Adian (1971c, Theorem 1) showed
every commutative subgroup is cyclic and consequently (Corollary 2) that every finite
subgroup is cyclic. This means that even the two-generator subgroups must be fairly

complicated.

PROBLEM 10. *Describe the two-generator subgroups of the infinite* $B(m, n)$.
For prime exponents are they also Burnside groups? For non-prime exponents are they
at least periodic products in the sense of Adian (1976)? More generally can one prove
for periodic products an analogue of the Kurosh Subgroup Theorem for free products?

8. Postscripts

a. Razmyslov (1978) has now shown that there are insoluble groups of exponent
four, eight and nine. This answers Problem 9 and the related questions. It also
implies that part of Higman's conjecture mentioned at the end of section 2 cannot be
retained; however the particular form stated remains unresolved.

b. The recognition of Ree groups has now been completed by E. Bombieri with
assistance from D.C. Hunt and A.M. Odlyzko.

c. Recently Ol'šanskii (1979) has proved the existence of an infinite (simple)
group such that every proper subgroup is finite cyclic and the existence of an
infinite group (of exponent zero) such that every proper subgroup has prime order.

d. The answer is no. Hihro (1978) has shown there is a group of exponent $2^3 3^2$
with nilpotent length 6 .

9. References

Graham Higman, B.H. Neumann and Hanna Neumann (1949), "Embedding theorems for groups",
    J. London Math. Soc. **24**, 247-254. MR11,322; Zbl.34,301.

А.Ю. Ольшанский [A.Ju. Ol'šanskiĭ] (1979), "Бесконечные группы с циклическими
    подгруппами" [Infinite groups with cyclic subgroups], *Dokl. Akad. Nauk SSSR* **245**,
    785-787; *Soviet Math. Dokl.* **20**, 343-346.

John H. Walter (1969), "The characterization of finite groups with abelian Sylow
    2-subgroups", *Ann. of Math.* (2) **89**, 405-514. MR40#2749; Zbl.184,46.

See the bibliography which follows for more details on other works to which
reference has been made.

BIBLIOGRAPHY

Compiled by M.F. Newman

The bibliography attempts to list all papers which claim to contribute to the Burnside questions and their relatives. Information about omissions would be welcome. In order to make the list more useful references to reviews in Mathematical Reviews (MR), Zentralblatt für Mathematik (Zbl), Referativnyĭ Žurnal (RZ) and the Jahrbuch über die Fortschritte der Mathematik (FdM) are given; also a key index is provided.

С.И. Адян [S.I. Adjan] (1970a), "Тождественные соотношения в группах" [Identical relations in groups], *Uspehi Mat. Nauk* 25, 263. RZ [1970], 7A202.

С.И. Адян [S.I. Adjan] (1970b), "Бесконечные неприводимые системы групповых тождеств" [Infinite irreducible systems of group identities], *Dokl. Akad. Nauk SSSR* 190, 499-501; *Soviet Math. Dokl.* 11, 113-115. MR41#1842; Zbl.232.20059; RZ [1970], 6A193.

С.И. Адян [S.I. Adjan] (1970c), "Бесконечные неприводимые системы групповых тождеств" [Infinite irreducible systems of group identities], *Izv. Akad. Nauk SSSR Ser. Mat.* 34, 715-734; *Math. USSR-Izv.* 4, 721-739 (1971). MR44#4078; Zbl.221.20047; RZ [1971], 2A181.

С.И. Адян [S.I. Adjan] (1971a), "О подгруппах свободных периодических групп нечетного показателя" [On subgroups of free periodic groups of odd exponent], *Trudy Mat. Inst. Steklov* 112, 64-72, 386; *Proc. Steklov Inst. Math.* 112, 61-69. MR49#7049; Zbl.259.20028; RZ [1972], 1A331.

S.I. Adjan (1971b), "Identités dans les groupes", *Actes Congrès Internat. des Math.* I, pp. 263-267. Russian Transl. ("Nauka", pp. 7-13 (1972)). MR55#8185; Zbl.238.20030; RZ [1973], 4A288.

The variations in the Roman versions of АДЯН reflect actual variations which occur in papers, translations or reviews.

С.И. Адян [S.I. Adjan] (1971c), "О некоторых группах без кручения" [On some torsion-free groups], *Izv. Akad. Nauk SSSR Ser. Mat.* **35**, 459-468; *Math. USSR-Izv.* **5**, 475-484 (1972). MR44#303; Zbl.271.20016; RZ [1973], 2A218.

S.I. Adjan (1973a), "Burnside groups of odd exponent and irreducible systems of group identities", *Word problems. Decision problems and the Burnside problem in group theory*, pp. 19-37. MR55#8195; Zbl.264.20027; RZ [1974], 6A271.

С.И. Адян [S.I. Adjan] (1973b), "О работах П.С. Новикова и его учеников по алгоритмическим вопросам алгебры" [On the work of P.S. Novikov and his graduate students in algorithmic questions of algebra], *Trudy Mat. Inst. Steklov* **133**, 23-32, 274; *Proc. Steklov. Inst. Math.* **133**, 21-30 (1977). MR54#10424; Zbl.351.02034; RZ [1974], 3A101.

S.I. Adyan (1974), "Periodic groups of odd exponent", *Proc. Second Internat. Conf. Theory of Groups*, pp. 8-12. MR51#3318; Zbl.306.20044; RZ [1975], 6A295.

С.И. Адян [S.I. Adjan] (1975), *Проблема Бернсайда и тождества в группах* (Izdat. "Nauka", Moscow). English translation: S.I. Adian (1979), *The Burnside problem and identities in groups* (translated by J. Lennox, J. Wiegold. Ergebnisse der Mathematik und ihrer Grenzgebiete, **95**. Springer-Verlag, Berlin, Heidelberg, New York). MR55#5753; Zbl.306.20045; RZ [1975], 12A241н.

С.И. Адян [S.I. Adjan] (1976), "Периодические произведения групп" [Periodical products of groups], *Trudy Mat. Inst. Steklov* **142**, 3-21. RZ [1977], 3A184.

С.И. Адян [S.I. Adyan] (1977), "Аксиоматический метод построения групп с заданными свойствами" [An axiomatic method of constructing groups with given properties], *Uspehi Mat. Nauk* **32** (193), no. 1, 3-15, 271; *Russian Math. Surveys* **32**, no. 1, 1-14. MR57#3269; Zbl.375.20025; RZ [1977], 8A247.

С.И. Адян [S.I. Adian] (1978), "О простоте периодических произведений групп" [On simple periodic products of groups], *Dokl. Akad. Nauk SSSR* **241**, 745-748. RZ [1978], 12A332.

S.I. Adian (1980), "Classifications of periodic words and their application in group theory", these proceedings, pp. 1-40.

С.В. Алешин [S.V. Aleŝin] (1972), "Конечные автоматы и проблема Бернсайда о периодических группах" [Finite automata and the Burnside problem for periodic groups], *Mat. Zametki* **11**, 319-328; *Math. Notes* **11**, 199-203. MR46#265; Zbl.246.20024; RZ [1972], 8A271.

William A. Alford, George Havas and M.F. Newman (1975), "Groups of exponent four", *Notices Amer. Math. Soc.* **22**, A.301.

W.A. Alford and Bodo Pietsch (1980), "An application of the nilpotent quotient program", these proceedings, pp. 47-48.

S. Bachmuth, H.A. Heilbronn, and H.Y. Mochizuki (1968), "Burnside metabelian groups",
   Proc. Roy. Soc. London Ser. A **307**, 235-250. MR38#4561; Zbl.167, 293;
   RZ [1969], 7A185.

S. Bachmuth and H.Y. Mochizuki (1968), "The class of the free metabelian group with
   exponent p^2 ", *Comm. Pure Appl. Math.* **21**, 385-399. MR38#234; Zbl.176,297;
   RZ [1969], 3A180.

S. Bachmuth and H.Y. Mochizuki (1971), "Third Engel groups and the Macdonald-Neumann
   conjecture", *Bull. Austral. Math. Soc.* **5**, 379-386. MR46#1917; Zbl.221.20053;
   RZ [1972], 5A229.

Seymour Bachmuth and Horace Y. Mochizuki (1973), "A criterion for non-solvability of
   exponent 4 groups", *Comm. Pure Appl. Math.* **26**, 601-608. MR50#13278;
   Zbl.267.20023; RZ [1974], 11A271.

Seymour Bachmuth, Horace Y. Mochizuki and David Walkup (1970), "A nonsolvable group of
   exponent 5 ", *Bull. Amer. Math. Soc.* **76**, 638-640. MR41#1862; Zbl.218,144;
   RZ [1971], 2A194.

S. Bachmuth, H.Y. Mochizuki and D.W. Walkup (1973), "Construction of a non-solvable
   group of exponent 5 ", *Word problems. Decision problems and the Burnside
   problem in group theory*, pp. 39-66. MR54#399; Zbl.262.20033; RZ [1974], 6A241.

S. Bachmuth, H.Y. Mochizuki and K. Weston (1973), "A group of exponent 4 with
   derived length at least 4 ", *Proc. Amer. Math. Soc.* **39**, 228-234. MR47#3492;
   Zbl.284.20044; RZ [1974], 2A220.

Reinhold Baer (1944), "The higher commutator subgroups of a group", *Bull. Amer. Math.
   Soc.* **50**, 143-160. MR5,227; Zbl.61,27.

A.J. Bayes, J. Kautsky and J.W. Wamsley (1974), "Computation in nilpotent groups
   (application)", *Proc. Second Internat. Conf. Theory of Groups*, pp. 82-89.
   MR50#7299; Zbl.288.20032; RZ [1975], 6A324.

E.D. Bolker [1972], "Groups whose elements are of order two or three", *Amer. Math.
   Monthly* **79**, 1007-1010. MR46#3616; Zbl.255.20027; RZ [1973], 5A189.

А.А. Боровков, П.Н. Голованов, В.Я. Козлов, А.И. Кострикин, Ю.В. Линник, П.С. Новиков,
   Д.К. Фадеев, Н.Н. Ченцов [A.A. Borovkov, P.N. Golovanov, V.Ya. Kozlov, A.I.
   Kostrikin, Yu.V. Linnik, P.S. Novikov, D.K. Faddeev, N.N. Čentsov] (1969), "Иван
   Николаевич Санов 1919-1968: Некролог" [Ivan Nikolaevič Sanov 1919-1968:
   Obituary], *Uspehi Mat. Nauk* **24** (148), no. 4, 177-179; *Russian Math. Surveys* **24**,
   no. 4, 159-161. MR41#1490; Zbl.199,294; RZ [1969], 12A32.

R. Bosbach (1977), "Concerning joins of equational classes of Burnside groups", *Acta
   Math. Acad. Sci. Hungar.* **30**, 19-20. MR57#12701; RZ [1978], 7A233.

J.L. Britton (1973), "The existence of infinite Burnside groups", *Word problems.
Decision problems and the Burnside problem in group theory*, pp. 67-348.
MR53#10945; Zbl.264.20028; RZ [1974], 12A173.

R.H. Bruck (1962), "On the restricted Burnside problem", *Arch. der Math.* 13, 179-186.
MR27#2552; Zbl.106,22; RZ [1963], 9A161.

R.H. Bruck (1963), *Engel conditions in groups and related questions* (Lecture Notes.
Third Summer Research Institute of the Australian Mathematical Society,
Canberra).

R.A. Bryce (1969), "On metabelian groups of prime-power exponent", *Proc. Roy. Soc.
London Ser. A* 130, 393-399. MR39#6982; Zbl.179,43; RZ [1969], 12A327.

W. Burnside (1902), "On an unsettled question in the theory of discontinuous groups",
Quart. J. Pure Appl. Math. 33, 230-238. FdM33,149.

W. Burnside (1905), "On criteria for the finiteness of the order of a group of linear
substitutions", *Proc. London Math. Soc.* (2) 3, 435-440. FdM36,199.

W. Burnside (1911), *Theory of groups of finite order* [2nd. ed.] (Cambridge University
Press, Cambridge; reprinted: Dover, New York, 1955). FdM42,151; MR16,1086.

Charlotte J. Chell (1973), "A programmed algorithm for a numerical solution of the
restricted Burnside problem for prime exponent", typescript.

H.S.M. Coxeter and W.O.J. Moser (1957), *Generators and relations for discrete groups*
(Ergebnisse der Mathematik und ihrer Grenzgebiete, 14. Springer-Verlag, Berlin,
Gottingen, Heidelberg. Third edition: 1972). MR19,527; Zbl.77,28;
RZ [1961], 11A222H.

J.K. Doyle, K.I. Mandelberg and M.R. Vaughan-Lee (1978), "On solvability of groups of
exponent four", *J. London Math. Soc.* (2) 18, 234-242. Zbl.396.20025;
RZ [1979], 5A168.

C.C. Edmunds and N.D. Gupta (1973), "On groups of exponent four IV", *Conference on
group theory*, pp. 57-70. MR51#10470; Zbl.256.20033; RZ [1973], 12A240.

Irwin Fischer and Ruth R. Struik (1968), "Nil algebras and periodic groups", *Amer.
Math. Monthly* 75, 611-623. MR38#179; Zbl.174,328; RZ [1969], 1A262.

A.R. Forsyth (1927), "Professor W. Burnside", *Nature* 120, 555-557. FdM53,34.

A.R. Forsyth (1928), "William Burnside", *J. London Math. Soc.* 3, 64-80. FdM54,38.

George Glauberman, Eugene F. Krause, and Ruth Rebekka Struik (1966), "Engel
congruences in groups of prime-power exponent", *Canad. J. Math.* 18, 579-588.
MR33#4138; Zbl.143,38; RZ [1967], 2A190.

Е.С. Голод [E.S. Golod] (1964), "О ниль-алгебрах и финитно-аппроксимируемых
    p-группах" [On nil-algebras and residually finite p-groups], *Izv. Akad. Nauk
    SSSR Ser. Mat.* 28, 273-276; *Amer. Math. Soc. Transl.* (2) 48 (1965), 103-106.
    MR28#5082; Zbl.215,392; RZ [1964], 9A220.

Е.С. Голод [E.S. Golod] (1968), "О некоторых проблемах Бернсайдовского типа" [Some
    problems of Burnside type], *Труды Международного Конгресса Математиков*, pp.
    284-289. English Translation: *Amer. Math. Soc. Transl.* (2) 70, 49; *Amer. Math.
    Soc. Transl.* (2) 84 (1969), 83-88. MR39#240; Zbl.206,324; RZ [1968], 12A127.

J.A. Green (1952), "On groups with odd prime-power exponent", *J. London Math. Soc.* 27,
    476-485. MR14,350; Zbl.47,25.

J.A. Green and D. Rees (1952), "On semi-groups in which $x^n = x$ ", *Proc. Cambridge
    Philos. Soc.* 48, 35-40. MR13,720; Zbl.46,19.

Fletcher Gross (1967), "On finite groups of exponent $p^m q^n$ ", *J. Algebra* 7, 238-253.
    MR36#273; Zbl.167,294; RZ [1969], 1A180.

O. Grün (1940), "Zusammenhang zwischen Potenzbildung und Kommutatorbildung", *J.
    reine angew. Math.* 182, 158-177. MR2,212; Zbl.25,300.

Fritz Grunewald - Jens Mennicke (1973), "Über eine Gruppe vom Exponenten acht"
    (Dissertation zur Erlangung des Doktorgrades der Fakultät für Mathematik der
    Universität Bielefeld, Bielefeld).

F.J. Grunewald and J. Mennicke (1980), "Finiteness proofs for groups of exponent 8 ",
    these proceedings, pp. 189-210.

Fritz J. Grunewald, George Havas, J.L. Mennicke, and M.F. Newman (1979), "Groups of
    exponent eight", *Bull. Austral. Math. Soc.* 20, 7-16.

Fritz J. Grunewald, George Havas, J.L. Mennicke, M.F. Newman (1980), "Groups of
    exponent eight", these proceedings, pp. 49-188.

C.K. Gupta and N.D. Gupta (1968), "Some groups of prime exponent", *J. Combinatorial
    Theory* 5, 397-407. MR38#1170; Zbl.197,19.

C.K. Gupta and N.D. Gupta (1972), "On groups of exponent four. II", *Proc. Amer. Math.
    Soc.* 31, 360-362. MR44#6819; Zbl.207,34; RZ [1972], 12A204.

N.D. Gupta (1968), "Polynilpotent groups of prime exponent", *Bull. Amer. Math. Soc.*
    74, 559-561. MR36#5221; Zbl.169,33; RZ [1970], 8A182.

N.D. Gupta (1969), "The free metabelian group of exponent p^2 ", *Proc. Amer. Math.
    Soc.* 22, 375-376. MR39#6984; Zbl.177,35; RZ [1970], 5A189.

Narain Gupta (1975-1976), *Burnside groups and related topics* (Notes from lectures
    given at Kurukshetra University and the Ruhr-Universität, Bochum).

Narain D. Gupta, Horace Y. Mochizuki, and Kenneth W. Weston (1974), "On groups of
     exponent four with generators of exponent two", *Bull. Austral. Math. Soc.* 10,
     135-142. MR49#2951; Zbl.277.20026, RZ [1974], 10A230.

N.D. Gupta and M.F. Newman (1968), "Engel congruences in groups of prime-power
     exponent", *Canad. J. Math.* 20, 1321-1323. MR38#1171; Zbl.194,36;
     RZ [1969], 7A186.

N.D. Gupta and M.F. Newman (1974), "The nilpotency class of finitely generated groups
     of exponent four", *Proc. Second Internat. Conf. Theory of Groups*, pp. 330-332.
     MR50#4752; Zbl.291.20034; RZ [1975], 6A320.

N.D. Gupta and M.F. Newman (1975), "Groups of finite exponent", *Bull. Austral. Math.
     Soc.* 12, 99. MR51#5781; Zbl.288.20053; RZ [1975], 12A235.

N.D. Gupta, M.F. Newman and S.J. Tobin (1968), "On metabelian groups of prime-power
     exponent", *Proc. Roy. Soc. London Ser. A* 302, 237-242. MR36#5222; Zbl.157,350;
     RZ [1969], 1A203.

N.D. Gupta and R.B. Quintana, Jr. (1972), "On groups of exponent four. III", *Proc.
     Amer. Math. Soc.* 33, 15-19. MR45#2000; Zbl.239,20034; RZ [1973], 1A206.

Narain D. Gupta and Seán J. Tobin (1967), "On certain groups with exponent four",
     Math. Z. 102, 216-226. MR36#5220; Zbl.157,350; RZ [1969], 1A230.

Narain D. Gupta and Kenneth W. Weston (1971), "On groups of exponent four", *J. Algebra*
     17, 59-66. MR42#3176; Zbl.206,308; RZ [1971], 7A247.

Marshall Hall, Jr. (1957), "Solution of the Burnside problem for exponent 6 ", *Proc.
     Nat. Acad. Sci. U.S.A.* 43, 751-753. MR19,728; Zbl.79,30; RZ [1958], 1839.

Marshall Hall, Jr. (1958), "Solution of the Burnside problem for exponent six",
     Illinois J. Math. 2, 764-786. MR21#1345; Zbl.83,248; RZ [1960], 4944.

Marshall Hall, Jr. (1959), *The theory of groups* (Macmillan, New York. 10th reprint,
     1968. Reprinted: Chelsea, New York, 1976). MR21#1996; Zbl.84,22;
     RZ [1960], 13617H.

Marshall Hall, Jr. (1964), "Generators and relations in groups - the Burnside
     problem", *Lectures on modern mathematics*, II, pp. 42-92. Zbl.123,20;
     RZ [1966], 1A220.

Marshall Hall, Jr. (1973), "Notes on groups of exponent four", *Conference on group
     theory*, pp. 91-118. MR52#579; Zbl.282.20024; RZ [1973], 12A242.

Marshall Hall, Jr. (1977), "Computers in group theory", *Topics in group theory and
     computation*, pp.1-37. MR57#9809; Zbl.382.20002.

P. Hall and Graham Higman (1956), "On the p-length of p-soluble groups and reduction
     theorems for Burnside's problem", *Proc. London Math. Soc.* (3) 6, 1-42. MR17,344;
     Zbl.73,255; RZ [1958], 4509.

George Havas (1974), "Computational approaches to combinatorial group theory" (PhD
    thesis, University of Sydney, Sydney). See also: Abstract, *Bull. Austral. Math.
    Soc.* 11, 475-476. RZ [1975], 8A224.

George Havas (to appear), "Commutators in groups expressed as products of powers",
    Comm. Algebra.

George Havas and M.F. Newman (1980), "Application of computers to questions like those
    of Burnside", these proceedings, pp. 211-230.

George Havas and G.E. Wall (1974), "The group $\overline{B}(5, 2)$ ", *Notices Amer. Math. Soc.* 21,
    A-364.

George Havas, G.E. Wall, and J.W. Wamsley (1974), "The two generator restricted
    Burnside group of exponent five", *Bull. Austral. Math. Soc.* 10, 459-470.
    MR51#3298; Zbl.277.20025; RZ [1975], 6A255.

Franz-Josef Hermanns (1976), "Eine metabelsche Gruppe vom Exponenten, 8 "
    Diplomarbeit, Fakultät für Mathematik, Universität Bielefeld, Bielefeld).

Franz-Josef Hermanns (1977), "Eine metabelsche Gruppe vom Exponenten 8 ", *Arch. der
    Math.* 29, 375-382. MR57#12697; Zbl.367.20034; RZ [1978], 6A217.

Franz-Josef Hermanns (1980), "On certain groups of exponent eight generated by three
    involutions", these proceedings, pp. 231-245.

I.N. Herstein (1968), *Noncommutative rings* (The Carus Mathematical Monographs, 15.
    The Mathematical Association of America). MR37#2790; Zbl.177,58;
    RZ [1970], 3A287н.

Marcel Herzog and Cheryl E. Praeger (1976), "On the order of linear groups of fixed
    finite exponent", *J. Algebra* 43, 216-220. MR54#12918; Zbl.343.20025;
    RZ [1977], 8A225.

K.K. Hickin and R.E. Phillips (1978), "Non-isomorphic Burnside groups of exponent
    p^2 ", *Canad. J. Math.* 30, 180-189. MR57#6208; Zbl.344.20029; RZ [1978], 9A223.

Kenneth K. Hickin and Richard E. Phillips (1979), "Joins of periodic groups", *Proc.
    London Math. Soc.* (3) 39, 176-192.

Graham Higman (1956a), "On a conjecture of Nagata", *Proc. Cambridge Philos. Soc.* 52,
    1-4. MR17,453; Zbl.72,25; RZ [1956], 8651.

Graham Higman (1956b), "On finite groups of exponent five", *Proc. Cambridge Philos.
    Soc.* 52, 381-390. MR18,377; Zbl.75,240; RZ [1958], 9557.

G. Higman (1957a), "Le problème de Burnside", *Colloque d'algèbre supérieure,*
    pp. 123-128. MR21#5679; Zbl.84,29; RZ [1960], 4945.

Graham Higman (1957b), "Finite groups in which every element has prime power order",
    J. London Math. Soc. 32, 335-342. MR19,633; Zbl.79,32; RZ [1958], 7497.

Graham Higman (1960), "Lie ring methods in the theory of finite nilpotent groups",
   Proceedings of the International Congress of Mathematicians (Edinburgh, 1958),
   pp. 307-312. MR22#6845; Zbl.122,274; RZ [1962], 3A172.

G. Higman (1962), "p-length theorems", 1960 *Institute on finite groups*, pp. 1-16.
   MR24#A3195; Zbl.115,253; RZ [1964], 5A143.

Graham Higman (1967), "The orders of relatively free groups", *Proc. Internat. Conf.
   Theory of Groups*, pp. 153-165. MR36#2676; Zbl.166,280.

W. Holenweg (1961), "Die Dimensionsdefekte der BURNSIDE-Gruppen mit zwei Erzeugenden",
   Comment. Math. Helv. **35**, 169-200. MR23#A2463; Zbl.102,19.

W. Holenweg (1962), "Über die Ordnung von BURNSIDE-Gruppen mit endlich vielen
   Erzeugenden", *Comment. Math. Helv.* **36**, 83-90. MR26#3771; Zbl.102,19;
   RZ [1963], 5A201.

Е.И. Хухро [E.I. Huhro] (1978), "О конечных группах периода $p^\alpha q^\beta$ " [On finite groups
   of exponent $p^\alpha q^\beta$], *Algebra i Logika* **17**, 727-740.

I.Д. Iванюта [Ī.D. Īvanjuta] (1969), "Про деякі груп експоненти чотири" [On some
   groups of exponent four], *Dopovīdī Akad. Nauk Ukrain. RSR Ser. A*, 787-789;   790;
   860; MR42#7762; Zbl.177,35; RZ [1970], 4A247.

N. Jacobson (1945), "Structure theory for algebraic algebras of bounded degree", *Ann.
   of Math.* (2) **46**, 695-707. MR7,238; Zbl.60,75.

S.C. Jeanes (1976), "The maximal condition on closed normal subgroups in the profinite
   topology" (Raleigh Prize Essay, Newnham College, Cambridge).

K.W. Johnson (1974), "Varietal generalisations of Schur multipliers, stem extensions
   and stem covers", *J. reine angew. Math.* **270**, 169-183. MR53#646; Zbl.291.18019;
   RZ [1975], 4A454.

Irving Kaplansky (1963), "Lie algebras", *Lectures on modern mathematics*, I,
   pp. 115-132. MR31#2355; Zbl.147,283; RZ [1966], 1A366.

Harry Kesten (1959), "Symmetric random walks on groups", *Trans. Amer. Math. Soc.* **92**,
   336-354. MR22#B253; Zbl.92,335; RZ [1960], 9235.

А.И. Кострикин [A.I. Kostrikin] (1955), "Решение ослабленной проблемы Бернсайда для
   показателя 5 " [Solution of the restricted Burnside problem for exponent 5],
   Izv. Akad. Nauk SSSR Ser. Mat. **19**, 233-244. MR17,126,1436; Zbl.66,12;
   RZ [1956], 5133.

А.И. Кострикин [A.I. Kostrikin] (1956а), *Нильпотентные группы и кольца Ли* [*Nilpotent
   groups and Lie rings* (Candidate in Physics-Mathematics. Math. Inst. Akad. Sci.
   USSR, Moscow). RZ [1957], 218д.

А.И. Костpикин [A.I. Kostrikin] (1956b), "Нильпотентные группы и кольца Ли" [Nilpotent groups and Lie rings], *Труды Третьего Всесоюзного Математического съезда.* I: *Секционные доклады*, p. 26. RZ [1957], 2105.

А.И. Костpикин [A.I. Kostrikin] (1956c), "О кольцах Ли удовлетворяющих условию Энгеля" [On Lie rings satisfying the Engel condition], *Dokl. Akad. Nauk SSSR (N.S.)* 108, 580-582. MR18,188; Zbl.72,14; RZ [1959], 2414.

А.И. Костpикин [A.I. Kostrikin] (1957a), "О связи между периодическими группами и кольцами Ли" [On the connection between periodic groups and Lie rings], *Izv. Akad. Nauk SSSR Ser. Mat.* 21, 289-310; *Amer. Math. Soc. Transl.* (2) 45 (1965), 165-189. MR20#898; Zbl.80,246; RZ [1959], 4483.

А.И. Костpикин [A.I. Kostrikin] (1957b), "Кольца Ли, удовлетворяющие условию Энгеля" [Lie rings satisfying the Engel condition], *Izv. Akad. Nauk SSSR Ser. Mat.* 21, 515-540; *Amer. Math. Soc. Transl.* (2) 45 (1965), 191-220. MR20#1701; Zbl.79,261; RZ [1959], 4484.

А.И. Костpикин [A.I. Kostrikin] (1958a), "О локальной нильпотентности колец Ли с условием Энгеля" [On local nilpotent Lie rings with Engel conditions], *Uspehi Mat. Nauk* 13, no. 3, 246. RZ [1959], 1282.

А.И. Костpикин [A.I. Kostrikin] (1958b), "О проблеме Бернсайда" [On Burnside's problem], *Dokl. Akad. Nauk SSSR (N.S.)* 119, 1081-1084. MR24#A1948; Zbl.84,255; RZ [1959], 4486.

А.И. Костpикин [A.I. Kostrikin] (1958c), "О проблеме Бернсайда" [On the Burnside problem] (Doctoral dissertation). See also: *Uspehi Mat. Nauk* 14 (90) (1959), no. 6, 237-240. Zbl.89,14; RZ [1960], 10046Д.

А.И. Костpикин [A.I. Kostrikin] (1959), "О проблеме Бернсайда" [The Burnside problem], *Izv. Akad. Nauk SSSR Ser. Mat.* 23, 3-34; *Amer. Math. Soc. Transl.* (2) 36 (1964), 63-99. MR24#A1947; Zbl.90,245; RZ [1960], 7311.

А.И. Костpикин [A.I. Kostrikin] (1960), "Об энгелевых свойствах групп с тождественным соотношением $x^{p^{\alpha}} = 1$ " [On Engel properties of groups with the identical relation $x^{p^{\alpha}} = 1$], *Dokl. Akad. Nauk SSSR* 135, 524-526; *Soviet Math. Dokl.* 1, 1282-1284. MR24#A766; Zbl.103,13; RZ [1962], 3A193.

А.И. Костpикин [A.I. Kostrikin] (1963), "Алгебры Ли и конечные группы" [Lie algebras and finite groups], *Proc. International Congress of Mathematicians* (Stockholm, 1962), pp. 264-269. *Amer. Math. Soc. Transl.* (2) 31, 40-46. MR32#4167; Zbl.122,273; RZ [1965], 5A217.

A.I. Kostrikin (1974), "Some related questions in the theory of groups and Lie algebras", *Proc. Second Internat. Conf. Theory of Groups*, pp. 409-416. MR51#5689; Zbl.297.20046; RZ [1975], 6A411.

А.И. Кострикин [A.I. Kostrikin] (1977), "Бёрнсайда проблема" [Burnside problems],
Математическая Энциклопедия, pp. 416-418.

L.G. Kovács (1967), "Varieties and the Hall-Higman paper", *Proc. Internat. Conf.
Theory of Groups*, pp. 217-219. MR35#6736; Zbl.189,309.

L.G. Kovács (1968), "Varieties of groups and Burnside's problem", *Bull. Amer. Math.
Soc.* **74**, 599-601. MR36#5195; Zbl.169,32; RZ [1970], 10A159.

L.G. Kovács (1969), "Varieties and finite groups", *J. Austral. Math. Soc.* **10**, 5-19.
MR40#1459; Zbl.205,31; RZ [1970], 5A190.

Eugene F. Krause and Kenneth W. Weston (1971), "On the Lie algebra of a Burnside group
of exponent 5 ", *Proc. Amer. Math. Soc.* **27**, 463-470. MR43#2034; Zbl.217,359;
RZ [1972], 3A253.

А.Г. Курош [A.G. Kuroš] (1941), "Проблемы теории колец, связанные с проблемой
Бернсайда о периодических группах" [Problems in the theory of rings, associated
with Burnside's problem about periodic groups], *Bull. Acad. Sci. URSS (Izv. Akad.
Nauk SSSR) Ser. Mat.* **5**, 233-240. MR3,194.

А.Г. Курош [A.G. Kuroš] (1944), *Теория групп [Theory of groups]* (OGIZ, Moscow-
Leningrad. 3rd edition: Izdat. "Nauka", Moscow, 1967). MR40#2740;
Zbl.189,308; RZ [1969], 2A223.

M. Lazard (1960), "Groupes, anneaux di Lie et problème de Burnside" (C.I.M.E. Gruppi,
Anelli di Lie e Teoria della Coomologia. Instituto Matematico dell'Universita,
Roma). Zbl.134,260.

John Leech (1963 - 1967), "Coset enumeration on digital computers", *Proc. Cambridge
Philos. Soc.* **59**, 257-267. (Privately reprinted 1967.) MR26#4513; Zbl.117,269;
RZ [1963], 12A176.

Friedrich Levi (1933), "Über die Untergruppen der freien Gruppen. II", *Math. Z.* **37**,
90-97. FdM59,142; Zbl.6,246.

Friedrich Levi und B.L. van der Waerden (1933), "Über eine besondere Klasse von
Gruppen", *Abh. Math. Sem. Univ. Hamburg* **9**, 154-158 (1932). FdM58,125;
Zbl.5,385.

Frank Levin and Gerhard Rosenberger (1980), "Generalized power laws", these
proceedings, pp. 246-248.

В.П. Лобыч, А.И. Скопин [V.P. Lobyč, A.I. Skopin] (1976), "О соотношениях в группах
экспоненты 8 " [Relations in groups of exponent 8], *Zap. Naučn. Sem.
Leningrad. Otdel. Mat. Inst. Steklov* **64**, 92-94, 161. MR57#3268; Zbl.366.20021;
RZ [1977], 4A159.

Alfred Loewy (1900), "Zur Theorie der Gruppen linearer Substitutionen", *Math. Ann.* **53**,
225-242. FdM31,130.

R.C. Lyndon (1954), "On Burnside's problem", *Trans. Amer. Math. Soc.* **77**, 202-215.
MR16,218; Zbl.58,17; RZ [1956], 1997.

R.C. Lyndon (1955), "On Burnside's problem. II", *Trans. Amer. Math. Soc.* **78**, 329-332.
MR16,792; Zbl.66,277; RZ [1956], 7894.

R.C. Lyndon (1959), "Burnside groups and Engel rings", *Finite groups*, pp. 4-14.
MR22#9522; Zbl.168,284; RZ [1963], 3A240.

I.D. Macdonald (1973), "Computer results on Burnside groups", *Bull. Austral. Math.
Soc.* **9**, 433-438. MR49#5165; Zbl.267.20011; RZ [1974], 10A201.

I.D. Macdonald (1974), "A computer application to finite *p*-groups", *J. Austral. Math.
Soc.* **17**, 102-112. MR51#13028; Zbl.277.20024; RZ [1974], 11A228.

Saunders Mac Lane (1963), "Some recent advances in algebra", *Studies in modern
algebra*, pp. 9-34.

Wilhelm Magnus (1937), "Neuere Ergebnisse über auflösbare Gruppen", *Jber. Deutsche
Math.-Verein.* **47**, 69-78. FdM63,69; Zbl.16,202.

Wilhelm Magnus (1950), "A connection between the Baker-Hausdorff formula and a problem
of Burnside", *Ann. of Math.* (2) **52**, 111-126. MR12,476; Zbl.37,304.

W. Magnus (1953), "Errata: A connection between the Baker-Hausdorff formula and a
problem of Burnside", *Ann. of Math.* (2) **57**, 606. MR14,723; Zbl.51,255.

Wilhelm Magnus, Abraham Karrass, Donald Solitar (1966), *Combinatorial group theory:
Presentations of groups in terms of generators and relations* (Pure and Applied
Mathematics, **13**. Interscience [John Wiley & Sons], New York, London, Sydney.
Second, revised edition: Dover, New York, 1976). MR34#7617; Zbl.138,256;
RZ [1967], 9A143к.

В.Д. Мазуров [V.D. Mazurov] (1969), "Ослабленная проблема Бернсайда для показателя
30 " [The restricted Burnside problem for exponent 30], *Algebra i Logika* **8**,
460-477; *Algebra and Logic* **8**, 264-273. MR43#3330; Zbl.213,303;
RZ [1970], 6A177.

John R. McMullen (1974), "Compact torsion groups", *Proc. Second Internat. Conf. Theory
of Groups*, pp. 453-462. MR50#13292; Zbl.286.22003; RZ [1975], 6A352.

Heinrich Meier-Wunderli (1950), "Über endliche *p*-Gruppen, deren Elemente der
Gleichung $x^p = 1$ genügen", *Comment. Math. Helv.* **24**, 18-45. MR11,579;
Zbl.38,164.

H. Meier-Wunderli (1951), "Metabelsche Gruppen", *Comment. Math. Helv.* **25**, 1-10.
MR12,671; Zbl.44,15.

H. Meier-Wunderli (1952), "Note on a basis of P. Hall for the higher commutators in
free groups", *Comment. Math. Helv.* **26**, 1-5. MR13,818.

Heinrich Meier-Wunderli (1956), "Über die Struktur der Burnsidegruppen mit zwei
    Erzeugenden und vom Primzahlexponenten $p > 3$ ", *Comment. Math. Helv.* **30**,
    144-174. MR18,376; Zbl.70,22; RZ [1957], 5360.

Thomas Meixner (1977), "Eine Bemerkung zu p-Gruppen vom Exponenten p ", *Arch. der*
    Math. **29**, 561-565. MR57#3263; Zbl.378.20020; RZ [1978], 7A247.

Jean Michel (1976), "Calculs dans les algèbres de Lie libres: la série de Hausdorff
    et le problème de Burnside", *Journées Algorithmiques*, pp. 139-148. MR57#6116;
    Zbl.351.17012; RZ [1977], 8A333.

Jay I. Miller (1977), "Center-by-metabelian groups of prime exponent", *Notices Amer.*
    Math. Soc. **24**, A-415.

Jay I. Miller (1979), "Center-by-metabelian groups of prime exponent", *Trans. Amer.*
    Math. Soc. **249**, 217-224.

Horace Y. Mochizuki (1974), "On groups of exponent four: a criterion for
    nonsolvability", *Proc. Second Internat. Conf. Theory of Groups*, pp. 499-503.
    MR50#7347; Zbl.291.20019; RZ [1975], 6A322.

M. Muzalewski (1976), "Burnside's problems, residual finiteness and finite
    reducibleness", *Bull. Acad. Polon. Sci. Sér. Sci. Math. Astronom. Phys.* **24**,
    1067-1068. MR55#8184; Zbl.355.20033; RZ [1977], 11A272.

B.H. Neumann (1935), "Identical relations in groups" (Dissertation, The University,
    Cambridge).

B.H. Neumann (unpublished), "Bericht über 'Identical relations in groups'".

B.H. Neumann (1937a), "Identical relations in groups. I", *Math. Ann.* **114**, 506-525.
    FdM63,64; Zbl.16,351.

B.H. Neumann (1937b), "Groups whose elements have bounded orders", *J. London Math.*
    Soc. **12**, 195-198. FdM63,66; Zbl.16,393.

B.H. Neumann (1960), *Lectures on topics in the theory of infinite groups* (Notes by
    M. Pavman Murthy. Tata Institute of Fundamental Research, Bombay. Reissued:
    1968). MR42#1881; Zbl.237.200001.

B.H. Neumann (1967), "Varieties of groups", *Bull. Amer. Math. Soc.* **73**, 603-613.
    MR35#2952; Zbl.149,267; RZ [1969], 1A210.

B.H. Neumann and Hanna Neumann (1959), "Embedding theorems for groups", *J. London*
    Math. Soc. **34**, 465-479. MR29#1267; Zbl.102,264; RZ [1969], 12519.

Hanna Neumann (1967), *Varieties of groups* (Ergebnisse der Mathematik und ihrer
    Grenzgebiete, **37**. Springer-Verlag, Berlin, Heidelberg, New York). MR35#6734;
    Zbl.251.200001; RZ [1968], 3A194 .

M.F. Newman (1964), "A theorem of Golod-Šafarevič and an application in group theory",
    typescript.

M.F. Newman (1976), "A computer aided study of a group defined by fourth powers",
Bull. Austral. Math. Soc. 14, 293-297. MR53#8266; Zbl.326.20001.
RZ [1977], 2B867. Addendum: *Bull. Austral. Math. Soc.* 15, 477-479. MR55#8153;
Zbl.354.20028; RZ [1977], 11A251.

M.F. Newman (1979), "Groups of exponent dividing seventy", *Math. Scientist* 4, 149-157.

M.F. Newman, K.W. Weston and Tah-Zen Yuan (1975), "Polynomials associated with groups
of exponent four", *Bull. Austral. Math. Soc.* 12, 81-87. MR51#704;
Zbl.292.20030; RZ [1975], 11A296.

П.С. Новиков [P.S. Novikov] (1959a), "Решение проблемы Бернсайда о периодических
группах" [Solution of Burnside's problem on periodic groups], *Uspehi Mat. Nauk
(N.S.)* 14 (89), 236-237.

П.С. Новиков [P.S. Novikov] (1959b), "О периодических группах" [On periodic groups],
Dokl. Akad. Nauk SSSR 127, 749-752; *Amer. Math. Soc. Transl.* (2) 45 (1965),
19-22. MR21#5680; Zbl.119,22; RZ [1960], 6197.

П.С. Новиков, С.И. Адян [P.S. Novikov, S.I. Adjan] (1968a), "О бесконечных
периодических группах. I" [Infinite periodic groups. I], *Izv. Akad. Nauk SSSR
Ser. Mat.* 32, 212-244; *Math. USSR-Izv.* 2, 209-236 (1969). MR39#1532a;
Zbl.194,33; RZ [1969], 4A169.

П.С. Новиков, С.И. Адян [P.S. Novikov, S.I. Adjan] (1968b), "О бесконечных
периодических группах. II" [Infinite periodic groups. II], *Izv. Akad. Nauk
SSSR Ser. Mat.* 32, 251-524; *Math. USSR-Izv.* 2, 241-479 (1969). MR39#1532b;
Zbl.194,33; RZ [1969], 4A170.

П.С. Новиков, С.И. Адян [P.S. Novikov, S.I. Adjan] (1968c), "О бесконечных
периодических группах. III" [Infinite periodic groups. III], *Izv. Akad. Nauk
SSSR Ser. Mat.* 32, 709-731; *Math. USSR-Izv.* 2, 665-685 (1969). MR39#1532c;
Zbl.194,33; RZ [1969], 4A171.

П.С. Новиков, С.И. Адян [P.S. Novikov, S.I. Adjan] (1968d), "Определяющие соотношения
и проблема тождества для свободных периодических групп нечетного порядка"
[Defining relations and the word problem for free periodic groups of odd order],
Izv. Akad. Nauk SSSR Ser. Mat. 32, 971-979; *Math. USSR-Izv.* 2, 935-942 (1969).
MR40#4344; Zbl.194,33.

П.С. Новиков, С.И. Адян [P.S. Novikov, S.I. Adjan] (1968e), "О коммутативных
подгруппах и проблеме сопряженности в свободных периодических группах нечетного
порядка" [Commutative subgroups and the conjugacy problem in free periodic groups
of odd order], *Izv. Akad. Nauk SSSR Ser. Mat.* 32, 1176-1190; *Math. USSR-Izv.* 2,
1131-1144 (1970). MR38#2197; Zbl.204,341; RZ [1969], 4A48.

П.С. Новиков [P.S. Novikov] (1971), "Петр Сергеевич Новиков. (К семидесятилетию со
    дня рождения)" [Petr Sergeevich Novikov (on his seventieth birthday)] (written by
    editors), *Uspehi Mat. Nauk* 26 (161), no. 5, 231-241; *Russian Math. Surveys* 26,
    no. 5, 165-176. MR52#5334; RZ [1972], 2A34.

Петр Сергеевич Новиков [Petr Sergeevic Novikov] (1975), "Некролог" [Obituary], *Izv.*
    Akad. Nauk SSSR Ser. Mat. 39, 469-470; Abridged Translation: *Fiz.-Mat. Spis.*
    B"lgar. Akad. Nauk 18 (51), no. 3, 227. MR51#12470; RZ [1975], 8A26.

Gianfranco Panella (1966), "Un teorema di Golod-Šafarevič e alcune sue conseguenze",
    Confer. Sem. Mat. Univ. Bari No. 104, 17pp. MR33#5673; Zbl.148,263;
    RZ [1968], 3A260.

J. Płonka (1976), "On direct products of some Burnside groups", *Acta Math. Acad. Sci.*
    Hungar. 27, 43-45. MR55#8190; Zbl.339.20010; RZ [1977], 4A216.

C. Procesi (1966), "The Burnside problem", *J. Algebra* 4, 421-425. MR35#2956;
    Zbl.152,3; RZ [1967], 9A145.

Claudio Procesi (1973), *Rings with polynomial identities* (Pure and Applied
    Mathematics, 17. Marcel Dekker, New York). MR51#3214; Zbl.262.16018;
    RZ [1974], 6A339.

Ricardo B. Quintana (1973), "An attack on the restricted Burnside problem for groups
    of exponent 8 on 2 generators", *Conference on group theory*, pp. 140-147.
    MR54#12909; Zbl.262.20040; RZ [1973], 12A201.

Ю.П. Размыслов [Ju.P. Razmyslov] (1971), "Об энгелевых алгебрах Ли" [On Engel Lie
    algebras], *Algebra i Logika* 10, 33-44; *Algebra and Logic* 10, 21-29. MR45#3498;
    Zbl.253.17005; RZ [1971], 11A307.

Ю.П. Размыслов [Ju.P. Razmyslov] (1972), "Об одном примере неразрешимых почти
    кроссовых многообразий групп" [A certain example of unsolvable almost Cross
    varieties of groups], *Algebra i Logika* 11, 186-205, 238; *Algebra and Logic* 11,
    108-120 (1973). MR47#332; Zbl.248.20033; RZ [1972], 11A139.

Ю.П. Размыслов [Ju.P. Razmyslov] (1978), "О проблеме Холла-Хигмена" [On a problem of
    Hall-Higman], *Izv. Akad. Nauk SSSR Ser. Mat.* 42, 833-847. Zbl.394.20030;
    RZ [1979], 1A246.

И.Н. Санов [I.N. Sanov] (1940), "Решение проблемы Бернсайда для показателя 4 "
    [Solution of Burnside's problem for exponent four], *Leningrad Gos. Univ. Ped.*
    Inst. Uč. Zap. Mat. Ser. 10, 166-170. MR2,212.

И.Н. Санов [I.N. Sanov] (1946), "Периодические группы с малыми периодами" [Periodic
    groups with small periods] (PhD Dissertation, Leningrad University).

И.Н. Санов [I.N. Sanov] (1947), "О проблеме Бернсайда" [On Burnside's problem], *Dokl.*
    Akad. Nauk SSSR (N.S.) 57, 759-761. MR9,224; Zbl.29,102.

И.Н. Санов [I.N. Sanov] (1951), "О некоторой системе соотношений в периодических группах с периодом степенью простого числа" [On a certain system of relations in periodic groups with period a power of a prime number], *Izv. Akad. Nauk SSSR Ser. Mat.* 15, 477-502. MR14,722; Zbl.45,302.

И.Н. Санов [I.N. Sanov] (1952), "Установление связи между периодическими группами с периодом простым числом и кольцами Ли" [Establishment of a connection between periodic groups with period a prime number and Lie rings], *Izv. Akad. Nauk SSSR Ser. Mat.* 16, 23-58. MR13,721; Zbl.46,32.

Eugene Schenkman (1954), "Two theorems on finitely generated groups", *Proc. Amer. Math. Soc.* 5, 497-498. MR16,671; Zbl.56,255; RZ [1956], 215.

I. Schur (1911), "Über Gruppen periodischer linearer Substitutionen", *S.-B. Preuss. Akad. Wiss. Phys.-Mat. Kl*, 619-627. See also: Issai Schur, *Gesammelte Abhandlungen* I, no. 17, pp. 442-450 (Springer-Verlag, Berlin, Heidelberg, New York, 1973). FdM42,155-156.

J.-A. de Séguier (1904), *Théorie des groupes finis. Éléments de la théorie des groupes abstraits* (Gauthier-Villars, Paris). FdM35,181.

G.B. Seligman (1967), *Modular Lie algebras* (Ergebnisse der Mathematik und ihrer Grenzgebiete, 40. Springer-Verlag, Berlin, Heidelberg, New York). MR39#6933; Zbl.189,32; RZ [1969], 5A231.

Stephen S. Shatz (1972), *Profinite groups, arithmetic, and geometry* (Annals of Mathematics Studies, 67. Princeton University Press, Princeton, New Jersey; University of Tokyo Press, Tokyo). MR50#279; Zbl.236.12002; RZ [1972], 9A311.

David Shield (1977), "The class of a nilpotent wreath product", *Bull. Austral. Math. Soc.* 17, 53-89. MR57#3266; Zbl.396.20015; RZ [1978], 9A232.

В.Л. Ширванян [V.L. Širvanjan] (1976), "Вложение группы $B(\infty, n)$ в группу $B(2, n)$" [Imbedding of the group $B(\infty, n)$ in the group $B(2, n)$], *Izv. Akad. Nauk SSSR Ser. Mat.* 40, 190-208, 223; *Math. USSR-Izv.* 10, 181-199 (1977). MR54#2821; Zbl.336.20027; RZ [1976], 7A262.

А.И. Скопин [A.I. Skopin] (1976), "О соотношениях в группах экспоненты 8 " [Relations in groups of exponent 8], *Zap. Naučn. Sem. Leningrad. Otdel. Mat. Inst. Steklov (LOMI)* 57, 129-170. MR57#3267; Zbl.366.20020; RZ [1976], 9A202.

А.И. Скопин [A.I. Skopin] (1978), "Об одной группе экспоненты 8 " [On a group of exponent 8], *Zap. Naučn. Sem. Leningrad. Otdel. Mat. Inst. Steklov (LOMI)* 75, 164-165. RZ [1978], 8A210.

J.P. Soublin (1974), "Problèmes de Burnside", *Comptes Rendus des Journées Mathématiques de la Société Mathématique de France*, pp. 151-156. MR51#8263.

Michio Suzuki (1962), "On a class of doubly transitive groups", *Ann. of Math.* (2) 75, 105-145. MR25#112; Zbl.106,247; RZ [1962], 10A132.

John Joseph Tobin (1954), "On groups with exponent 4 " (PhD thesis, University of
    Manchester).

Seán Tobin (1960), "Simple bounds for Burnside p-groups", *Proc. Amer. Math. Soc.* 11,
    704-706. MR23#A202; Zbl.96,16; RZ [1962], 8A149.

S. Tobin (1975), "On groups with exponent four", *Proc. Roy. Irish Acad. Sect. A* 75,
    115-120. MR52#572; Zbl.319.20043; RZ [1976], 2A262.

А.И. Токаренко [A.I. Tokarenko] (1968), "О линейных группах над кольцами" [Linear
    groups over rings], *Sibirsk. Mat. Z.* 9, 951-959; *Siberian Math. J.* 9, 708-713.
    MR39#325; Zbl.256.20061; RZ [1969], 2A281.

A.L. Tritter (1970), "A module-theoretic computation related to the Burnside problem",
    Computational problems in abstract algebra, pp. 189-198. MR41#6981;
    Zbl.217,358.

M.R. Vaughan-Lee (1979), "Derived lengths of Burnside groups of exponent 4 ", *Quart.
    J. Math. Oxford* (2), 30, 495-504.

Б.Б. Венков [B.B. Venkov] (1973), "О некоторых гомологических свойствах групп
    Бернсайда" [Certain homological properties of Burnside groups], *Zap. Naučn. Sem.
    Leningrad. Otdel. Mat. Inst. Steklov (LOMI)* 31, 38-54. MR51#5798;
    RZ [1973], 8A337.

Э.Б. Винберг [È.B. Vinberg] (1965), "К теореме о бесконечномерности ассоциативной
    алгебры" [On the theorem concerning the infinite dimensionality of an associative
    algebra], *Izv. Akad. Nauk SSSR Ser. Mat.* 29, 209-214. MR30#3108; Zbl.171,294;
    RZ [1965], 7A191.

Ascher Wagner and Verity Mosenthal (1978), "A bibliography of William Burnside",
    Historia Math. 5, 307-312.

G.E. Wall (1974), "On the Lie ring of a group of prime exponent", *Proc. Second
    Internat. Conf. Theory of Groups*, pp. 667-690. MR50#10098; Zbl.286.20050;
    RZ [1975], 7A389.

G.E. Wall (1978a), "Lie methods in group theory", *Topics in algebra*, pp. 137-173.

G.E. Wall (1978b), "On the Lie ring of a group of prime exponent II", *Bull. Austral.
    Math. Soc.* 19, 11-28.

James Wiegold (1965), *Kostrikin's proof of the restricted Burnside conjecture for
    prime exponent* (Lectures given at the Institute of Advanced Studies, Australian
    National University, Canberra).

C.R.B. Wright (1960), "On groups of exponent four with generators of order two",
    Pacific J. Math. 10, 1097-1105. MR22#6844; Zbl.91,26; RZ [1971], 5A222.

C.R.B. Wright (1961), "On the nilpotency class of a group of exponent four", *Pacific
    J. Math.* 11, 387-394. MR23#A927; Zbl.126,47; RZ [1962], 3A170.

Rodney I. Yager (1977), "The Burnside problem" (Honours Essay, University of Sydney).

Giovanni Zacher (1956), "Sull'ordine di un gruppo finito risolubile somma dei suoi
     sottogruppi di Sylow", *Atti Accad. Naz. Lincei. Rend. Cl. Sci. Fis. Mat. Nat.*
     (8) **20**, 171-174. MR18,377; Zbl.75,239; RZ [1957], 1152.

Giovanni Zacher (1957), "Sui gruppi finiti somma dei loro sottogruppi di Sylow",
     Rend. Sem. Mat. Univ. Padova **27**, 267-275. MR20#4594.

KEY INDEX

Actes du Congrès International des Mathématiciens (Nice, 1970), 3 volumes (Gauthier-
   Villars, Paris, 1971). MR54#3,4,5; Zbl.219.00001.

Colloque d'algèbre supérieure (Bruxelles, 1956) (Centre Belge de Recherches
   Mathématiques. Etablissements Ceuterick, Louvain, 1957).

Comptes rendus des journées mathématiques de la société mathématique de France [Yves
   Cesari et Artibano Micali, Eds] (L'Université des Sciences et Techniques du
   Languedoc, Montpellier, 1974) (Cahiers Mathématiques Montpellier, 3. U.E.R. de
   Mathématiques, Université des Sciences et Techniques du Languedoc, Montpellier,
   1974). MR51#2822.

Computational problems in abstract algebra [John Leech, Ed.] (Proc. Conf. Oxford,
   1967) (Pergamon, Oxford, London, New York, 1970). MR40#5374; Zbl.186,299.

Conference on group theory [R.W. Gatterdam and K.W. Weston, Eds] (held at University
   of Wisconsin, Parkside, Kenosha, 1972) (Lecture Notes in Mathematics, 319.
   Springer-Verlag, Berlin, Heidelberg, New York, 1973.) MR51#658; Zbl.253.00005;
   RZ [1973], 10A145.

Finite Groups [A.A. Albert, Irving Kaplansky, Eds] (Proc. Sympos. Pure Math. 1, New
   York, 1959. American Mathematical Society, Providence, Rhode Island, 1959).
   Zbl.97,256.

Journées Algorithmiques [Maurice Nivat, Gérard Viennot, Organisateurs] (L'école
   Normale Supérieure, Paris, 1975), Astérisque 38-39 (Société Mathématique de
   France, Paris, 1976). MR54#12421.

Коуровская тетрадь. Нерешенные вопросы теории групп [*Kourovka notebook. Unsolved
   problems in the theory of groups*] (1st edition, 1965. 6th augmented edition:
   В.Д. Мазуров, Ю.И. Мерзляков, В.А. Чуркин, Редакторы [V.D. Mazurov, Ju.I.
   Merzljakov, V.A. Čurkin, editors]. Akad. Nauk SSSR Sibirsk. Otdel. Inst. Mat.,
   Novosibirisk, 1978).

Lectures on Modern Mathematics I [T.L. Saaty, Ed.] (John Wiley & Sons, New York,
   London, 1963). MR31#2104a; Zbl.124,241; RZ [1964], 11A64н.

Lectures on Modern Mathematics II [T.L. Saaty, Ed.] (John Wiley & Sons, New York,
   London, Sydney, 1964). MR31#2104b; Zbl.129,243.

Математическая Энциклопедия [*Mathematical Encyclopedia*] [И.М. Виноградов, Главный
    редактор] [I.M. Vinogradov, Editor-in-Chief] (Izdatel'stvo Sovetskaja
    Enciklopedija, Moscow, 1977).

Proceedings of the International Conference on Theory of Groups (Canberra, 1965)
    [L.G. Kovács and B.H. Neumann, Eds] (Gordon and Breach, New York, London, Paris,
    1967). MR35#6732; Zbl.158,2.

Proceedings of the International Congress of Mathematicians (Edinburgh, 1958) [J.A.
    Todd, Ed.] (Cambridge University Press, Cambridge, 1960). MR22#5537;
    Zbl.119,242; RZ [1961], 10A34н.

Proceedings of the International Congress of Mathematicians (Stockholm, 1962)
    (Institute Mittag-Leffler, Djursholm, Sweden, 1963). MR28#1; Zbl.114,2.

Труды Международного Конгресса Математиков (Москва, 1966) [*Proceedings of the
    International Congress of Mathematicians* (Moscow, 1966)] (Izdat. "Mir", Moscow,
    1968. Amer. Math. Soc. Transl. (2) 70 (1968)). MR38#960; Zbl.185,1;
    RZ [1968], 12A3.

Труды Третьего Всесоюзного Математического с'езда (Москва, 1956). Том 1: *Секционные
    Доклады* [*Proceedings of the Third All-Union Mathematical Conference* (Moscow,
    1956). Volume 1: *Sectional reports*] (Izdat. Akad. Nauk SSSR, Moscow, 1956).
    MR30#6973a.

Proceedings of the Second International Conference on the Theory of Groups (Canberra,
    1973) [M.F. Newman,Ed.](Lecture Notes in Mathematics, 372. Springer-Verlag,
    Berlin, Heidelberg, New York, 1974). MR49#9054; Zbl.282.00012;
    RZ [1975], 5A165н.

Studies in modern algebra, Volume 2 [A.A. Albert, Ed.] (Mathematical Association of
    America and Prentice-Hall, Englewood Cliffs, New Jersey, 1963). MR26#3750.

Topics in algebra [M.F. Newman, Ed.] (Proceedings, 18th Summer Research Institute of
    the Australian Mathematical Society, Australian National University, Canberra,
    1978) (Lecture Notes in Mathematics, 697. Springer-Verlag, Berlin, Heidelberg,
    New York, 1978).

Topics in group theory and computation [Michael P.J. Curran, Ed.] (Proceedings of
    the Summer School, University College, Galway, 1973) (Academic Press [Harcourt
    Brace Jovanovich], London, New York, San Francisco, 1977). MR57#421;
    Zbl.361.00004.

Word Problems. Decision problems and the Burnside problem in group theory [William W.
    Boone, Frank B. Cannonito and Roger C. Lyndon, Eds] (Studies in Logic and the
    Foundations of Mathematics, 71. North-Holland, Amsterdam, London, 1973).
    MR49#10789; Zbl.254.00004; RZ [1974], 5A264н.

1960 *Institute on finite groups* [Marshall Hall, Jr., Ed.] (California Institute of
      Technology, Pasadena, California, 1960) (Proc. Sympos. Pure Math. 6.   American
      Mathematical Society, Providence, Rhode Island, 1962).   MR24#A1942;   Zbl.112,260;
      RZ [1969], 1A176и.

Department of Mathematics,
Institute of Advanced Studies,
Australian National University,
Canberra, ACT, Australia.